Metagenomics and Microbial Ecology

Metagenomics and Microbial Ecology

Techniques and Applications

Edited by
Surajit De Mandal, Amrita Kumari Panda,
Nachimuthu Senthil Kumar, Satpal Singh Bisht,
and Fengliang Jin

CRC Press is an imprint of the
Taylor & Francis Group, an **informa** business

First edition published 2022
by CRC Press
6000 Broken Sound Parkway NW, Suite 300, Boca Raton, FL 33487–2742

and by CRC Press
2 Park Square, Milton Park, Abingdon, Oxon, OX14 4RN

Library of Congress Cataloging-in-Publication Data

Names: De Mandal, Surajit, editor. | Panda, Amrita Kumari, editor. | Senthil Kumar, N., editor. | Bisht, Satpal Singh, editor. | Jin, Fengliang, editor.
Title: Metagenomics and microbial ecology : techniques and applications / edited by Surajit De Mandal, Amrita Kumari Panda, N. Senthil Kumar, Satpal Singh Bisht, Fengliang Jin.
Description: First edition. | Boca Raton : CRC Press, 2022. | Includes bibliographical references and index.
Identifiers: LCCN 2021031758 | ISBN 9780367487348 (hardback) | ISBN 9781032140209 (paperback) | ISBN 9781003042570 (ebook)
Subjects: MESH: Metagenomics | Environmental Microbiology | Microbiological Phenomena
Classification: LCC QH434 | NLM QU 460 | DDC 572.8/629—dc23
LC record available at https://lccn.loc.gov/2021031758

ISBN: 9780367487348 (hbk)
ISBN: 9781032140209 (pbk)
ISBN: 9781003042570 (ebk)

DOI: 10.1201/9781003042570

Typeset in Times LT Std
by Apex CoVantage, LLC

Contents

Preface

Natural environments have unique ecosystems with a large diversity of microbial dark matter. Understanding and characterizing these microorganisms will be an important step toward discovering novel products or understanding complex biological systems. Researchers working in the field of microbial taxonomy are well aware of the remarkable changes that have occurred over the last decade through next-generation sequencing (NGS) technology. The introduction of high-throughput cost-effective NGS technology has expanded the possibilities of genomic research in various biological systems, allowing parallel sequencing analyses of multiple genes effectively at any desired depth of coverage. These methods are currently considered an important tool for studying the taxonomic and functional characteristics of microorganisms in various ecosystems. This also led to the discovery of rare microorganisms present in low abundance. However, it is often difficult for young researchers to fully understand and apply NGS in their research due to the sophisticated instrumentation, large amounts of generated data, and complex bioinformatics tools for annotation. Metagenomics completely revolutionized microbial dark matter research by exploring environmental samples. Metagenomics diverges from genomics study, as it focuses on intricate biological interactions such as horizontal gene transfer, population dynamics, and nutritional complementation among microbial consortia. Metagenomics projects mainly use two types of methods: amplicon sequencing and shotgun sequencing, but the experimental strategy, sequencing technology, and analytical bioinformatics tools vary per the requirements to facilitate microbial ecology research.

The purpose of this book is to present these advances to better understand the status and application of NGS technology in microbial ecology projects. This book will also help young researchers to understand and learn new scientific tools and their subsequent applications to address future challenges. In this book, *Metagenomics and Microbial Ecology: Techniques and Applications*, renowned scientists and professors explore the most recent advances in the landscape of next-generation sequencing technologies for studying microorganisms and their communities in different environments. The whole book is divided into five sections. Section I deals with an overview of metagenomics. Section II describes various metagenomic tools used to assess microbial diversity. Section III describes metagenomic studies conducted under extreme conditions, including halophiles in the coastal area, desert ecosystems, and xenobiotics. Section IV of this book describes the metagenomic studies in various ecotypes. Finally, the recent application of metagenomics is discussed in Section V.

Editors

Surajit De Mandal, Ph.D., is a post-doctoral researcher at the College of Agriculture, South China Agricultural University, Guangzhou, P. R. China. He has eight years of research experience and has published several research articles in international journals. He also acts as an editorial board member/reviewer for various international journals. His fields of interest include microbial diversity and metagenomics, molecular phylogeny, bioinformatics, and microbial control of insect pests. He is presently working on microbial community analysis using next-generation sequencing methods.

Amrita Kumari Panda, Ph.D., is currently working as an assistant professor of biotechnology at the Department of Biotechnology, Sant Gahira Guru Vishwavidyalaya Sarguja, Chhattisgarh, India. She obtained her doctoral degree in biotechnology from Berhampur University, Odisha, India. Dr. Panda received a Fast Track Post Doctoral fellowship from the Science and Engineering Research Board, Government of India, New Delhi. She has more than eight years of research experience in the field of molecular microbiology and microbial diversity and has published several articles in national and international peer-reviewed journals. She received the Federation of European Microbiological Society (FEMS) Meeting grant award in the year 2014. Her research interests include metagenomics and microbial diversity.

Nachimuthu Senthil Kumar, Ph.D., is Professor in the Department of Biotechnology, Mizoram University, with 20 years of teaching and research experience. His fields of interest include genomics, molecular phylogeny, and bioinformatics. He is presently working on metagenomics, cancer DNA markers, and mutation analysis. During his post-doctoral stint at Sun Yat-sen University, China, he was involved in RNAi studies with the chitin synthase gene. He has authored more than 120 research papers, and he is the investigator of various national research projects.

Satpal Singh Bisht, D.Sc., has been an inspiring professor and researcher in the field of life sciences for 20 years. Dr. Bisht is an internationally acclaimed biologist/academician and an administrator. He completed his doctoral degree in 1992 from Kumaun University, Nainital Uttarakhand, India. Dr. Bisht presently works as Professor of Zoology and Dean of Biomedical Sciences at Kumaun University. He has published many research papers and articles in journals of international repute and visited many countries on academic assignments. Prof. Bisht is a member of various academic bodies at many universities and colleges. His major research interests include exploring the gut microbiome of earthworms using next-generation sequencing and DNA barcoding of mountain earthworms. He has served as an editor and reviewer for many journals and has been nominated for and received many professional awards from many academic organizations.

Fengliang Jin, Ph.D., is presently working as a professor in the Department of Entomology, College of Agriculture, South China Agricultural University, Guangzhou, P.R. China. His major research interest is in the role of non-coding RNAs in the regulation of insect immune signal transduction, exploring the interaction mechanism of insects to entomopathogens using next-generation sequencing, bioinformatic analysis, and RNAi-based functional analysis of immune-related genes and their role in the regulation of antimicrobial peptides of insects. Prof. Jin served as a principal investigator for several research projects related to the development of biopesticides at various agencies. He has several patents and scientific articles and is currently serving as a reviewer and editor for many international journals.

Contributors

Totan Adak
ICAR–National Rice Research Institute, Crop Protection Division, Cuttack, India

S. Al Khodor
Sidra Medicine, Ar-Rayyan, Doha, Qatar

Athira CH
Department of Genomic Science, Central University of Kerala, Tejaswini Hills, Kerala, India

Monu Bala
Department of Zoology, P.G. College, Syalde Almora, Uttarakhand, India

Ramon Alberto Batista-García
Universidad Autónoma del Estado de Morelos, Cuernavaca, México

Bharat Bhushan
ICAR–Indian Institute of Maize Research, Ludhiana, Punjab, India

Satpal Singh Bisht
Laboratory of Earthworm Biotechnology and Microbial Metagenomics, Department of Zoology, D.S.B Campus, Kumaun University, Nainital, India

Hugo G. Castelán-Sánchez
Centro de Investigación en Dinámica Celular, Instituto de Investigación en Ciencias Básicas y Aplicadas, Universidad Autónoma del Estado de Morelos, Cuernavaca, Mexico

Natividad Castro-Alarcón
Universidad Autónoma de Guerrero, México

Sara Cuadros-Orellana
Universidad Católica del Maule, Chile

Sonia Dávila-Ramos
Centro de Investigación en Dinámica Celular, Instituto de Investigación en Ciencias Básicas y Aplicadas, Universidad Autónoma del Estado de Morelos, Cuernavaca, Mexico

María del Rayo Sánchez-Carbente
Centro de Investigación en Biotecnología, Universidad Autónoma del Estado de Morelos, Cuernavaca, Mexico

Surajit De Mandal
Laboratory of Bio-Pesticide Innovation and Application of Guangdong Province, College of Agriculture, South China Agricultural University, Guangzhou, China

Vinay Shankar Dubey
Yoga and Psychotherapy Association of India

Joystu Dutta
Sant Gahira Guru University, Ambikapur, Chhattisgarh, India

Jorge Luis Folch-Mallol
Universidad Autónoma del Estado de Morelos, Cuernavaca, México

Sampat Ghosh
Agriculture Science and Technology Research Institute, Andong National University, Republic of Korea

Adhikesavan Harikrishnan
Department of Chemistry, School of Arts and Sciences, Vinayaka Mission Research Foundation-Aarupadai Veedu (VMRF-AV) campus, Paiyanoor, Chennai, India

Madangchanok Imchen
Department of Genomic Science, Central University of Kerala, Tejaswini Hills, Kerala, India

Chuleui Jung
Agriculture Science and Technology Research Institute, Andong National University, Republic of Korea and Department of Plant Medicals, Andong National University, Republic of Korea

Ankit Kumar
Department of Pharmaceutical Sciences, Kumaun University Bhimtal Campus Bhimtal, India

Mahesh Kumar
ICAR–National Institute of Abiotic Stress Management, Baramati, Pune, India

Satish Kumar
ICAR–National Institute of Abiotic Stress Management, Baramati, Pune, India

Ranjith Kumavath
Department of Genomic Science, Central University of Kerala, Tejaswini Hills, Kerala, India

Basavegowda Lakshmi
Department of Chemistry, REVA University, Kattigenahalli, Yelahanka, Bengaluru, India

Srikanta Lenka
ICAR–National Rice Research Institute, Crop Protection Division, Cuttack, India

Xiaolong Liang
Department of Earth and Planetary Sciences, Washington University in St. Louis, Missouri

Arabinda Mahanty
ICAR–National Rice Research Institute, Crop Protection Division, Cuttack, India

Kamlesh K. Meena
ICAR–National Institute of Abiotic Stress Management, Baramati, Pune, India

Rashi Miglani
Laboratory of Earthworm Biotechnology and Microbial Metagenomics Department of Zoology, D.S.B Campus, Kumaun University, Nainital, India

Jamseel Moopantakath
Department of Genomic Science, Central University of Kerala, Tejaswini Hills, Kerala, India

Saeed Mohamadzade Namin
Agriculture Science and Technology Research Institute, Andong National University, Republic of Korea

Alma Delia Nicolás-Morales
Universidad Autónoma de Guerrero, México

Oshin K
Department of Biotechnology, REVA University, Kattigenahalli, Yelahanka, Bengaluru, India

Amrita Kumari Panda
Department of Biotechnology, Sant Gahira Guru University, Ambikapur, Chhattisgarh, India

Koustav Kumar Panda
Department of Biochemistry, Biotechnology and Physiology, Centurion University, Paralakhemundi, India

Nagma Parveen
Laboratory of Earthworm Biotechnology and Microbial Metagenomics Department of Zoology, D.S.B Campus, Kumaun University, Nainital, India

D. Patel
Department of Bioinformatics, Bharati Vidyapeeth (Deemed to be University), Pune, Maharashtra, India

P. Chunarkar Patil
Department of Bioinformatics, Bharati Vidyapeeth (Deemed to be University), Pune, Maharashtra, India

Yordanis Pérez-Llano
Universidad Autónoma del Estado de Morelos, Cuernavaca, México

Prabhukarthikeyan SR
ICAR–National Rice Research Institute, Crop Protection Division, Cuttack, India

Raghu S.
ICAR–National Rice Research Institute, Crop Protection Division, Cuttack, India

Jagadish Rane
ICAR–National Institute of Abiotic Stress Management, Baramati, Pune, India

Prakash Chandra Rath
ICAR–National Rice Research Institute, Crop Protection Division, Cuttack, India

A. Rawat
Sidra Medicine, Ar-Rayyan, Doha, Qatar

Tirthankar Sen
Indian Institute of Technology-Guwahati, India

Ramasamy Shanmugavalli
Department of Chemistry, School of Arts and Sciences, Vinayaka Mission Research Foundation-Aarupadai Veedu (VMRF-AV) campus, Paiyanoor, Chennai, India

Ashok Kumar Sharma
University of Minnesota, Twin Cities, St Paul, Minnesota

Ajinath Shridhar Dukare
ICAR—Central Institute of Post-Harvest Engineering and Technology, Ludhiana, India

Ajay Kumar Singh
ICAR–National Institute of Abiotic Stress Management, Baramati, Pune, India

P. Singh
Sidra Medicine, Ar-Rayyan, Doha, Qatar

Jyoti Upadhyay
School of Health Science, University of Petroleum and Studies, Dehradun, India

G.C. Wakchaure
ICAR–National Institute of Abiotic Stress Management, Baramati, Pune, India

Vijaykumar Veena
Department of Biotechnology, REVA University, Kattigenahalli, Yelahanka, Bengaluru, India

Section I

An Overview of Metagenomics

1

Principles and Analysis of Metagenomics Data

A. Rawat, P. Singh, S. Al Khodor*
Sidra Medicine, Ar-Rayyan, Doha, Qatar

CONTENTS

1.1 Introduction

The ongoing developments in high-throughput sequencing and large-scale endeavors such as the Human Microbiome Project (HMP) (Group et al. 2009), METAgenomics of the Human Intestinal Tract (MetaHIT) (Qin et al. 2010), Metagenomics and Metadesign of Subways & Urban Biomes (MetaSUB) (Meta 2016), Extreme Microbiome Project (XMP) (Tighe et al. 2017), and the Earth Microbiome Project (EMP) (Gilbert, Jansson, and Knight 2018) have helped unfold the complex genomic composition of the 'microbiome.' The two methodologies generally used for microbial identification and classification are amplicon/marker genes (e.g., 16S rRNA) and whole metagenome shotgun (WMGS) metagenomics.

The advancement of sequencing technologies both in terms of the quantity and the length of sequences has allowed the generation of large full-genome datasets. However, it has been observed that the sequencing yield consists of large amounts of 'dark matter' or unknown sequences (Sedlazeck et al. 2018). Additionally, there are challenges with the difficulty of sorting reads from the different microbial entities and exploring microbial diversity (Nayfach and Pollard 2016). Furthermore, downstream analysis is confounded by biological and technical variability (Wooley and Ye 2009). Technical variation can arise during different stages like sampling size/storage/handling and DNA extraction methods that can introduce biases toward certain taxa (Boers, Jansen, and Hays 2019). Biological variance is significantly contributed by the high level of diversity and complexity of microbial communities (Delmont et al. 2011). Microbial species composition is known to vary considerably inter-sample (David et al. 2014) as well as intra-sample in a population (Kashtan et al. 2014). All these factors, combined with the choice of bioinformatics tools for microbial diversity analysis (e.g., quality filtering tools and assemblers), contribute to complexities in the statistical inference of metagenomic data. Standardization of wet-lab and bioinformatics protocols is necessary to reduce various biases, which otherwise might lead to non-reproducible

DOI: 10.1201/9781003042570-2

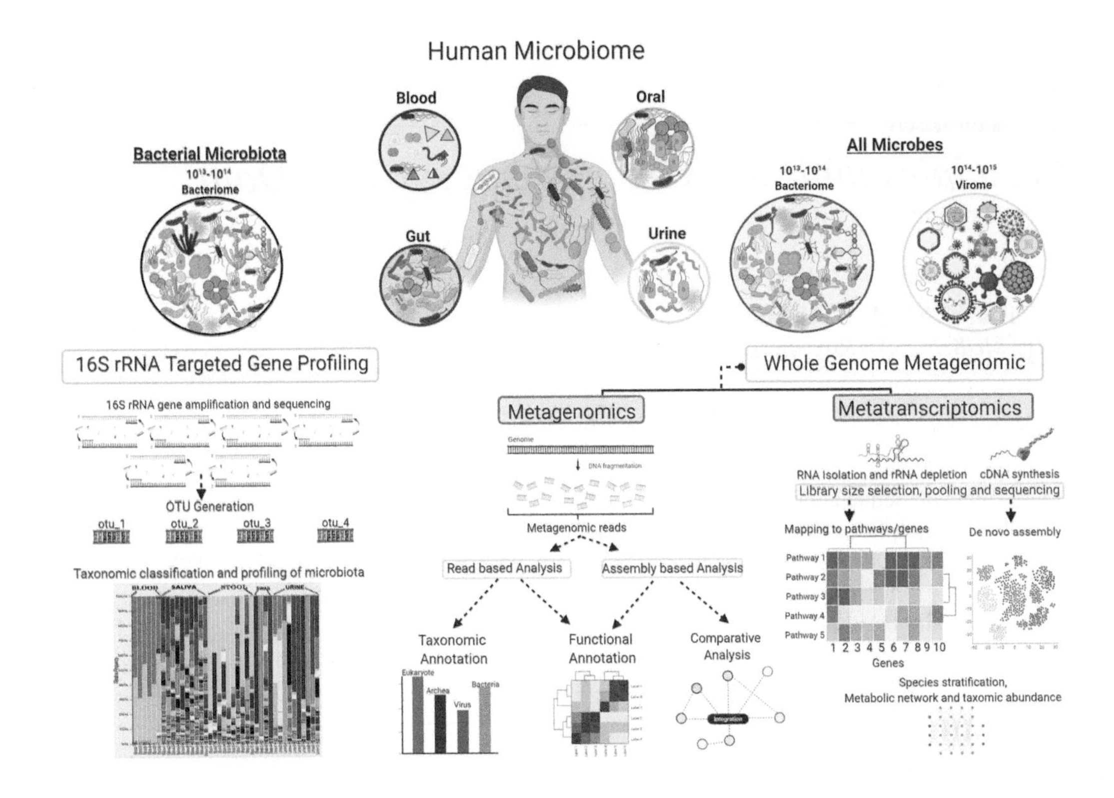

FIGURE 1.1 Overview of the metagenomics workflow.

results. This chapter discusses best-practice bioinformatics analysis workflows in an effort to achieve robustness and reproducibility.

Microbial metagenomic data analysis typically encompasses two sequencing strategies: amplicon sequencing targeting the bacterial 16S rRNA marker gene to gain an overview of community structure and the rapidly developing WMGS and metatranscriptomics to capture the genes and genomes of all microorganisms from a sample along with their functional profiles (Figure 1.1). Novel development in analytical methods, for example, PICRUSt that allows for the imputation of functional content based on taxon abundances with 16s rRNA (Langille et al. 2013), has allowed amplicon sequencing to remain an efficient and cost-effective strategy for microbiome analysis. Numerous statistical/computational tools and databases have been developed in order to allow for the analysis of the huge amount of data; however, the selection of the correct tool depends largely on the biological question and available computational resources. We will discuss the data analysis of these two dichotomous technologies in separate sections.

1.2 Amplicon Sequencing

High-throughput genomic approaches have effectively transformed microbial research throughout the last decade, bypassing onerous culture-based microbiological, molecular, and serological tests for identifying, typing, and characterizing microbes. Amplicon sequencing targeting the universal 16S rRNA gene in bacteria allows segregation between organisms at the genus or even species level across various bacteria phyla and is considered highly informative for bacterial communities/strains that are phenotypically aberrant, less known, or difficult to isolate (Clarridge 2004). The 16S rRNA gene sequence is approximately 1,550 bp long, and though the gene is highly conserved across prokaryotes, it comprises nine hypervariable regions with sufficient interspecific polymorphisms for taxonomic distinction that can be validated statistically.

16S rRNA amplicon sequencing performs well with regard to the classification of bacteria; however, the possibility of lateral or horizontal gene transfer can pose challenges with the use of a single marker gene to assess diversity; another factor to consider is the low taxonomic resolution at the level of species and within closely related species (Janda and Abbott 2007; Bosshard et al. 2006; Mignard and Flandrois 2006). Amplicon sequencing is also known to be constrained by short read lengths and sequencing errors (Quince et al. 2009) and taxonomic accuracies of different hypervariable regions (Youssef et al. 2009); reports have also highlighted difficulties in clustering sequences into the appropriate operational taxonomic units (OTUs) (Kunin et al. 2010).

Targeted amplicon analyses use specific primers designed to amplify the nine different hypervariable regions of choice (V1–V9) in the 16S rRNA gene. Based upon the similarity (above 97%), the sequences are clustered into OTU bins, following which a single sequence is selected as a representative sequence that is annotated using a 16S rRNA classification method, and the same annotation is applied to all sequences within the OTU. Though, as mentioned, historically, OTUs are typically constructed using an identity threshold of 97%, more recent studies have shown that the optimal clustering threshold might be ~99% for full-length sequences (Edgar 2018; Nguyen et al. 2016).

Additionally, downstream analysis of 16S rRNA data is also confounded by variables, as highlighted by Weiss et al. (2017). First, differential sampling and sequencing throughput (i.e., library sizes) may represent the community in each biological sample, which may be reflective of the sequencing process rather than a true biological variation. Second, a large number of OTUs are sparse, meaning that they contain a high proportion of zero counts (up to ~90%). This sparsity can cause various hindrances, from interpretation (of rare OTUs) to analytical issues, and assigning pseudocount values (replacing zeros with an insignificant number) is considered questionable by many. Third, the number of sequencing reads in no way represents the actual microbes in the community, and each sample is a fractional representation of the original community. Despite these challenges, 16S rRNA amplicon-based next-generation sequencing (NGS) has been widely used in the characterization of the bacterial communities associated with various samples from major human body sites such as the gastrointestinal tract and skin (Turnbaugh et al. 2007), as well as environmental samples such as soil, water sources (Marotz et al. 2017), and so on. In the following, we summarize the typical workflow for amplicon sequencing and highlight common approaches to overcome the associated challenges.

1.2.1 Amplicon Sequencing Workflow Overview

Illumina sequencing platforms are widely used for amplicon libraries and allow the assessment of multiple samples via a multiplexed PCR approach; however, the short read lengths limit resolution. The Oxford Nanopore and Pacific Biosciences sequencing platforms overcome this limitation but are prone to higher error rates. Depending on methodology, amplicon sequencing can result in single- or paired-end sequences. The single-end sequences are read from one end to another (unidirectional), while paired sequencing runs are read from one end to the other end and then start another round of reading from the opposite end (bidirectional), resulting in nearly twice the read length. These pairs are separated by an insert, and insert size is usually variable, dependent on technology. Usually paired reads are preferred for interrogation of bacterial communities, while their counterpart mate-pair (in different orientation) allows for larger distance between reads and is more suitable for genome finishing, structural variation, and so on. A workflow for processing sequence data is heavily dependent on commonly available tools; we have categorized the 16S rRNA amplicon sequencing data analysis into two stages:

1. Pre-OTU
2. Post-OTU

The pre-OTU stage attempts to generate high-quality OTUs, and these resulting OTUs are then utilized in the analysis during the post-OTU stage. As the post-OTU stage addresses the biological question relevant to the individual project, we will focus mainly on the pre-OTU stage, involving preprocessing (quality control) and quantification (taxonomic profiling) (Figure 1.2).

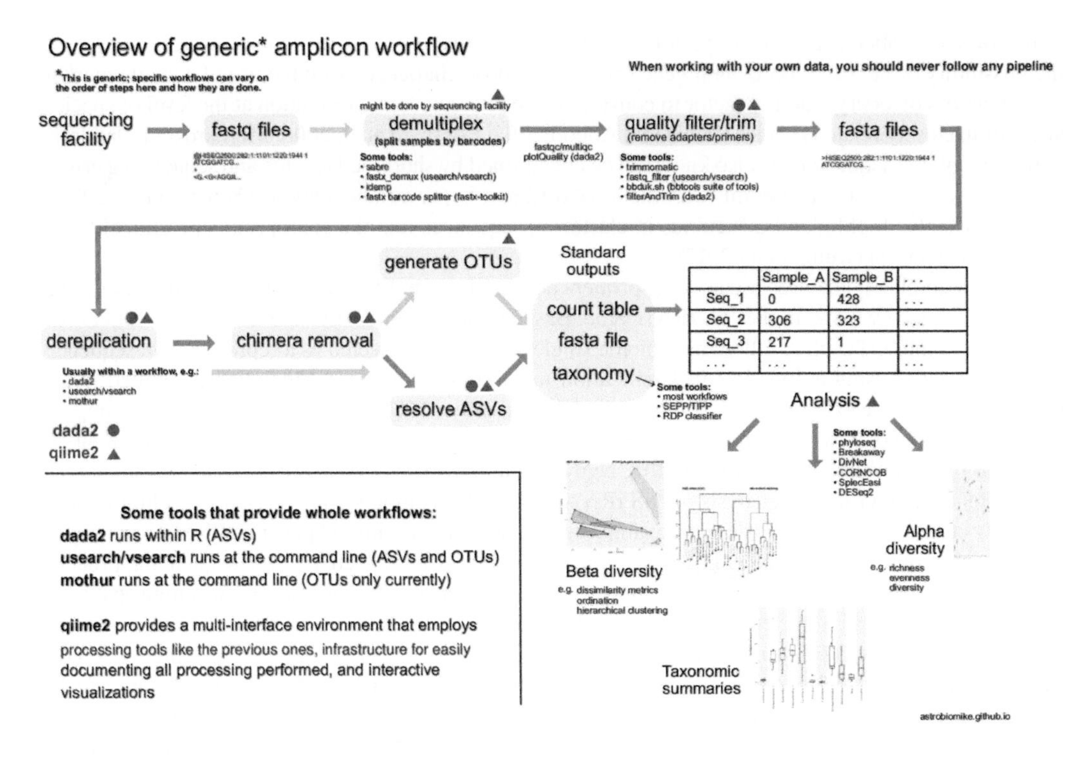

FIGURE 1.2 The generic workflow associated with amplicon data analysis.

Source: Lee (2019)

1.2.1.1 Pre-Processing

Pre-processing of the sequences is done with the goal of allowing only high-quality reads to be evaluated for downstream analysis and involves the following:

1. Demultiplexing: Barcode information is used to identify which sequences came from which samples after they all have been sequenced together. Demultiplexing can be done for the Illumina platform by its software bcl2fastq (https://support.illumina.com/) or uploaded to basespace (basespace.illumina.com), or independent tools (for all sequencing platforms) like fastx_demux (https://drive5.com/usearch/manual/pipe_demux.html) and fastx barcode splitter (http://hannonlab.cshl.edu/fastx_toolkit/) can be utilized.
2. Rarefaction Curves: These are usually employed to assess sampling sensitivity. These are usually plotted with a number of OTUs on the *y*-axis and a number of (subsampled) sequences on the *x*-axis. As the sequencing yield increases, it is a common observation that number of OTUs will increase in differing proportions across samples. However, it is expected that the curve will flatten or taper, representing that no new OTUs or only rare OTUs remain to be generated given a sequencing yield. Therefore, these curves are helpful in the assessment of sample diversity vis-à-vis different samples of the community sequenced in a study.
3. Trimming Adapters, PCR Primers, and Removal of Low-Quality Bases: There are numerous tools designed to address and augment overall sequence quality. FastQC (www.bioinformatics.babraham.ac.uk/projects/fastqc/) presents the quality/Phred score. The Phred score defines sequencing error in terms of Q-score, where Q10 represents 1 error expected out of every 10 bases; Q20 represents 1 error expected out of every 100 bases, and so forth (Ewing et al. 1998; Ewing and Green 1998). Addressing sequences with low-quality scores can improve the overall

accuracy of bioinformatics analyses. Based on the overall sequencing quality derived by different evaluation metrics, sequencing reads can be trimmed for adapters and low-quality regions. Some notable published tools for sequence trimming are Trimmomatic (Bolger, Lohse, and Usadel 2014), PRINSEQ (Schmieder and Edwards 2011), and Cutadapt (Martin 2011).

After preprocessing, the reads are subjected to quantification with processes like denoising.

1.2.1.2 Quantification

The landscape of 16S rRNA amplicon sequencing analysis has seen huge transmutation. On the one hand, highly implemented tools like MOTHUR (Schloss et al. 2009) and QIIME (Caporaso et al. 2010) exist. On the other hand, newer approaches like denoising have been developed. This has so far not led to any general consensus regarding the best strategy for quantification. It has been demonstrated that denoising generates amplicon sequence variants (ASVs), oftentimes referred as sub-OTUs (sOTUs), and might provide better resolution in contrast to the OTU-level approach (Prodan et al. 2020). The main aim of the quantification stage is to output high-quality OTUs/sOTUs which can be used for downstream analysis and consists of stages like chimera removal and clustering.

1. Chimera Removal: Chimeras are artifact sequences formed by two or more biological sequences that are incorrectly joined together, and removal of these artifacts with tools like ChimeraSlayer (Haas et al. 2011), UCHIME (Edgar et al. 2011), or Bellerophon (Huber, Faulkner, and Hugenholtz 2004) is helpful.
2. Clustering: Clustering sequences requires huge computational resources and was considered an important/integrated step along with dereplication and chimera removal, but its importance is now considered debatable (Callahan, McMurdie, and Holmes 2017). We discuss this more in Section 2.2.2. The recent introduction of Dada2 (Callahan et al. 2016) and Deblur (Amir et al. 2017) introduces the concept of ASV/sOTU, which has also been recently added to QIIME. In a recent study, six commonly used tools/pipelines were described, three at the OTU level: USEARCH-UPARSE, MOTHUR, and QIIME-UCLUST, and the other three at the ASV level: USEARCH-UNOISE3, Qiime2-Deblur, and DADA2. The study on the efficiency of these pipelines reported that DADA2 has the best resolution and sensitivity, followed by USEARCH-UNOISE3 (Prodan et al. 2020).
3. Merge Paired-Read: Previous workflows involved an additional step to merge or assemble paired reads to a consensus sequence using tools like PANDAseq (Masella et al. 2012), PEAR (Zhang et al. 2014), FLASH (Magoc and Salzberg 2011), and COPE (Liu et al. 2012). Newer algorithms like DADA2 (Callahan, McMurdie, and Holmes 2017), Deblur (Amir et al. 2017), and UNOISE (Edgar and Flyvbjerg 2015) have changed the arena of integrating many of these multiple steps in a single workflow.

1.2.2 Types of Quantification Platforms for Amplicon Sequencing

Quantification platforms can be segregated into three platforms:

1. Workflow/pipeline
2. Standalone programs
3. Web interfaces

    1. Workflow/Pipeline: Multiple workflows like QIIME (Caporaso et al. 2010), Mothur (Schloss et al. 2009), LotuS (Hildebrand et al. 2014), and BioMaS (Fosso et al. 2015) are available. The development of two highly cited pipelines (QIIME and MOTHUR) primarily came from the need to cluster sequences (OTUs) based on sequence similarities (Schloss 2020). Clustering sequences requires huge computational resources, and these two workflows recommend a dichotomous approach. QIIME recommends open, closed,

and *de novo* clustering with the help of UPARSE (Edgar 2013) and VSEARCH (Rognes et al. 2016). Meanwhile, MOTHUR has *de novo* clustering built into the application itself. Both pipelines otherwise follow nearly the same workflow of 'cleaning' reads, followed by clustering and assignment of taxonomy. Many studies demonstrate the sensitivity of *de novo* clustering (Westcott and Schloss 2015; Schloss 2020). However, the given sensitivity comes with a computational cost, and huge projects should be mindful of such bottlenecks. The integration of DADA2 and Deblur introduces the concept of ASV/sOTU and has recently been added as a functionality in QIIME2.

2. Standalone Tools: With the recent availability of DADA2 in R, a statistical platform, free to use (GNU General Public License) is rapidly becoming the preferred platform of analysis. Also, the availability of several downstream statistical analysis tools like Phyloseq (McMurdie and Holmes 2013), MetagenomeSeq (Paulson et al. 2013), the microbiome R package (Lahti et al. 2012–2017), and Vegan (Oksanen et al. 2013), among others, allow users to perform data analysis in a common platform. The Biological Observation Matrix (BIOM) file format, developed by QIIME, can be generated with output analysis and imported into other platforms/tools for further downstream statistical analysis. The power of R lies in easy sharing of analysis script codes, exploring different aspects of data like taxonomic trees, and sample metadata aiding reproducibility. R is an incredibly flexible tool commonly utilized by data scientists; however, learning R can be challenging for clinicians and wet-lab researchers.
3. Web Interface: Platforms such as QIIME and MOTHUR have incorporated the flexibility of command line independence and user-defined exploration of sequencing data suitable for clinicians and bench researchers. QIIME2 is the next-generation microbiome bioinformatics platform with a shiny derivative that is graphical user interface (GUI) based, where users can explore taxonomic information and also generate figures. QIIME users are, however, limited to taxonomic interrogation and therefore lack the capacity to perform correlative analysis with other datasets such as gene expression. In addition to these widely used resources, there is constant development of new tools and methods to deal with the challenges of analyzing and visualizing these complex datasets. Earlier, there were online interfaces from well-known reference databases like SILVA (Quast et al. 2013), Greengenes (DeSantis et al. 2006), and the Ribosomal Database Project (RDP) (Cole et al. 2014). These online resources helped users identify their datasets against the collated database. However, these resources lacked the flexibility of using different (for example, 18S rRNA) or even custom databases. Also, oftentimes, huge computational resources were required, especially for large projects. Keeping this in context, QIIME and Mothur were developed and were discussed in more detail previously. Other highly cited web servers, MG-RAST (Meyer et al. 2008) and MEGAN (Huson et al. 2007), allow for higher flexibility in terms of overall utility to perform 16S rRNA analysis. Recently there have been efforts like MicrobiomeAnalyst (Dhariwal et al. 2017), Metaviz (Wagner et al. 2018), Calypso (Zakrzewski et al. 2017), and PUMA (Mitchell et al. 2018) to help users with limited programming skills. An OTU/sOTU table with sample details can be uploaded directly and visualized for different comparisons. However, like most web interfaces, these are limited to only methods provided by the web host.

Metadata Association: The OTU generated can be associated with metadata for any meaningful biological interpretation.

1. Diversity Measurement: Community composition in microbial ecology can be measured with Alpha and Beta diversity. Alpha diversity measures within-sample diversity, while Beta diversity is the divergence across samples in the community composition. Alpha diversity is usually measured by Shannon, inverse Simpson, observed OTU, and Chao1. Shannon and inverse Simpson take into account the diversity within a sample, while observed OTU and Chao1

consider the richness of the community. Depending on the study involved, additional metrics like evenness and phylogeny can also be utilized. Beta diversity can be explored with different methods like NMDS, MDS, and PCoA with different distance metrics like weighted Unifrac, unweighted Unifrac, Bray-Curtis, and Jaccard.

2. Taxonomy: Taxonomic resolution (phylum to species level) is dependent on factors like database and similarity level for OTUs (Chen and Li 2013). OTU sequences can be mapped against any of the well-known databases like SILVA (Quast et al. 2013), Greengenes (DeSantis et al. 2006), the Ribosomal Database Project (Cole et al. 2014), or Unite (Koljalg et al. 2013).
3. Functional Annotation: Tax4fun or PICRUSt can be utilized to explore the functional profile of microbial communities. Tax4fun is available in R, but its counterpart PICRUSt is an elaborate pipeline that first normalizes the data on 16S rRNA gene copy number and then annotates sequences based on KEGG and/or COG protein families.

1.3 Shotgun Metagenomics Approach

1.3.1 Overview

Metagenomics catalogs the gene content of the microbial community (Bharti and Grimm 2019) utilizing either long-read (500–4000 bp) with PacBio and Oxford Nanopore or short-read (250–300 bp) with Illumina. One of the prominent features that separate amplicon sequencing from the metagenomics approach is that while amplicon technology sequences only marker-genes (e.g., 16S rRNA, 18S rRNA, ITS), WMGS sequences the entire DNA content for all genes. Previous studies have reported that the 16S rRNA gene captures sufficient biodiversity and community ecology for different taxonomic levels (Tessler et al. 2017), and the characterization of microbial communities is consistent between 16S rRNA and metagenomic sequencing (Rausch et al. 2019). Nevertheless, WMGS is generally considered preferable to 16S amplicon sequencing in the human microbiome, providing increased resolution of bacterial species, enhanced coverage of diversity, and gene prediction (Ranjan et al. 2016). A few drawbacks associated with WMGS are that the host DNA often overwhelm the sequencing output compared to the small fraction of microbial DNA, and extensive profiling can lead to cost escalation. These conflicting challenges must be overcome through continued research and technological advancements in NGS. In the following, we provide an overview of computational workflows, along with commonly used pipelines/platforms for WMGS analysis.

1.3.2 WMGS Application

One of the prominent features of WMGS is that it allows for better characterization of the microbial population than 16S rRNA, including subtypes, strains, and pathogenic gene load. Therefore, the role of WMGS as a powerful tool in medical microbiology is increasingly being demonstrated by typing of pathogens with the highest possible discriminatory power (Oude Munnink et al. 2020). Keeping the clinical objective in mind, the workflows can be specific for DNA or RNA. For DNASeq, the commonly used workflows are PathSeq (Kostic et al. 2011), Metagenie (Rawat et al. 2014), Clinical Pathoscope (Byrd et al. 2014; Francis et al. 2013), RINS (Bhaduri et al. 2012), and SURPI (Naccache et al. 2014). For RNA-Seq, tools like IMSA+A (Cox et al. 2017) and nf-rnaSeqMetagen (Mpangase et al. 2020) are the commonly used ones. A few workflows/tools that support a metataxonomics framework include GOTTCHA (Freitas et al. 2015), VirusFinder (Wang, Jia, and Zhao 2013), VirusSeq (Chen et al. 2013), IMSA+A (Cox et al. 2017), MEGAN-CE (Huson et al. 2016), Kraken (Wood and Salzberg 2014), and MetaPhlAn2 (Truong et al. 2015).

1.3.3 WMGS Data Analysis

There are several analytical approaches that can be performed in lieu of 'one size fits all' tools or established best practice solutions for analysis of WMGS datasets. Various approaches have their pros and

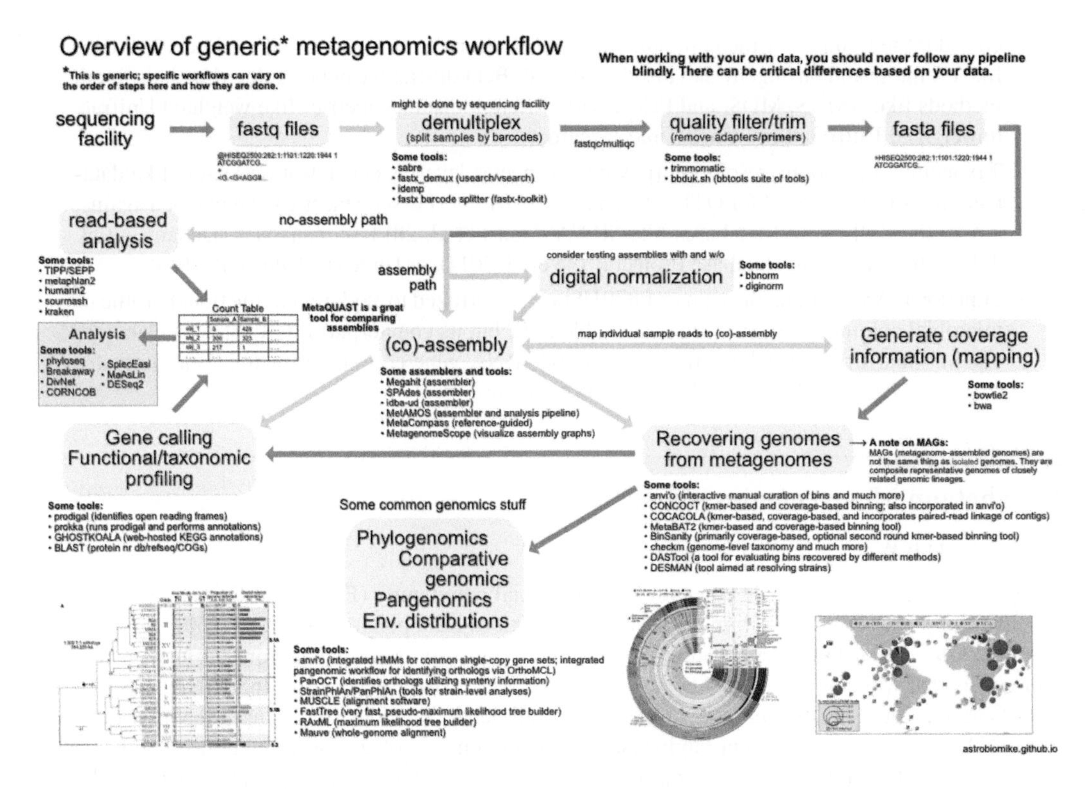

FIGURE 1.3 A generic workflow associated with metagenomic data analysis.

Source: Lee (2019)

cons and are constantly evolving. Typically, a computational pipeline for WGS can broadly follow three steps: preprocessing, assembling, and finally taxonomic and phylogenetic resolution (Figure 1.3). The preprocessing step hugely overlaps with amplicon sequencing and includes steps such as filtering data for adapters, host contaminants, and low-quality bases. For instance, MG-RAST (Meyer et al. 2008) employs a normalization step in order to remove duplicate sequences and sequencing artifacts. Users are empowered to choose the best parameter that suits their dataset; they can select filtering for contaminants and obtain a quality estimate for each sample. Tools like BBtools (Bushnell et al. 2017) with a repair function and Sickle (Joshi and Fass 2011) with an adaptive sliding window might be helpful both for long and short reads. The methodological approaches can be summarized broadly into two areas: read-based approaches and assembly-based approaches.

1.3.3.1 Read-Based Analysis

One of the first methods to be utilized for metagenomic interrogation was read based, which analyzes unassembled reads and is still valuable for quantitative analysis, provided there is availability of appropriate references. Each metagenomic read from a sample is explored that can be utilized as a machine learning feature and used for prediction. A typical workflow involves the following analysis:

a. Taxonomy: Community profiling for metagenomic reads can be performed with MetaPhlAn2 (Truong et al. 2015), which involves comparing each read from a sample to the marker database (~17K reference genomes consisting of unique clade-specific marker genes) to identify high-confidence matches, Kraken (Wood and Salzberg 2014) is an ultra-fast and sensitive sequence classifier that assigns taxonomic ranks to short DNA reads using exact alignments of k-mers.

b. Functional Annotation: The functional profile of metagenomic reads can be performed with the Blast (Altschul et al. 1990) suite against a database (like NCBI or Silva) or pipeline like HUMAnN2 (Franzosa et al. 2018) or Megan (Huson et al. 2007).

1.3.3.2 Assembly-Based Analysis

Assembling is usually an iterative process whereby sequencing reads are first assembled into contigs, which are then merged into *scaffolds.* Assembling can be performed with *de novo* or reference-based assemblers. Some of the well-known assemblers utilized for WMGS are MetaSPAdes (Nurk et al. 2017), MetaIDBA (Peng et al. 2011), and MegaHit (Li et al. 2015). The assembled reads are then subjected to characterization, as discussed in the following.

a. Taxonomy and Phylogeny: Metataxonomic profiles can be generated using Kraken and other tools. Phylogenomics comparative analysis for pangenomes can also be performed on assembled reads utilizing Anvi'o (Eren et al. 2015), PanOCT (Fouts et al. 2012), and PanPhlAn (Scholz et al. 2016), among others.
b. Functional Annotation: Sequencing data consist of a high percentage of unknown reads, often referred to as 'dark matter' (Qin et al. 2010; van Bakel et al. 2010). Gene prediction is quite helpful to annotate these unknown sequences. Prokka (Seemann 2014) is one such helpful tool that performs functional annotation on the gene predicted by Prodigal (Hyatt et al. 2010).

Binning: Oftentimes, binning might be performed on sequences or assembled reads to generate taxonomic information and functional annotation and reconstruct/recover individual genomes from metagenomes. This is performed with a supervised classifier like PhyloPythiaS (Patil, Roune, and McHardy 2012; Gregor et al. 2016) or Phylosift (Darling et al. 2014) or an unsupervised classifier for new association without *a priori* information, for example, CONCOCT (Alneberg et al. 2014) or Maxbin (Wu, Simmons, and Singer 2016; Wu et al. 2014). Other binning tools are based on a hybrid approach with both read similarity and composition, like PhymmBL (Brady and Salzberg 2009, 2011), and MetaCluster (Wang et al. 2012a, 2012b) can also be utilized.

1.4 Conclusion

As next-generation sequencing becomes more affordable, the microbial metagenomic approach has become popular for clinical and biomedical research. Both target-gene and metagenomic sequencing approaches are vital for a comprehensive understanding of microbial communities. Here, we gave an overview of optimally tested tools available for the interrogation of microbial diversity, abundance, assembly, and functional annotations. However, there are bottlenecks and biases associated with both sequencing and computational methods. Therefore, it is critical to enable biomedical researchers to explore datasets using efficient, user-friendly, whole-pipeline tools. Furthermore, investigation into the unexplored multiple aspects of metagenomic data is much needed and could present interesting topics for research.

REFERENCES

Alneberg, J., B. S. Bjarnason, I. de Bruijn, M. Schirmer, J. Quick, U. Z. Ijaz, L. Lahti, N. J. Loman, A. F. Andersson, and C. Quince. 2014. "Binning metagenomic contigs by coverage and composition." *Nat Methods* 11 (11):1144–6. doi: 10.1038/nmeth.3103.

Altschul, S. F., W. Gish, W. Miller, E. W. Myers, and D. J. Lipman. 1990. "Basic local alignment search tool." *J Mol Biol* 215 (3):403–10. doi: 10.1016/S0022-2836(05)80360-2.

Amir, A., D. McDonald, J. A. Navas-Molina, E. Kopylova, J. T. Morton, Z. Zech Xu, E. P. Kightley, L. R. Thompson, E. R. Hyde, A. Gonzalez, and R. Knight. 2017. "Deblur rapidly resolves single-nucleotide community sequence patterns." *mSystems* 2 (2). doi: 10.1128/mSystems.00191-16.

Bhaduri, A., K. Qu, C. S. Lee, A. Ungewickell, and P. A. Khavari. 2012. "Rapid identification of non-human sequences in high-throughput sequencing datasets." *Bioinformatics* 28 (8):1174–5. doi: 10.1093/bioinformatics/bts100.

Bharti, R., and D. G. Grimm. 2019. "Current challenges and best-practice protocols for microbiome analysis." *Brief Bioinform*. doi: 10.1093/bib/bbz155.

Boers, S. A., R. Jansen, and J. P. Hays. 2019. "Understanding and overcoming the pitfalls and biases of next-generation sequencing (NGS) methods for use in the routine clinical microbiological diagnostic laboratory." *Eur J Clin Microbiol Infect Dis* 38 (6):1059–70. doi: 10.1007/s10096-019-03520-3.

Bolger, A. M., M. Lohse, and B. Usadel. 2014. "Trimmomatic: A flexible trimmer for Illumina sequence data." *Bioinformatics* 30 (15):2114–20. doi: 10.1093/bioinformatics/btu170.

Bosshard, P. P., R. Zbinden, S. Abels, B. Boddinghaus, M. Altwegg, and E. C. Bottger. 2006. "16S rRNA gene sequencing versus the API 20 NE system and the VITEK 2 ID-GNB card for identification of nonfermenting Gram-negative bacteria in the clinical laboratory." *J Clin Microbiol* 44 (4):1359–66. doi: 10.1128/JCM.44.4.1359-1366.2006.

Brady, A., and S. L. Salzberg. 2009. "Phymm and PhymmBL: Metagenomic phylogenetic classification with interpolated Markov models." *Nat Methods* 6 (9):673–6. doi: 10.1038/nmeth.1358. BBTools Bioinformatics Tools, Including BBMap.

Brady, A., and S. L. Salzberg. 2011. "PhymmBL expanded: Confidence scores, custom databases, parallelization and more." *Nat Methods* 8 (5):367. doi: 10.1038/nmeth0511-367.

Bushnell, B., J. Rood, and E. Singer. 2017. "BBMerge–Accurate paired shotgun read merging via overlap." *PloS One* 12 (10):e0185056.

Byrd, A. L., J. F. Perez-Rogers, S. Manimaran, E. Castro-Nallar, I. Toma, T. McCaffrey, M. Siegel, G. Benson, K. A. Crandall, and W. E. Johnson. 2014. "Clinical PathoScope: Rapid alignment and filtration for accurate pathogen identification in clinical samples using unassembled sequencing data." *BMC Bioinformatics* 15:262. doi: 10.1186/1471-2105-15-262.

Callahan, B. J., P. J. McMurdie, and S. P. Holmes. 2017. "Exact sequence variants should replace operational taxonomic units in marker-gene data analysis." *ISME J* 11 (12):2639–43. doi: 10.1038/ismej.2017.119.

Callahan, B. J., P. J. McMurdie, M. J. Rosen, A. W. Han, A. J. Johnson, and S. P. Holmes. 2016. "DADA2: High-resolution sample inference from Illumina amplicon data." *Nat Methods* 13 (7):581–3. doi: 10.1038/nmeth.3869.

Caporaso, J. G., J. Kuczynski, J. Stombaugh, K. Bittinger, F. D. Bushman, E. K. Costello, N. Fierer, A. G. Pena, J. K. Goodrich, J. I. Gordon, G. A. Huttley, S. T. Kelley, D. Knights, J. E. Koenig, R. E. Ley, C. A. Lozupone, D. McDonald, B. D. Muegge, M. Pirrung, J. Reeder, J. R. Sevinsky, P. J. Turnbaugh, W. A. Walters, J. Widmann, T. Yatsunenko, J. Zaneveld, and R. Knight. 2010. "QIIME allows analysis of high-throughput community sequencing data." *Nat Methods* 7 (5):335–6. doi: 10.1038/nmeth.f.303.

Chen, J., and H. Li. 2013. "Variable selection for sparse Dirichlet-multinomial regression with an application to microbiome data analysis." *Ann Appl Stat* 7 (1). doi: 10.1214/12-AOAS592.

Chen, Y., H. Yao, E. J. Thompson, N. M. Tannir, J. N. Weinstein, and X. Su. 2013. "VirusSeq: Software to identify viruses and their integration sites using next-generation sequencing of human cancer tissue." *Bioinformatics* 29 (2):266–7. doi: 10.1093/bioinformatics/bts665.

Clarridge, J. E., 3rd. 2004. "Impact of 16S rRNA gene sequence analysis for identification of bacteria on clinical microbiology and infectious diseases." *Clin Microbiol Rev* 17 (4):840–62, table of contents. doi: 10.1128/CMR.17.4.840-862.2004.

Cole, J. R., Q. Wang, J. A. Fish, B. Chai, D. M. McGarrell, Y. Sun, C. T. Brown, A. Porras-Alfaro, C. R. Kuske, and J. M. Tiedje. 2014. "Ribosomal Database Project: Data and tools for high throughput rRNA analysis." *Nucleic Acids Res* 42 (Database issue):D633–42. doi: 10.1093/nar/gkt1244.

Cox, J. W., R. A. Ballweg, D. H. Taft, P. Velayutham, D. B. Haslam, and A. Porollo. 2017. "A fast and robust protocol for metataxonomic analysis using RNAseq data." *Microbiome* 5 (1):7. doi: 10.1186/s40168-016-0219-5.

Darling, A. E., G. Jospin, E. Lowe, F. A. Matsen, H. M. Bik, and J. A. Eisen. 2014. "PhyloSift: Phylogenetic analysis of genomes and metagenomes." *PeerJ* 2:e243. doi: 10.7717/peerj.243.

David, L. A., A. C. Materna, J. Friedman, M. I. Campos-Baptista, M. C. Blackburn, A. Perrotta, S. E. Erdman, and E. J. Alm. 2014. "Host lifestyle affects human microbiota on daily timescales." *Genome Biol* 15 (7):R89. doi: 10.1186/gb-2014-15-7-r89.

Delmont, T. O., C. Malandain, E. Prestat, C. Larose, J. M. Monier, P. Simonet, and T. M. Vogel. 2011. "Metagenomic mining for microbiologists." *ISME J* 5 (12):1837–43. doi: 10.1038/ismej.2011.61.

DeSantis, T. Z., P. Hugenholtz, N. Larsen, M. Rojas, E. L. Brodie, K. Keller, T. Huber, D. Dalevi, P. Hu, and G. L. Andersen. 2006. "Greengenes, a chimera-checked 16S rRNA gene database and workbench compatible with ARB." *Appl Environ Microbiol* 72 (7):5069–72. doi: 10.1128/AEM.03006-05.

Dhariwal, A., J. Chong, S. Habib, I. L. King, L. B. Agellon, and J. Xia. 2017. "MicrobiomeAnalyst: A web-based tool for comprehensive statistical, visual and meta-analysis of microbiome data." *Nucleic Acids Res* 45 (W1):W180–W188. doi: 10.1093/nar/gkx295.

Edgar, R. C. 2013. "UPARSE: Highly accurate OTU sequences from microbial amplicon reads." *Nat Methods* 10 (10):996–8. doi: 10.1038/nmeth.2604.

Edgar, R. C. 2018. "Updating the 97% identity threshold for 16S ribosomal RNA OTUs." *Bioinformatics* 34 (14):2371–5. doi: 10.1093/bioinformatics/bty113.

Edgar, R. C., and H. Flyvbjerg. 2015. "Error filtering, pair assembly and error correction for next-generation sequencing reads." *Bioinformatics* 31 (21):3476–82. doi: 10.1093/bioinformatics/btv401.

Edgar, R. C., B. J. Haas, J. C. Clemente, C. Quince, and R. Knight. 2011. "UCHIME improves sensitivity and speed of chimera detection." *Bioinformatics* 27 (16):2194–200. doi: 10.1093/bioinformatics/btr381.

Eren, A. M., O. C. Esen, C. Quince, J. H. Vineis, H. G. Morrison, M. L. Sogin, and T. O. Delmont. 2015. "Anvi'o: an advanced analysis and visualization platform for 'omics data." *PeerJ* 3:e1319. doi: 10.7717/peerj.1319.

Ewing, B., and P. Green. 1998. "Base-calling of automated sequencer traces using Phred. II. Error probabilities." *Genome Res* 8 (3):186–94.

Ewing, B., L. Hillier, M. C. Wendl, and P. Green. 1998. "Base-calling of automated sequencer traces using Phred. I. Accuracy assessment." *Genome Res* 8 (3):175–85. doi: 10.1101/gr.8.3.175.

Fosso, B., M. Santamaria, M. Marzano, D. Alonso-Alemany, G. Valiente, G. Donvito, A. Monaco, P. Notarangelo, and G. Pesole. 2015. "BioMaS: A modular pipeline for Bioinformatic analysis of Metagenomic AmpliconS." *BMC Bioinformatics* 16:203. doi: 10.1186/s12859-015-0595-z.

Fouts, D. E., L. Brinkac, E. Beck, J. Inman, and G. Sutton. 2012. "PanOCT: Automated clustering of orthologs using conserved gene neighborhood for pan-genomic analysis of bacterial strains and closely related species." *Nucleic Acids Res* 40 (22):e172. doi: 10.1093/nar/gks757.

Francis, O. E., M. Bendall, S. Manimaran, C. Hong, N. L. Clement, E. Castro-Nallar, Q. Snell, G. B. Schaalje, M. J. Clement, K. A. Crandall, and W. E. Johnson. 2013. "Pathoscope: Species identification and strain attribution with unassembled sequencing data." *Genome Res* 23 (10):1721–9. doi: 10.1101/gr.150151.112.

Franzosa, E. A., L. J. McIver, G. Rahnavard, L. R. Thompson, M. Schirmer, G. Weingart, K. S. Lipson, R. Knight, J. G. Caporaso, N. Segata, and C. Huttenhower. 2018. "Species-level functional profiling of metagenomes and metatranscriptomes." *Nat Methods* 15 (11):962–8. doi: 10.1038/s41592-018-0176-y.

Freitas, T. A., P. E. Li, M. B. Scholz, and P. S. Chain. 2015. "Accurate read-based metagenome characterization using a hierarchical suite of unique signatures." *Nucleic Acids Res* 43 (10):e69. doi: 10.1093/nar/gkv180.

Gilbert, J. A., J. K. Jansson, and R. Knight. 2018. "Earth microbiome project and global systems biology." *mSystems* 3 (3). doi: 10.1128/mSystems.00217-17.

Gregor, I., J. Droge, M. Schirmer, C. Quince, and A. C. McHardy. 2016. "PhyloPythiaS+: A self-training method for the rapid reconstruction of low-ranking taxonomic bins from metagenomes." *PeerJ* 4:e1603. doi: 10.7717/peerj.1603.

Group, Nih Hmp Working, J. Peterson, S. Garges, M. Giovanni, P. McInnes, L. Wang, J. A. Schloss, V. Bonazzi, J. E. McEwen, K. A. Wetterstrand, C. Deal, C. C. Baker, V. Di Francesco, T. K. Howcroft, R. W. Karp, R. D. Lunsford, C. R. Wellington, T. Belachew, M. Wright, C. Giblin, H. David, M. Mills, R. Salomon, C. Mullins, B. Akolkar, L. Begg, C. Davis, L. Grandison, M. Humble, J. Khalsa, A. R. Little, H. Peavy, C. Pontzer, M. Portnoy, M. H. Sayre, P. Starke-Reed, S. Zakhari, J. Read, B. Watson, and M. Guyer. 2009. "The NIH human microbiome project." *Genome Res* 19 (12):2317–23. doi: 10.1101/gr.096651.109.

Haas, B. J., D. Gevers, A. M. Earl, M. Feldgarden, D. V. Ward, G. Giannoukos, D. Ciulla, D. Tabbaa, S. K. Highlander, E. Sodergren, B. Methe, T. Z. DeSantis, Consortium Human Microbiome, J. F. Petrosino, R. Knight, and B. W. Birren. 2011. "Chimeric 16S rRNA sequence formation and detection in Sanger and 454-pyrosequenced PCR amplicons." *Genome Res* 21 (3):494–504. doi: 10.1101/gr.112730.110.

Hildebrand, F., R. Tadeo, A. Y. Voigt, P. Bork, and J. Raes. 2014. "LotuS: An efficient and user-friendly OTU processing pipeline." *Microbiome* 2 (1):30. doi: 10.1186/2049-2618-2-30.

Huber, T., G. Faulkner, and P. Hugenholtz. 2004. "Bellerophon: A program to detect chimeric sequences in multiple sequence alignments." *Bioinformatics* 20 (14):2317–9. doi: 10.1093/bioinformatics/bth226.

Huson, D. H., A. F. Auch, J. Qi, and S. C. Schuster. 2007. "MEGAN analysis of metagenomic data." *Genome Res* 17 (3):377–86. doi: 10.1101/gr.5969107.

Huson, D. H., S. Beier, I. Flade, A. Gorska, M. El-Hadidi, S. Mitra, H. J. Ruscheweyh, and R. Tappu. 2016. "MEGAN community edition—Interactive exploration and analysis of large-scale microbiome sequencing data." *PLoS Comput Biol* 12 (6):e1004957. doi: 10.1371/journal.pcbi.1004957.

Hyatt, D., G. L. Chen, P. F. Locascio, M. L. Land, F. W. Larimer, and L. J. Hauser. 2010. "Prodigal: Prokaryotic gene recognition and translation initiation site identification." *BMC Bioinformatics* 11:119. doi: 10.1186/1471-2105-11-119.

Janda, J. M., and S. L. Abbott. 2007. "16S rRNA gene sequencing for bacterial identification in the diagnostic laboratory: Pluses, perils, and pitfalls." *J Clin Microbiol* 45 (9):2761–4. doi: 10.1128/JCM.01228-07. A Sliding-Window, Adaptive, Quality-Based Trimming Tool for Fastq Files.

Joshi, N. A., and J. N. Fass. 2011. "Sickle: A sliding-window, adaptive, quality-based trimming tool for FastQ files (version 1.33)." https://github.com/najoshi/sickle.

Kashtan, N., S. E. Roggensack, S. Rodrigue, J. W. Thompson, S. J. Biller, A. Coe, H. Ding, P. Marttinen, R. R. Malmstrom, R. Stocker, M. J. Follows, R. Stepanauskas, and S. W. Chisholm. 2014. "Single-cell genomics reveals hundreds of coexisting subpopulations in wild prochlorococcus." *Science* 344 (6182):416–20. doi: 10.1126/science.1248575.

Koljalg, U., R. H. Nilsson, K. Abarenkov, L. Tedersoo, A. F. Taylor, M. Bahram, S. T. Bates, T. D. Bruns, J. Bengtsson-Palme, T. M. Callaghan, B. Douglas, T. Drenkhan, U. Eberhardt, M. Duenas, T. Grebenc, G. W. Griffith, M. Hartmann, P. M. Kirk, P. Kohout, E. Larsson, B. D. Lindahl, R. Lucking, M. P. Martin, P. B. Matheny, N. H. Nguyen, T. Niskanen, J. Oja, K. G. Peay, U. Peintner, M. Peterson, K. Poldmaa, L. Saag, I. Saar, A. Schussler, J. A. Scott, C. Senes, M. E. Smith, A. Suija, D. L. Taylor, M. T. Telleria, M. Weiss, and K. H. Larsson. 2013. "Towards a unified paradigm for sequence-based identification of fungi." *Mol Ecol* 22 (21):5271–7. doi: 10.1111/mec.12481.

Kostic, A. D., A. I. Ojesina, C. S. Pedamallu, J. Jung, R. G. Verhaak, G. Getz, and M. Meyerson. 2011. "PathSeq: Software to identify or discover microbes by deep sequencing of human tissue." *Nat Biotechnol* 29 (5):393–6. doi: 10.1038/nbt.1868.

Kunin, V., A. Engelbrektson, H. Ochman, and P. Hugenholtz. 2010. "Wrinkles in the rare biosphere: Pyrosequencing errors can lead to artificial inflation of diversity estimates." *Environ Microbiol* 12 (1):118–23. doi: 10.1111/j.1462-2920.2009.02051.x. microbiome R package.

Lahti, L., S. Shetty, T. Blake, and J. Salojarvi. 2012–2017. Microbiome R package. http://microbiome.github.io.

Langille, M. G., J. Zaneveld, J. G. Caporaso, D. McDonald, D. Knights, J. A. Reyes, J. C. Clemente, D. E. Burkepile, R. L. Vega Thurber, R. Knight, R. G. Beiko, and C. Huttenhower. 2013. "Predictive functional profiling of microbial communities using 16S rRNA marker gene sequences." *Nat Biotechnol* 31 (9):814–21. doi: 10.1038/nbt.2676.

Lee, M. D. 2019. "Happy belly bioinformatics: An open-source resource dedicated to helping biologists utilize bioinformatics." *J Open Source Ed*. doi: 10.21105/jose.00053.

Li, D., C. M. Liu, R. Luo, K. Sadakane, and T. W. Lam. 2015. "MEGAHIT: An ultra-fast single-node solution for large and complex metagenomics assembly via succinct de Bruijn graph." *Bioinformatics* 31 (10):1674–6. doi: 10.1093/bioinformatics/btv033.

Liu, B., J. Yuan, S. M. Yiu, Z. Li, Y. Xie, Y. Chen, Y. Shi, H. Zhang, Y. Li, T. W. Lam, and R. Luo. 2012. "COPE: An accurate k-mer-based pair-end reads connection tool to facilitate genome assembly." *Bioinformatics* 28 (22):2870–4. doi: 10.1093/bioinformatics/bts563.

Magoc, T., and S. L. Salzberg. 2011. "FLASH: Fast length adjustment of short reads to improve genome assemblies." *Bioinformatics* 27 (21):2957–63. doi: 10.1093/bioinformatics/btr507.

Marotz, C., A. Amir, G. Humphrey, J. Gaffney, G. Gogul, and R. Knight. 2017. "DNA extraction for streamlined metagenomics of diverse environmental samples." *Biotechniques* 62 (6):290–293. doi: 10.2144/000114559.

Martin, M. 2011. "Cutadapt removes adapter sequences from high-throughput sequencing reads." *EMBnet Journal* 17.

Masella, A. P., A. K. Bartram, J. M. Truszkowski, D. G. Brown, and J. D. Neufeld. 2012. "PANDAseq: Paired-end assembler for illumina sequences." *BMC Bioinformatics* 13:31. doi: 10.1186/1471-2105-13-31.

McMurdie, P. J., and S. Holmes. 2013. "phyloseq: An R package for reproducible interactive analysis and graphics of microbiome census data." *PLoS One* 8 (4):e61217. doi: 10.1371/journal.pone.0061217.

Meta, S. U. B. International Consortium. 2016. "The metagenomics and metadesign of the subways and urban biomes (MetaSUB) international consortium inaugural meeting report." *Microbiome* 4 (1):24. doi: 10.1186/s40168-016-0168-z.

Meyer, F., D. Paarmann, M. D'Souza, R. Olson, E. M. Glass, M. Kubal, T. Paczian, A. Rodriguez, R. Stevens, A. Wilke, J. Wilkening, and R. A. Edwards. 2008. "The metagenomics RAST server—A public resource for the automatic phylogenetic and functional analysis of metagenomes." *BMC Bioinformatics* 9:386. doi: 10.1186/1471-2105-9-386.

Mignard, S., and J. P. Flandrois. 2006. "16S rRNA sequencing in routine bacterial identification: A 30-month experiment." *J Microbiol Methods* 67 (3):574–81. doi: 10.1016/j.mimet.2006.05.009.

Mitchell, K., C. Dao, A. Freise, S. Mangul, and J. M. Parker. 2018. "PUMA: A tool for processing 16S rRNA taxonomy data for analysis and visualization." *bioRxiv*:482380.

Mpangase, P. T., J. Frost, M. Ramsay, and S. Hazelhurst. 2020. "nf-rnaSeqMetagen: A nextflow metagenomics pipeline for identifying and characterizing microbial sequences from RNA-seq data." *Med Microecol* 4.

Naccache, S. N., S. Federman, N. Veeraraghavan, M. Zaharia, D. Lee, E. Samayoa, J. Bouquet, A. L. Greninger, K. C. Luk, B. Enge, D. A. Wadford, S. L. Messenger, G. L. Genrich, K. Pellegrino, G. Grard, E. Leroy, B. S. Schneider, J. N. Fair, M. A. Martinez, P. Isa, J. A. Crump, J. L. DeRisi, T. Sittler, J. Hackett, Jr., S. Miller, and C. Y. Chiu. 2014. "A cloud-compatible bioinformatics pipeline for ultrarapid pathogen identification from next-generation sequencing of clinical samples." *Genome Res* 24 (7):1180–92. doi: 10.1101/gr.171934.113.

Nayfach, S., and K. S. Pollard. 2016. "Toward accurate and quantitative comparative metagenomics." *Cell* 166 (5):1103–1116. doi: 10.1016/j.cell.2016.08.007.

Nguyen, N. P., T. Warnow, M. Pop, and B. White. 2016. "A perspective on 16S rRNA operational taxonomic unit clustering using sequence similarity." *NPJ Biofilms Microbiomes* 2:16004. doi: 10.1038/npjbiofilms.2016.4.

Nurk, S., D. Meleshko, A. Korobeynikov, and P. A. Pevzner. 2017. "metaSPAdes: A new versatile metagenomic assembler." *Genome Res* 27 (5):824–834. doi: 10.1101/gr.213959.116. Package 'Vegan'. Community Ecology Package 2.

Oksanen, J., F. G. Blanchet, R. Kindt, P. Legendre, P. R. Minchin, R. O'Hara, G. L. Simpson, P. Solymos, M. H. H. Stevens, E. Szoecs, and H. Wagner. 2013. "Package 'Vegan'. Community Ecology Package (Version 2)." http://CRAN.R-project.org/package=vegan.

Oude Munnink, B. B., D. F. Nieuwenhuijse, M. Stein, A. O'Toole, M. Haverkate, M. Mollers, S. K. Kamga, C. Schapendonk, M. Pronk, P. Lexmond, A. van der Linden, T. Bestebroer, I. Chestakova, R. J. Overmars, S. van Nieuwkoop, R. Molenkamp, A. A. van der Eijk, C. GeurtsvanKessel, H. Vennema, A. Meijer, A. Rambaut, J. van Dissel, R. S. Sikkema, A. Timen, M. Koopmans, and Team Dutch-Covid-19 response. 2020. "Rapid SARS-CoV-2 whole-genome sequencing and analysis for informed public health decision-making in the Netherlands." *Nat Med.* doi: 10.1038/s41591-020-0997-y.

Patil, K. R., L. Roune, and A. C. McHardy. 2012. "The PhyloPythiaS web server for taxonomic assignment of metagenome sequences." *PLoS One* 7 (6):e38581. doi: 10.1371/journal.pone.0038581.

Paulson, J. N., N. D. Olson, D. J. Braccia, J. Wagner, H. Talukder, M. Pop, and H. C. Bravo. 2013. "MetagenomeSeq: Statistical analysis for sparse high-throughput sequencing." http://www.cbcb.umd.edu/software/metagenomeSeq.

Peng, Y., H. C. Leung, S. M. Yiu, and F. Y. Chin. 2011. "Meta-IDBA: A de novo assembler for metagenomic data." *Bioinformatics* 27 (13):i94–101. doi: 10.1093/bioinformatics/btr216.

Prodan, A., V. Tremaroli, H. Brolin, A. H. Zwinderman, M. Nieuwdorp, and E. Levin. 2020. "Comparing bioinformatic pipelines for microbial 16S rRNA amplicon sequencing." *PLoS One* 15 (1):e0227434. doi: 10.1371/journal.pone.0227434.

Qin, J., R. Li, J. Raes, M. Arumugam, K. S. Burgdorf, C. Manichanh, T. Nielsen, N. Pons, F. Levenez, T. Yamada, D. R. Mende, J. Li, J. Xu, S. Li, D. Li, J. Cao, B. Wang, H. Liang, H. Zheng, Y. Xie, J. Tap, P. Lepage, M. Bertalan, J. M. Batto, T. Hansen, D. Le Paslier, A. Linneberg, H. B. Nielsen, E. Pelletier, P. Renault, T. Sicheritz-Ponten, K. Turner, H. Zhu, C. Yu, S. Li, M. Jian, Y. Zhou, Y. Li, X. Zhang, S. Li, N. Qin, H. Yang, J. Wang, S. Brunak, J. Dore, F. Guarner, K. Kristiansen, O. Pedersen, J. Parkhill, J. Weissenbach, H. I. T. Consortium Meta, P. Bork, S. D. Ehrlich, and J. Wang. 2010. "A human gut microbial gene catalogue established by metagenomic sequencing." *Nature* 464 (7285):59–65. doi: 10.1038/nature08821.

Quast, C., E. Pruesse, P. Yilmaz, J. Gerken, T. Schweer, P. Yarza, J. Peplies, and F. O. Glockner. 2013. "The SILVA ribosomal RNA gene database project: Improved data processing and web-based tools." *Nucleic Acids Res* 41 (Database issue):D590–6. doi: 10.1093/nar/gks1219.

Quince, C., A. Lanzen, T. P. Curtis, R. J. Davenport, N. Hall, I. M. Head, L. F. Read, and W. T. Sloan. 2009. "Accurate determination of microbial diversity from 454 pyrosequencing data." *Nat Methods* 6 (9):639–41. doi: 10.1038/nmeth.1361.

Ranjan, R., A. Rani, A. Metwally, H. S. McGee, and D. L. Perkins. 2016. "Analysis of the microbiome: Advantages of whole genome shotgun versus 16S amplicon sequencing." *Biochem Biophys Res Commun* 469 (4):967–77. doi: 10.1016/j.bbrc.2015.12.083.

Rausch, P., M. Ruhlemann, B. M. Hermes, S. Doms, T. Dagan, K. Dierking, H. Domin, S. Fraune, J. von Frieling, U. Hentschel, F. A. Heinsen, M. Hoppner, M. T. Jahn, C. Jaspers, K. A. B. Kissoyan, D. Langfeldt, A. Rehman, T. B. H. Reusch, T. Roeder, R. A. Schmitz, H. Schulenburg, R. Soluch, F. Sommer, E. Stukenbrock, N. Weiland-Brauer, P. Rosenstiel, A. Franke, T. Bosch, and J. F. Baines. 2019. "Comparative analysis of amplicon and metagenomic sequencing methods reveals key features in the evolution of animal metaorganisms." *Microbiome* 7 (1):133. doi: 10.1186/s40168-019-0743-1.

Rawat, A., D. M. Engelthaler, E. M. Driebe, P. Keim, and J. T. Foster. 2014. "MetaGeniE: Characterizing human clinical samples using deep metagenomic sequencing." *PLoS One* 9 (11):e110915. doi: 10.1371/journal.pone.0110915.

Rognes, T., T. Flouri, B. Nichols, C. Quince, and F. Mahe. 2016. "VSEARCH: A versatile open source tool for metagenomics." *PeerJ* 4:e2584. doi: 10.7717/peerj.2584.

Schloss, P. D. 2020. "Reintroducing Mothur: 10 years later." *Appl Environ Microbiol* 86 (2). doi: 10.1128/AEM.02343-19.

Schloss, P. D., S. L. Westcott, T. Ryabin, J. R. Hall, M. Hartmann, E. B. Hollister, R. A. Lesniewski, B. B. Oakley, D. H. Parks, C. J. Robinson, J. W. Sahl, B. Stres, G. G. Thallinger, D. J. Van Horn, and C. F. Weber. 2009. "Introducing Mothur: Open-source, platform-independent, community-supported software for describing and comparing microbial communities." *Appl Environ Microbiol* 75 (23):7537–41. doi: 10.1128/AEM.01541-09.

Schmieder, R., and R. Edwards. 2011. "Quality control and preprocessing of metagenomic datasets." *Bioinformatics* 27 (6):863–4. doi: 10.1093/bioinformatics/btr026.

Scholz, M., D. V. Ward, E. Pasolli, T. Tolio, M. Zolfo, F. Asnicar, D. T. Truong, A. Tett, A. L. Morrow, and N. Segata. 2016. "Strain-level microbial epidemiology and population genomics from shotgun metagenomics." *Nat Methods* 13 (5):435–8. doi: 10.1038/nmeth.3802.

Sedlazeck, F. J., H. Lee, C. A. Darby, and M. C. Schatz. 2018. "Piercing the dark matter: Bioinformatics of long-range sequencing and mapping." *Nat Rev Genet* 19 (6):329–346. doi: 10.1038/s41576-018-0003-4.

Seemann, T. 2014. "Prokka: Rapid prokaryotic genome annotation." *Bioinformatics* 30 (14):2068–9. doi: 10.1093/bioinformatics/btu153.

Tessler, M., J. S. Neumann, E. Afshinnekoo, M. Pineda, R. Hersch, L. F. M. Velho, B. T. Segovia, F. A. Lansac-Toha, M. Lemke, R. DeSalle, C. E. Mason, and M. R. Brugler. 2017. "Large-scale differences in microbial biodiversity discovery between 16S amplicon and shotgun sequencing." *Sci Rep* 7 (1):6589. doi: 10.1038/s41598-017-06665-3.

Tighe, S., E. Afshinnekoo, T. M. Rock, K. McGrath, N. Alexander, A. McIntyre, S. Ahsanuddin, D. Bezdan, S. J. Green, S. Joye, S. Stewart Johnson, D. A. Baldwin, N. Bivens, N. Ajami, J. R. Carmical, I. C. Herriott, R. Colwell, M. Donia, J. Foox, N. Greenfield, T. Hunter, J. Hoffman, J. Hyman, E. Jorgensen, D. Krawczyk, J. Lee, S. Levy, N. Garcia-Reyero, M. Settles, K. Thomas, F. Gomez, L. Schriml, N. Kyrpides, E. Zaikova, J. Penterman, and C. E. Mason. 2017. "Genomic methods and microbiological technologies for profiling novel and extreme environments for the extreme microbiome project (XMP)." *J Biomol Tech* 28 (1):31–39. doi: 10.7171/jbt.17-2801-004.

Truong, D. T., E. A. Franzosa, T. L. Tickle, M. Scholz, G. Weingart, E. Pasolli, A. Tett, C. Huttenhower, and N. Segata. 2015. "MetaPhlAn2 for enhanced metagenomic taxonomic profiling." *Nat Methods* 12 (10):902–3. doi: 10.1038/nmeth.3589.

Turnbaugh, P. J., R. E. Ley, M. Hamady, C. M. Fraser-Liggett, R. Knight, and J. I. Gordon. 2007. "The human microbiome project." *Nature* 449 (7164):804–10. doi: 10.1038/nature06244.

van Bakel, H., C. Nislow, B. J. Blencowe, and T. R. Hughes. 2010. "Most 'dark matter' transcripts are associated with known genes." *PLoS Biol* 8 (5):e1000371. doi: 10.1371/journal.pbio.1000371.

Wagner, J., F. Chelaru, J. Kancherla, J. N. Paulson, A. Zhang, V. Felix, A. Mahurkar, N. Elmqvist, and H. Corrada Bravo. 2018. "Metaviz: Interactive statistical and visual analysis of metagenomic data." *Nucleic Acids Res* 46 (6):2777–2787. doi: 10.1093/nar/gky136.

Wang, Q., P. Jia, and Z. Zhao. 2013. "VirusFinder: Software for efficient and accurate detection of viruses and their integration sites in host genomes through next generation sequencing data." *PLoS One* 8 (5):e64465. doi: 10.1371/journal.pone.0064465.

Wang, Y., H. C. Leung, S. M. Yiu, and F. Y. Chin. 2012a. "MetaCluster 4.0: A novel binning algorithm for NGS reads and huge number of species." *J Comput Biol* 19 (2):241–9. doi: 10.1089/cmb.2011.0276.

Wang, Y., H. C. Leung, S. M. Yiu, and F. Y. Chin. 2012b. "MetaCluster 5.0: A two-round binning approach for metagenomic data for low-abundance species in a noisy sample." *Bioinformatics* 28 (18):i356–i362. doi: 10.1093/bioinformatics/bts397.

Weiss, S., Z. Z. Xu, S. Peddada, A. Amir, K. Bittinger, A. Gonzalez, C. Lozupone, J. R. Zaneveld, Y. Vazquez-Baeza, A. Birmingham, E. R. Hyde, and R. Knight. 2017. "Normalization and microbial differential abundance strategies depend upon data characteristics." *Microbiome* 5 (1):27. doi: 10.1186/s40168-017-0237-y.

Westcott, S. L., and P. D. Schloss. 2015. "De novo clustering methods outperform reference-based methods for assigning 16S rRNA gene sequences to operational taxonomic units." *PeerJ* 3:e1487. doi: 10.7717/peerj.1487.

Wood, D. E., and S. L. Salzberg. 2014. "Kraken: Ultrafast metagenomic sequence classification using exact alignments." *Genome Biol* 15 (3):R46. doi: 10.1186/gb-2014-15-3-r46.

Wooley, J. C., and Y. Ye. 2009. "Metagenomics: Facts and artifacts, and computational challenges." *J Comput Sci Technol* 25 (1):71–81. doi: 10.1007/s11390-010-9306-4.

Wu, Y. W., B. A. Simmons, and S. W. Singer. 2016. "MaxBin 2.0: An automated binning algorithm to recover genomes from multiple metagenomic datasets." *Bioinformatics* 32 (4):605–7. doi: 10.1093/bioinformatics/btv638.

Wu, Y. W., Y. H. Tang, S. G. Tringe, B. A. Simmons, and S. W. Singer. 2014. "MaxBin: An automated binning method to recover individual genomes from metagenomes using an expectation-maximization algorithm." *Microbiome* 2:26. doi: 10.1186/2049-2618-2-26.

Youssef, N., C. S. Sheik, L. R. Krumholz, F. Z. Najar, B. A. Roe, and M. S. Elshahed. 2009. "Comparison of species richness estimates obtained using nearly complete fragments and simulated pyrosequencing-generated fragments in 16S rRNA gene-based environmental surveys." *Appl Environ Microbiol* 75 (16):5227–36. doi: 10.1128/AEM.00592-09.

Zakrzewski, M., C. Proietti, J. J. Ellis, S. Hasan, M. J. Brion, B. Berger, and L. Krause. 2017. "Calypso: A user-friendly web-server for mining and visualizing microbiome-environment interactions." *Bioinformatics* 33 (5):782–783. doi: 10.1093/bioinformatics/btw725.

Zhang, J., K. Kobert, T. Flouri, and A. Stamatakis. 2014. "PEAR: A fast and accurate Illumina paired-end reAd mergeR." *Bioinformatics* 30 (5):614–20. doi: 10.1093/bioinformatics/btt593.

Section II

Metagenomics Tools to Access Microbial Diversity

2

Metagenomic Tools for Taxonomic and Functional Annotation

Yordanis Pérez-Llano*, Ramon Alberto Batista-García, María del Rayo Sánchez-Carbente, Jorge Luis Folch-Mallol
Universidad Autónoma del Estado de Morelos, México
Sara Cuadros-Orellana
Universidad Católica del Maule, Chile
Alma Delia Nicolás-Morales, Natividad Castro-Alarcón,
Universidad Autónoma de Guerrero, México

CONTENTS

2.1 Introduction

The rise of next-generation sequencing (NGS) technologies and their application to describe microbial communities have reshaped our understanding of biology. Metagenomics allowed the uncovering of a microscopic universe that is not only part of our environment but also a key player in ecosystem interactions that were until recently elusive. Microbes are involved in processes ranging from worldwide geochemical carbon cycling to changing human physiology and behavior. Metagenomic analysis of environments such as deep-sea or extreme ecosystems has significantly expanded and restructured the tree of life, rewriting our current knowledge about the evolution of life on Earth (Parks et al. 2019). Though it is a powerful technology, it is not free of pitfalls and limitations that arise both from the technical processing of samples and the computational analysis of the large amount of data it generates.

Traditionally, metagenomics comprises a group of analytical and microbiological techniques that allow the identification of the microorganisms present in an environment and their relative abundance (Fosso et al. 2018). Although functional screening of environmental DNA libraries has also been considered as a metagenomic approach, currently the term is preferentially applied to the analysis of datasets obtained from next-generation sequencing platforms. There are two major technological approaches to obtaining metagenomic data: amplicon-based sequencing (or metabarcoding) and whole-genome shotgun (WGS) sequencing. In the former, the DNA extracted from the biological sample is used as a

DOI: 10.1201/9781003042570-4

template for the amplification of marker genes. The amplicons are then sequenced by NGS technology, allowing, for example, identification of the individual members of a microbial community by phylogeny-based approaches. Shotgun metagenomics, on the contrary, relies on the untargeted sequencing of all available genome fragments after DNA extraction from the biological sample. A pivotal procedure in understanding the type of data generated by both technologies is sequence annotation. In the field of 'omics' bioinformatics, annotation is the process of finding biologically relevant features to genomic elements, and it consists of gene prediction and taxonomy or function assignment.

Other steps, such as sequence quality control and assembly, usually precede the annotation of genes in genomes and metagenomes. Analyzing microbial community datasets follows the same general principles of genome analysis. For instance, sequence assembly (often described with the analogy of putting together the pieces of a puzzle) is a general process of concatenation of overlapping fragmented sequences. For some time, it was intuitive to think that taxonomic and functional assignment profited from having the full gene or even several contiguous genes (as in contigs) rather than just short reads. However, most current methods are assembly indifferent and instead rely on the direct annotation of raw reads.

In this chapter, we intend to cover recent advances in the annotation of metagenomic sequencing data. The aim is not to present an exhaustive description of all available tools, as this is a very dynamic field. We also do not intend to make tool recommendations, as the best tool to use probably depends on the sample and data characteristics. Instead, this chapter aims to review some of the general strategies and methods used for metagenome taxonomic and functional annotation and their strengths and limitations and refer our readers to recently published benchmarking studies for the selection of tools that are appropriate for their research purpose.

2.2 General Methods for Taxonomic Annotation of Sequencing Data

Taxonomic annotation methods essentially aim to solve a classification problem in which the raw or processed sequencing data must be categorized using a reference taxonomy. This generally implies that the elements of the query set (sample data) must be matched to labeled elements on a reference database. Although it might seem simple, it becomes a cumbersome task because the reference databases are generally incomplete and contain sparse or even skewed data (Balvočiute and Huson 2017). Another particularity of this classification problem is that the biological taxonomy system follows a hierarchical structure by which the organisms are grouped into different ranks or taxa. Hence, the classification can be performed at different depths on the hierarchy, and in doing so, the classification will be more accurate for high-ranking taxa (e.g., kingdom or phylum) than for low-ranking ones (e.g., genus or species). To further complicate the scenario, the discovery of new microorganisms and the generation of new phylogenomic data impose changes in the reference taxonomy. This often leads to specimens being renamed to accommodate the new taxonomic system. The available databases should be updated every time this renaming occurs, but in practice, this is generally not the case. Therefore, achieving a highly confident annotation depends on the target taxonomy rank and relies on the matching algorithm and the reference database quality.

Figure 2.1 shows the common annotation pipelines that are followed for amplicon-based and shotgun sequencing data from metagenomic samples. The raw metagenomic data consist of many relatively short DNA fragment sequences, called reads. Whether these reads come from a metabarcoding or a WGS experiment, the annotation can be performed on the raw sequence reads or assembled contigs. The assembly step is often resource intensive, especially in terms of memory usage, though modern assemblers have somewhat reduced this limitation. Assembly tools use various algorithms and heuristics that may produce different results and thus directly affect taxonomy annotation. Probably due to these issues, assembly is not always performed, and alternatively, the taxonomic annotation is obtained directly from the raw reads, especially if the aim is to characterize the taxonomic composition of the microbiota (Tamames, Cobo-Simón, and Puente-Sánchez 2019). For some other purposes, such as functional annotation or genome reconstruction, assembly could be a required step.

Another common stage in data processing is the clustering of elements of the same apparent genomic origin. In the case of amplicon-based sequencing, the goal is to identify either the reads derived from the

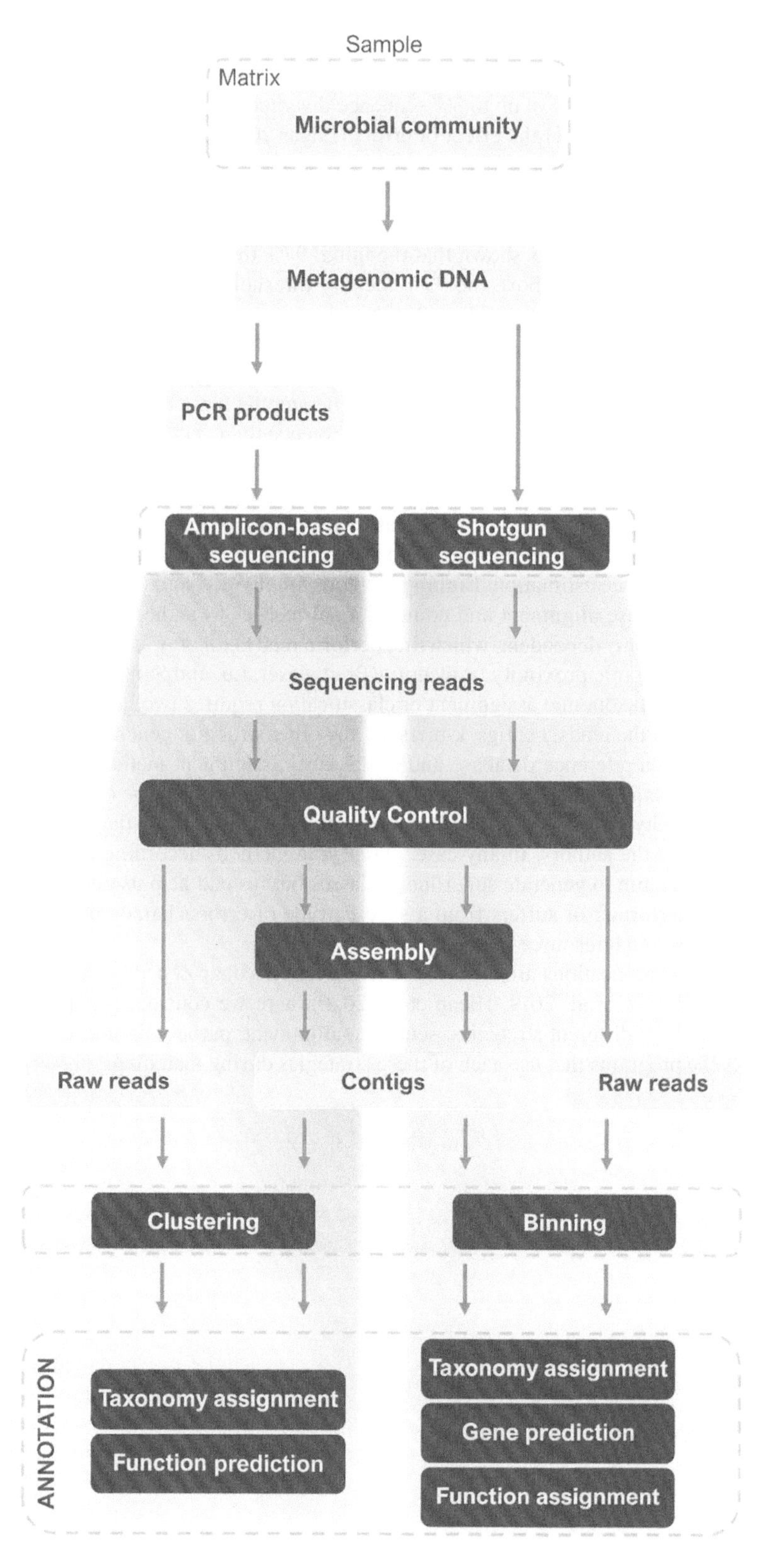

FIGURE 2.1 General workflows of metagenome taxonomic annotation of amplicon-based sequencing and WGS metagenomics.

same amplicon or from different amplicons. This identification is hampered by sequencing errors, which can be misinterpreted as biological variation. To overcome this sequencing platform limitation, assembled contigs are clustered if they have up to 3% sequence divergence (calculated for 16S rRNA variable region fragments), aiming to cancel the effect of errors (Edgar 2018b). These clusters of sequences are known as operational taxonomic units (OTUs) and are represented by only one sequence in the ensuing annotation and post-processing steps. The 'operational' part in the term emphasizes the notion that these clusters are not an accurate representation of the biological variation in the sample. Current evidence coming from assembled genomes has shown that the initial 97% threshold does not adequately capture species variation, as this must be above the 99% identity threshold for the entire 16S rRNA gene to achieve accurate species classification (Edgar 2018b). More recent methods (e.g., DADA2, Callahan et al. 2016; Deblur, Amir et al. 2017) take into consideration the lower error rates in current sequencing platforms and perform a clustering (sometimes referred to as denoising) that aims to differentiate technical variation from biological variation. The clusters generated by these methods, known as amplicon sequence variants (ASVs), tend to have less sequence divergence than OTUs.

The equivalent clustering step for WGS data is known as binning, that is, separating reads or contigs into bins that should ideally contain all the elements derived from a single genome. Instead, in practice, the bins usually contain sequences from closely related strains or species within a community. Binning might serve various purposes, such as genome reconstruction, functional classification of microbial entities, and microbial abundance estimation. Binning methods are divided into the ones that are taxonomy independent, that is, that use alignment and compositional metrics to gather all elements with shared characteristics, and taxonomy dependent, which first perform read taxonomy assignment and then cluster sequences based on taxonomic proximity or identity (Breitwieser, Lu, and Salzberg 2018).

As initially discussed, taxonomic assignment or classification requires two general steps: 1) a matching step that implies that the reads, contigs, k-mers, or any other structure generated from these must be matched to elements on a reference database and 2) the label assignment method based on the matching results. For some methods, discerning between the two processes might be challenging, sometimes because of the complexity of the methods or because both processes are not delineated in the method description provided by the authors. In any case, classifying methods according to their internal logic or functioning does not aim to generate superfluous classes but instead help us determine if a group of methods is the best performer or suffers from a specific type of errors/characteristics (e.g., high false positive rate, low recall, high resource consumption).

Although different classifications are used in the literature (Escobar-Zepeda et al. 2018; Breitwieser, Lu, and Salzberg 2018; Ye et al. 2019; Hleap et al. 2020), here we consider that the matching algorithms are based on three different strategies: sequence alignment, pseudoalignment, and composition. Figure 2.2 shows the programs that use each of these strategies during matching.

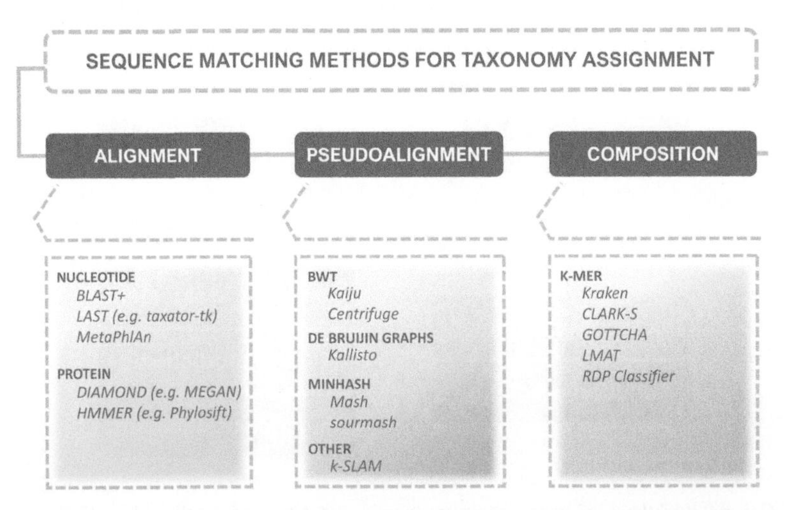

FIGURE 2.2 Types of sequence-matching algorithms used by different tools for metagenome taxonomy assignment.

Alignment-based matching algorithms perform local or global alignment of reads to a reference database. The alignment can be performed using nucleotide sequences against nucleotide databases or translated sequences against protein databases. BLAST is among the best performers in terms of accuracy and is the best-known alignment-based algorithm. The major drawback of these methods is that the alignment is resource intensive, so all the tools that rely on them are burdened with high memory and time consumption.

Once the alignment is performed, taxonomy assignment occurs by several methods. In the case of BLAST (or MegaBLAST, the most-used version in metagenomics), assignment is achieved by the best scoring hit of the sequence search. Other methods like MEGAN, which also uses BLAST, perform the assignment by a method called lowest common ancestor (LCA) (Huson et al. 2011). The LCA taxonomy assignment method is currently used by several tools, independent of what matching algorithm they use. The idea behind this method is to assign the most confident label based on annotations placed on distinct levels of the taxonomical hierarchy. In the case of MEGAN, the alignment of sequences is performed against various databases, and the taxonomic label is that of the lower-ranked taxa that unambiguously accommodates the annotations obtained from all databases (Huson et al. 2011). A third algorithm (or metric) to assign taxonomic labels is the average nucleotide identity (ANI), a measure of overall genome relatedness that is inspired in the DNA-to-DNA hybridization technique. As ANI is more related to complete genomes, its application is focused on the taxonomic annotation of metagenome-assembled genomes (MAGs) or metagenomic bins. A recent implementation of this algorithm is the tool OrthoANI, in which the ANI is calculated from the BLAST alignment of orthologous gene sequences from different genomes (Lee et al. 2016). The currently accepted boundary that delineates species is around 95–96% ANI. This method performs well for species classification where closely related species are available in the database, whereas it underperforms when comparing genomes from different genera (Lee et al. 2016; Yoon et al. 2017; Jain et al. 2018).

The pseudo-alignment term was used by the authors of kallisto to describe their method of matching reads to genomes based on k-mers represented in de Bruijn graphs (Yi et al. 2018; Bray et al. 2016). Here we chose the same term in a different context to include all the methods that perform taxonomic classification based on an alignment conducted in a space other than the sequence space of the original reads or their direct translation into peptides. In this group of methods, the reads or contigs are generally transformed into a different space using some form of convolution process, achieving a new way to represent the sequences. These methods aim to take advantage of the properties of the new space to optimize the matching step by one parameter (e.g., speed, RAM requirement, database storage, etc.). The most-used method for sequence transformation is the Burrow-Wheelers transform (BWT), a compression system that speeds up the matching to a reference database and at the same time reduces the database storage space. Some tools that use BWT in their algorithm are Kaiju (Menzel, Ng, and Krogh 2016) and Centrifuge (D. Kim et al. 2016).

Sequence composition methods are generally based on k-mer composition, although some methods might use k-mers to perform alignment or pseudoalignment as a primary matching algorithm, such as in the case of kallisto. In k-mer composition algorithms, the reads or contigs are split into small fragments of defined length (called k-mers) that are mapped by alignment to a reference database. The matching is performed by assessing the k-mer composition of the query sequence against the defined k-mer composition in the database. In some tools, this matching is not performed explicitly, as the taxonomy assignment occurs by using this composition information instead. For example, tools like the RDP classifier use probabilistic methods (in this case, naïve Bayes classifier) to determine the probability that a query sequence belongs to a set of known sequences within a genus (Wang et al. 2007; Bacci et al. 2015).

The current database growth due to metagenome-derived genomes, particularly in the case of RefSeq, has led to an expansion in species and genera that has reshaped taxonomical structures (Nasko et al. 2018). One direct repercussion of this growth is the reduction in the species classification accuracy of k-mer based methods, especially when several closely related genomes are found in the database (Nasko et al. 2018). Another confounding factor for species classification is the poorly defined species boundaries in taxa where high horizontal gene transfers occur (Nasko et al. 2018). This implies that, in the foreseeable future, k-mer–based methods will be increasingly prone to false positives for the classification of lower-ranked taxa (e.g., species or strain), so other alternative methods (even if more resource intensive) should be adopted.

2.3 Benchmarking Studies of Metagenomic Annotation Software

The pursuit of optimal methods for metagenomic analysis has resulted in an explosion of tools, all oriented to accomplish the same set of tasks but pledging to outperform their predecessors by some metric (Marx 2020). New tools are generally evaluated within a specific context or with a particular data type or application in mind. Assessing whether these tools can be applied broadly and how they compare to alternative software is generally carried out in so-called benchmarking studies. When conducted by relatively neutral authors (e.g., they are not involved in the development of tools evaluated in the study) and using appropriate test data and metrics, these studies are of vital importance to guide software users in the selection of the most accurate and/or fast tools for a specific task.

Referring to benchmarking studies for pipeline selection should be a common practice for software users and bioinformaticians, and we highly encourage it. A detailed discussion of the many reasons used by researchers to select bioinformatic tools can be found in (Gardner et al. 2017). Among these reasons, we can find the notions that:

1. Recently published software should be better or have improvements over older software.
2. Highly cited tools are widely accepted by the community (including potential reviewers of your work) and therefore more desirable.
3. The reputation of the authors or the journal is a guarantee of the quality of the tool.
4. Software tools often trade accuracy for speed, and therefore slower software should be more accurate.

Unfortunately, researchers might select tools only because they are user friendly or extensively documented. By correlating software accuracy with speed, age (i.e., to test the effect of recency), citation number (to test the wide use of the tools), and commonly accepted merit indexes such as journal impact or author reputation (H-index), Gardner et al. (2017) demonstrated that these metrics are not reliable predictors of accuracy. Basing research decisions on these widely accepted preconceptions could affect the reliability of our results and should therefore be avoided.

Systematic and standardized guidelines for benchmarking omics tools have recently been proposed (Mangul et al. 2019). Ensuring a scientifically rigorous comparison of different tools requires a gold standard dataset that serves as ground truth and a general set of metrics that can score the performance of any given method. Thus, as gold standard tools are not yet available for many bioinformatics applications, resorting to benchmarking studies is highly recommended for the selection of appropriate tools that fit our experimental setup and data.

Several benchmark studies assess the taxonomic annotation accuracy and recall of bioinformatics tools or pipelines (Bazinet and Cummings 2012; Peabody et al. 2015; Lindgreen, Adair, and Gardner 2016; Siegwald et al. 2017; McIntyre et al. 2017; Sczyrba et al. 2017; Almeida et al. 2018; Escobar-Zepeda et al. 2018; Gardner et al. 2019; Ye et al. 2019; Velsko et al. 2018). Some of these also evaluate the effect of databases on classification accuracy and recall (Escobar-Zepeda et al. 2018; Velsko et al. 2018). There are even benchmarking studies oriented to evaluate tools for specific tasks or datatypes, such as in rumen microbiome analysis (López-García et al. 2018) or ancient microbiome samples (Velsko et al. 2018). Some of these studies will be covered further in this chapter.

2.4 Tools for Taxonomic Annotation

The study of the microbial diversity of different natural environments can be carried out by targeting specific genes that can inform us about the taxonomic composition of the ecosystem. The 16S ribosomal RNA (rRNA) gene is commonly used for diversity analysis of Bacteria and Archaea communities, as it is present in all members of both domains and contains enough variability (up to nine hypervariable regions) to make it a very reliable marker for genus and species identification (Yang, Wang, and Qian 2016). For eukaryotes, the internal transcribed spacer (ITS), the 18S rRNA, and 28S rRNA genes are used.

Although marker-based metagenomic sequencing is still a standard procedure for the taxonomical description of microbial communities, various studies have shown that this technique has biases that can be alleviated by WGS sequencing (Khachatryan et al. 2020). Although WGS experiments are considerably more expensive and require computationally intensive analyses, higher accuracy in terms of identified taxa and abundance estimation is achieved by this methodology (Khachatryan et al. 2020). An affordable yet more accurate alternative is to perform experiments with little sequencing effort (approximately 0.5 million sequences for gut microbiomes), which are known as shallow sequencing, and that have been shown to achieve similar diversity estimation to ultradeep WGS at a price comparable to marker-based sequencing experiments (Hillmann et al. 2018). Shallow- or moderate-depth shotgun sequencing may be used by researchers to obtain species-level taxonomic and functional data at approximately the same cost as amplicon sequencing.

With the increasing size of metagenomic projects, biological databases are also growing exponentially in size. Under this scenario, the computational cost of sequence alignment needs to be considered. The development of faster tools, such as UBLAST and USEARCH (Edgar 2010), LAST (Kiełbasa et al. 2011), RAPSearch2 (Zhao, Tang, and Ye 2012), and DIAMOND (Buchfink, Xie, and Huson 2015), to cite a few, represents a new trend in bioinformatics.

The USEARCH algorithm is mostly used for high-identity searches. Conversely, UBLAST is used to search for more divergent protein or translated nucleotide sequences. LAST improves seed-and-extend heuristic methods by using an adaptive seed. RAPSearch (reduced alphabet based protein similarity search) uses an optimized suffix array data structure to accelerate the identification of alignment seeds, together with multi-thread modes (available in RAPSearch2 only) to further speed the process.

Among these tools, DIAMOND is probably the fastest one. It aligns short reads to the complete National Center for Biotechnology Information (NCBI)'s nr protein database. For this alignment, both the database and the query sequences are indexed, making the matching process more computationally feasible (Buchfink, Xie, and Huson 2015). Like BLASTX, DIAMOND is an 'all mapper' that attempts to exhaustively determine all significant alignments for the query (Buchfink, Xie, and Huson 2015). Nevertheless, it outperforms BLASTX by an impressive 20,000 times on short reads while maintaining a similar degree of sensitivity.

For the alignment of long reads, the most-used method is LAST, which allows a quick and sensitive comparison of sequences with an arbitrarily non-uniform composition. The LAST aligner is tolerant to single-base insertions and deletions, therefore outperforming non-gapped aligners (Kiełbasa et al. 2011).

Kraken uses the k-mer composition and LCA methods to perform taxonomy classification. Initially, a reference database is constructed by finding unique k-mers within genomic sequences and assigning a taxid (taxonomy label) based on the lowest common ancestor that shares that k-mer.

A meta-analysis of classifier benchmarking studies revealed that, when performed without strict guidelines, these studies can lead to contradictory results (Gardner et al. 2019). Part of the findings of this meta-analysis are shown in Figure 2.3. The first issue raised by the authors is that many of the tools available for taxonomic classification have not been independently evaluated (see Figure 2.3a). The reasons for this are many, but the authors underscore as more likely causes that the unevaluated tools are the most recently published, are no longer available/functional, or provide results that are not suited for comparison (Gardner et al. 2019). One other cause could be the popularity of the tools, but among the unevaluated set are some highly cited tools (e.g., UPARSE, PathSeq, Phylosift), which can be identified by the black labels in Figure 2.3b. In any case, this hidden group may contain promising tools that did not make it to the spotlight.

A more concerning observation in (Gardner et al. 2019) was that the number of citations per year does not correlate with the tool accuracy estimates, with less accurate tools being highly cited (see Figure 2.3c). To normalize comparison across different studies, these authors transformed the reported F-measures (a measure of accuracy) into a Z-score metric that eliminated the methodological and data differences. According to this metric, less accurate tools score negatively, while a positive score is given to those tools that consistently have a higher F-measure. As can be observed in Figure 2.3c, the classifiers in tools such as QIIME and mothur are highly used but are significantly less accurate than other tools such as Kraken, NBC, CLARK, or OneCodex (Gardner et al. 2019).

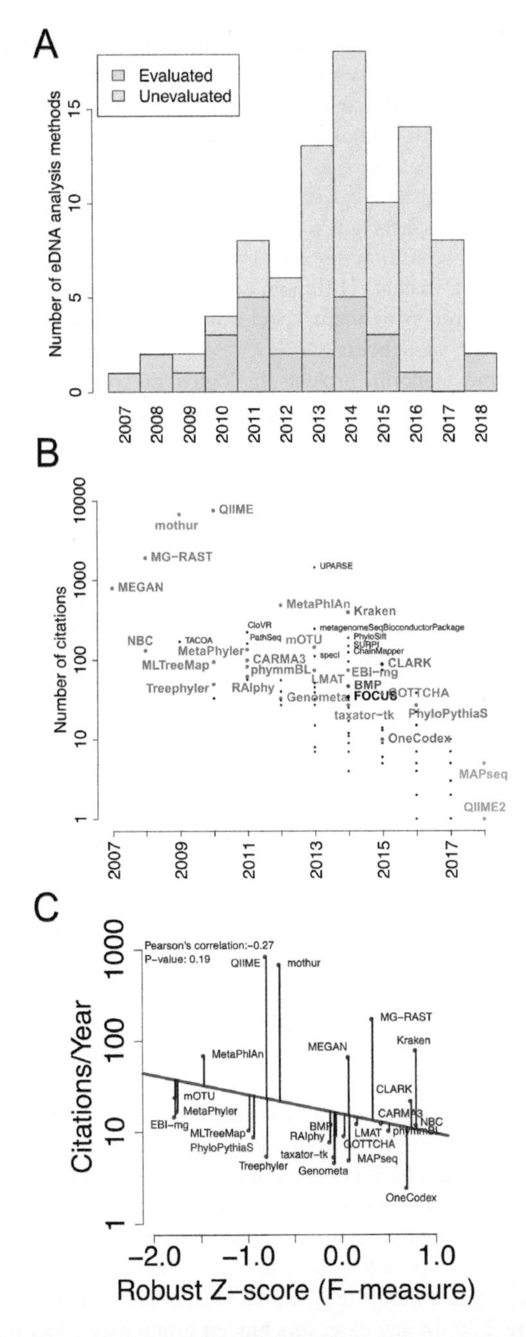

FIGURE 2.3 Analysis of software tool benchmarking studies, publication date, and citations as part of a meta-analysis of software accuracy conducted by Gardner et al. (2019). The results show (a) many of the available tools have not been independently evaluated as part of a benchmark study; (b) the number of citations for each software tool versus the year it was published, where tools that have been evaluated are colored and labeled, while highly cited tools that have not been independently evaluated are labeled in black, and (c) the number of citations per year does not correlate with the tool accuracy estimates, with less accurate tools being highly cited. (Modified from Gardner et al.[2019] with permission.)

A recent comprehensive benchmarking of metagenomic sequence classifiers was conducted to assess the species-level classification precision and recall of 20 different tools using a uniform database (Ye et al. 2019). This study considers tools that were not previously evaluated, such as PathSeq, MMseqs2, taxMaps, and mOTUs2. The authors of this study make a distinction of methods according to the type of sequence they handle during the matching algorithm. Hence, they assess DNA-to-DNA classifiers such as Kraken, k-SLAM, MegaBLAST, metaOthello, CLARK, GOTTCHA, taxMap, prophyle, PathSeq, Centrifuge, and Karp. DIAMOND, Kaiju, and MMseqs2 were evaluated as DNA-to-protein classifiers, and marker-based classifiers were MetaPhlAn2 and mOTUs2. All the tools were executed with default parameters as would be used by the average user of these packages. The precision of the abundance estimation of each of the methods was also assessed. Bracken, one of the tested tools, is an add-on to Kraken that allows more accurate abundance estimation, which is originally not performed by Kraken. According to their results, and in agreement with previous findings, the marker-based methods significantly underperformed compared to the shotgun-based methods. From the rest of the tested methods, Centrifuge, MegaBLAST, PathSeq, prophyle, DIAMOND, Kaiju, and MMseqs2 performed poorly in one or several metrics compared to Kraken, Kraken2, KrakenUniq, Bracken, CLARK, CLARK-S, k-SLAM, and taxMaps. From the latter methods, given that precision, recall, and abundance estimation are fairly similar, the selection for the optimal tool relies on the computing performance (i.e., CPU and memory requirements, execution time). For this reason, the combination of Kraken2/Bracken seems to be the most attractive among the tested classifiers, with execution time around 1 minute and memory consumption of ~36 Gb for Kraken2 and less than 1 Gb for Bracken (Ye et al. 2019).

As new tools or improvements on existing tools are likely to emerge in the future and no standardize pipeline applicable to most case scenarios can be envisioned, benchmarking studies will remain necessary. The Live Evaluation of Computational Methods for the Metagenome Investigation (LEMMI) tool has deployed a platform for the automatic benchmarking of taxonomic classification software (Seppey, Manni, and Zdobnov 2020). Results obtained from the assessment of the more popular tools can be obtained from the site. But, more importantly, users can set up evaluations of new algorithms with general or custom datasets. This platform will streamline future benchmarking studies and serves as an initiative to standardize research within the metagenomic field.

2.5 Databases for Taxonomic Annotation

Taxonomic assignment, as discussed earlier, depends on the availability of a reference database containing sequences labeled with known taxonomic information. As barcoding genes such as the 16S rRNA gene and ITS are well established in microbiome profiling, several databases collect taxonomic information based on these markers. Widely used RNA databases include GreenGenes (DeSantis et al. 2006), Ribosomal Database Project/RDP (Cole et al. 2014), SILVA (Quast et al. 2013), UNITE (Kõljalg et al. 2005), the NCBI Taxonomy database (Schoch et al. 2020), and the Genome Taxonomy Database (GTDB) (Parks et al. 2018).

The GreenGenes database stores only 16S (rRNA small subunit, SSU) sequences recovered from different sources (other databases, mainly NCBI). With this information, the taxonomy of Bacteria and Archaea was automatically constructed (DeSantis et al. 2006). The database was constructed with ~90,000 sequences in 2006, and its last release was in August 2013. Of the databases covered here, GreenGenes has the smallest number of taxonomic nodes and is the least supported. On the other hand, unlike the rest of the databases, all nodes in the GreenGene database phylogeny are assigned to a defined taxonomic rank (i.e., they have a corresponding domain, phylum, class, order, family, genus or species) (Balvočiute and Huson 2017). It is also a small database in terms of memory requirements and therefore easy to load and query.

The RDP database provides 16S (SSU) sequences from Bacteria and Archaea and 28S (rRNA large subunit, LSU) from Fungi. Its last documented release (September 30, 2016) contained ~3,356,000 aligned and annotated 16S rRNA sequences and ~125,000 28S rRNA sequences. The taxonomy assignment in this database is not phylogeny based but instead relies on a naïve Bayes classifier trained on a smaller subset of the database. RDP developed the RDPipeline program, an online complimentary service designed to perform several common processing steps for taxonomy-dependent analysis (using the RDP classifier) and for taxonomy-independent analysis (using hierarchical clustering) of large datasets (Cole et al. 2005; Bacci et al. 2015).

SILVA is a comprehensive, up-to-date quality-controlled database of rRNA gene sequences from Bacteria, Archaea, and Eukaryota (Glöckner et al. 2017; Woloszynek et al. 2018). This database is updated in a timely manner. Its last release (version 138, December 16, 2019) contains ~9,500,000 SSU sequences, and from these, ~510,000 are represented in the phylogenetic guide tree. Although the taxonomy is based on Bergey's Taxonomic Outlines and the List of Prokaryotic Names with Standing in Nomenclature (LPSN), its taxonomy classification process is phylogeny based, as it uses guide trees to resolve inconsistencies in nomenclature (Quast et al. 2013; Yilmaz et al. 2014). The taxonomy assignment is manually curated, which is the main difference from the previously mentioned databases. More recently, this database adopted the GTDB taxonomy (Parks et al. 2018), which implied a major rearrangement of the taxonomy, mainly in Bacteria and Archaea. Like RDP, the SILVA database has a web service (SILVAngs) that provides a fully automated analysis of rRNA gene amplicon sequencing data.

A comparison searching for inconsistencies in the taxonomies of these three databases estimated that the annotation error rate in the RDP database is ~10%, while for GreenGenes and SILVA, this value is around 17% (Edgar 2018a). As the author of this study points out, it is striking that the RDP database, in which taxonomy is not assigned explicitly from phylogeny, can be more accurate than the phylogeny-based and/or manually curated alternatives. This again reinforces the notion that 'more is not necessarily better' and that selecting any of these databases for a study should be taken under careful consideration.

These databases all contain marker-based taxonomic associations, which have been useful in the investigation of metagenomes. However, as discussed earlier, using exclusively amplicon sequence and OTU-based methods for taxonomic enumeration can substantially underestimate species diversity. To overcome this limitation, other taxonomies are constructed on more general sequence information. Such are the cases of the Microbial Genome Atlas (MiGA) (Rodriguez-R et al. 2018), the Genome Taxonomy Database (Parks et al. 2018, 2020), and the National Center for Biotechnology Information Taxonomy (Balvočiute and Huson 2017).

The GTDB database taxonomy was constructed using a phylogenomic alignment of 120 concatenated gene markers from more than 150,000 bacterial genomes, followed by the resolution of polyphyletic groups and taxonomy rank normalization by relative evolutionary divergence (Parks et al. 2018, 2019, 2020). One of its most relevant features is the phylogenetical consistency in the sense that it is highly congruent with the relative evolutionary divergence among species and, at the same time, eliminates phyletic conflicts. Also, the taxonomy required the reclassification of ~58% of genomes in the database, as well as the definition of new phyla (Parks et al. 2018). This database also includes several genomes of uncultured species that have been assembled from shotgun metagenomic data and are therefore unnamed. The total amount of unnamed species in the database was estimated around 40%, and more recently, the authors utilized the ANI and alignment fraction (AF) metrics to determine a species-level cluster of genomes that implied a major reordering and renaming at the genus and species rank level (Parks et al. 2019, 2020; Rinke et al. 2020). In our opinion, the GTDB seems to be the most accurate (i.e., closest to the true evolutionary tree) among the currently available taxonomic structures, but an independent formal evaluation of this statement has not been provided so far.

Finally, the largest sequence-associated taxonomy structure used for metagenomic annotation is the National Center for Biotechnology Information Taxonomy (Balvočiute and Huson 2017). There are larger taxonomy structures (e.g., the Open Tree of life Taxonomy; Hinchliff et al. 2015; Rees and Cranston 2017), but these have no sequence information available and are therefore not relevant for NGS sequence analysis.

The NCBI database is a collection of resources that contains nucleotides and protein sequence entries from different experimental origins, and its manually curated taxonomy covers all the sequences in the database (Wheeler et al. 2008; Schoch et al. 2020). The taxonomy assembles taxa naming information contained in 23 external resources, and it is updated weekly. The number of taxa in the NCBI taxonomy is currently over 460,000, which roughly represents a quarter of the described species so far, although it also contains sequence information on about 1.34 million species without formal names that are commonly regarded as 'dark taxa' (Schoch et al. 2020).

A comparison of the taxonomic structures of GreenGenes, RDP, SILVA, NCBI, and the Open Tree of Life Taxonomy (OTT) revealed that the 16S rRNA databases map well onto the larger databases (NCBI and OTT), but these do not map well onto the smaller taxonomies (Balvočiute and Huson 2017). This indicates that NCBI and OTT taxonomies are more explanatory than the 16S rRNA databases, although SILVA, as mentioned earlier, recently adopted the GDBT taxonomy, which could have significantly improved the content of this database. These authors also stated that, while larger, the OTT is not significantly more diverse than the NCBI taxonomy (Balvočiute and Huson 2017). To the best of our knowledge, more recent updates of independent database benchmarking including the latest genome-centered databases have not been published.

2.6 Microbiome Functional Inference from Community Structure

Ecological inferences from marker-based sequencing experiments are needed to fully understand the ecological niches of microbial communities in any environment. These inferences rely on verified metagenomic and metatranscriptomic profiles in several habitats. They allow correlating the taxonomic community structure with different functions in any ecosystem. The microbiome function is difficult to interpret when complete microbial profiling—bacteria, fungi, algae, protozoa—is obtained. Then robust algorithms to predict the ecological functions in nature are needed, and further efforts should be made to get bioinformatics tools to generate new insights related to the role of microorganisms in different ecosystems.

Microbial communities have been routinely studied using different molecular markers to provide powerful taxonomic interpretations. For example, as previously mentioned, 16S and 28S ribosomal RNA genes have been extensively used to assess bacterial and fungal biodiversity, respectively. However, these molecular markers are not useful in establishing the ecological roles of microbes. In this sense, metabolic and ecological functions should be predicted by bioinformatic algorithms using associations between the taxonomic profiles and experimentally demonstrated metabolic capabilities in well-studied species. Thus, methods based on ancestral state reconstruction should be developed, but many challenges are frequently related to these methods. For example, these strategies assign functional annotations by extrapolation of marker gene sequences with those known species. In this scenario, a database with a huge amount of complete microbial genomes properly annotated is necessary. That is an important bottleneck because a few thousand complete microbial genomes are currently available, and the annotation is deficient in many cases. For these reasons, interpretation of the microbial ecological functionality could be limited, and experimental validation should be performed to demonstrate that functional predictions are accurate and realistic.

SINAPS is a method that predicts microbial function from marker gene sequences (Edgar 2017). This algorithm was successfully validated to predict functions related to energy metabolism, Gram-positive staining, presence of flagella, 16S copy number, and number of V4 primer mismatches from 16S V4 ribosomal sequences. The validation of this method demonstrated that a large number of functions were correctly assigned.

Several bioinformatics tools have been developed to infer functional roles of taxa within a community: PICRUSt (Langille et al. 2013), Tax4Fun (Aßhauer et al. 2015), Piphillin (Iwai et al. 2016), Faprotax (Louca, Parfrey, and Doebeli 2016), and PAPRICA (Bowman and Ducklow 2015). Tax4Fun2 (improving on its predecessor Tax4Fun) is a tool for the prediction of functional profiles and redundancy of a metagenomic sequencing based on 16S ribosomal markers (Wemheuer et al. 2020). The accuracy and robustness of this tool were notably enhanced by the incorporation of habitat-specific genomic information.

These methods can predict functional capabilities from prokaryotic ribosomal sequences but not from metagenomic shotgun sequencing. These algorithms have been used to interpret ecological niches of microbial communities inhabiting soil, marine seawater, microbial mats, and so on (Wemheuer et al. 2020). However, the robustness of all these methods is based on the genomes available in public databases, and these only represent a very limited fraction of the functional diversity in nature. Thus, the accuracy and reliability of these methods could be unsatisfactory. To solve this inconvenience, new tools for specific habitats such as the rumen (Wilkinson et al. 2018) or marine ecosystems (Louca, Parfrey, and Doebeli 2016) have been developed.

A major question in environmental microbiology reflects the need to know if microbial communities contain redundant functional members. This is an urgent question, and its answer could probably allow understanding of how microbial communities provide functional stability to ecosystems. For example, this has importance in polluted habitats where environmental changes could produce ecological successions in microbial communities with unknown impacts on biodegradative functionalities. To address this need, Tax4Fun2 now offers a robust algorithm based on a functional redundancy index that reflects the proportion of species with the capabilities to perform a particular metabolic function and their phylogenetic relationship with others (Wemheuer et al. 2020). Although the authors stated that this tool is also available for fungi, to date, it is only validated for 16S rRNA gene data. Thus, further research is needed to generate new and robust tools useful to predict functions from eukaryotic gene data (18S or 28S rRNA).

PICRUSt2 is another predictive tool to provide functional inferences based on marker gene sequencing (Douglas et al. 2019). This method allows the analysis of eukaryotic communities and is compatible with any OTU-based algorithm. This method was successfully used to identify functional signatures in bacterial communities from humans with inflammatory diseases (Douglas et al. 2019).

Although there is some progress in the development of diverse tools to generate functional microbial inferences, there are important challenges associated with predicting metabolic networks from marker-based sequencing. This aids in designing robust methodologies with a positive impact on bioremediation, for example (Faust 2019). The generation of functional networks could allow optimizing the fitness of microbial consortia to enhance the biodegradation of certain pollutants, such as atrazine (Xu et al. 2019).

2.7 Functional Annotation of Shotgun Metagenome Data

The tools used to annotate metagenomic data often consist of new developments over previously existing tools designed to annotate isolated genomes. In both cases, the first step in the functional annotation is gene prediction. Once the candidate open reading frames (ORFs) are found, they can be linked to biological information based on current knowledge. This means that the functional annotation of the same dataset can be improved over time as our knowledge of biological systems increases.

Accurate gene prediction is a fundamental step in most metagenomics pipelines. Methods for gene prediction are classified as extrinsic (e.g., homology search) or intrinsic (e.g., sequence composition analysis).

Probably the most reliable way to predict a gene is to find a close homolog from another organism, and this is precisely what homology-based methods do: They perform pairwise alignments between metagenomic reads and a given database of known proteins. However, the main drawback of methods based exclusively on homology evidence is that they can only annotate previously known genes.

So, it soon became clear that computational methods that score the coding region using intrinsic sequence features are required for those genes lacking a significant homology to known genes. Intrinsic features can include signal sensors such as start/stop codons and promoters, as well as content sensors, such as patterns of codon usage, k-mer frequency profiles, or any other statistically inferable feature. These *ab initio* approaches increase the possibility of detecting novel genes, as they use linguistic or pattern recognition algorithms to detect specific sequence motifs or global statistical patterns that can consistently help in the process of gene finding. However, these methods are suitable for annotating assembled contigs rather than short reads, as they require a large number of genes for model training.

Methods based on intrinsic evidence include MetaGeneMark (Zhu, Lomsadze, and Borodovsky 2010), FragGeneScan (Rho, Tang, and Ye 2010), Glimmer-MG (Kelley et al. 2012), and MetaProdigal (Hyatt et al. 2012). MetaGeneMark, for instance, is a gene-prediction tool based on GeneMark-HMM. It uses a heuristic method to compute the parameters as functions of intrinsic features of individual sequences, which makes it efficient in predicting genes in metagenomic datasets, as such features (e.g., G+C content, codons, and oligomer frequencies) will probably vary widely among reads.

FragGeneScan is designed to find complete and fragmented genes in short reads (Rho, Tang, and Ye 2010). It combines codon usage information, sequencing error models, and start/stop codon patterns in a hidden Markov model (HMM) to find the most likely path of hidden states from a given input sequence. It accepts as inputs both short reads or assembled contigs, and it represents a suitable tool for gene prediction in incomplete metagenome assemblies.

Other popular tools are Glimmer-MG (Kelley et al. 2012; Salzberg et al. 1998) and MetaProdigal (Hyatt et al. 2012). The former incorporates classification and clustering of sequences prior to gene prediction using the Glimmer framework and uses a probabilistic model for prediction of gene length and start/stop codon presence in the case of truncated genes that are typical of shotgun metagenome sequencing. The latter is a metagenomic version of the gene prediction program Prodigal (Hyatt et al. 2010), which provides enhanced translation initiation site identification, the ability to identify sequences that use alternate genetic codes, and confidence values for each gene prediction.

Some pipelines use a combination of evidence-based (extrinsic) and *ab initio* (intrinsic) methods. This is the case of an in-house pipeline developed by researchers from the Max Planck Institute for Marine Microbiology, named meta-ORF-finder or mORFind (unpublished), which uses a combination of Orpheus (Frishman et al. 1998), CRITICA (Badger and Olsen 1999), and the previously mentioned Glimmer framework. Both CRITICA and Orpheus are BLAST-based tools that aim to identify coding regions in genomes invoking comparative analysis. CRITICA first considers the observed amino acid identity for the translated aligned sequences once their percentage nucleotide identity is known; if it is higher than expected, this is taken as evidence for coding. Next, it incorporates information of the relative hexanucleotide frequencies in coding frames versus other contexts, and this feature makes it less dependent on the accuracy of sequence annotation in databases and thus well suited for the analysis of novel genomes. Orpheus uses a very similar approach, in which the similarity-derived seed ORFs have their coding potential parameters calculated and scored. Those features, combined with those included in Glimmer, make mORFind a versatile tool suitable for metagenome gene prediction.

Identifying eukaryotic protein-coding genes in metagenomes is more challenging than identifying prokaryotic ones due to the exon-intron architecture of eukaryotic genes. A tool specially designed to meet this challenge is MetaEuk (Levy Karin, Mirdita, and Söding 2020). This toolkit allows high-throughput reference-based discovery and annotation of protein-coding genes in eukaryotic metagenomic contigs. Instead of doing a spliced alignment, which would be computationally costly, it takes as input a set of assembled contigs and scans each contig in all six reading frames to extract putative protein fragments between stop codons in each frame. Then it uses a sensitive and efficient method called MMseqs2 to perform an iterative search through any target database.

GeneMark-ES is an *ab initio* algorithm iterative unsupervised training to identify protein-coding genes in eukaryotic genomes. Augustus is one of the most accurate tools for eukaryotic gene prediction and is based on a generalized hidden Markov model (Stanke and Waack 2003). A web interface (WebAugustus) allows users to train their own gene structures or upload a training gene structure file or genome file and then perform eukaryotic gene predictions (Hoff and Stanke 2013).

Predicting protein function is the next step in metagenome functional annotation. It can be based on different sources of information, such as sequence similarity (mapping to databases), phylogenetic profiles, protein–protein interactions, and protein complexes.

A common approach is to perform a translated BLAST search to determine the annotation that will be assigned to a read based on its alignment scores, and this criterion can vary according to the protocol. For instance, the final annotation can consider only optimal alignments, including suboptimal alignments, or even use the average of multiple high-scoring hits. However, it is important to consider that the efficiency of the alignment methods is influenced by sequencing errors, read length, the phylogenetic coverage of the reference database, and the differences in annotation accuracy across the clades.

Also, the alignment of large datasets can be a computationally intensive and limiting step in the analysis of metagenomes. Thus, some tools were developed to achieve faster and accurate functional annotation. One example is Woods, an orthology-based functional classifier that uses a combination of machine learning (random forest) and similarity-based classification (RAPsearch2).

Methods based on protein interactions include interolog mapping. Interologs are interacting pairs of proteins that have homologs with conserved interaction in another organism. In interolog mapping, a known interolog in an organism is extended to a second organism assuming that the homologous proteins in different organisms maintain their interaction properties. A tool that uses the interolog concept is

STRING. It uses hierarchically arranged orthologous group relations, as defined in eggNOG, to transfer associations between organisms (prokaryotes and eukaryotes) where applicable.

Function prediction based on multilayer protein networks (FP-MPN) is a method that integrates protein–protein interaction (PPI) networks, protein domain content, and protein complex subunit information to predict protein function. This method assumes that diverse types of connections between groups of proteins reflect distinct roles and importance.

2.8 Main Databases Used for Functional Annotation

The main resources for functional annotation are currently the National Center for Biotechnology Information databases. The nr database is a protein database that contains non-identical sequences from GenBank CDS translations, Protein Data Bank (PDB), Swiss-Prot, Protein Information Resource (PIR), and Protein Research Foundation (PRF). RefSeq is an open-access, curated, and non-redundant database of publicly available genomes, transcripts, and protein products.

The UniProt database is a large collection of protein sequences and annotations from all domains of life. It contains more than 120 million protein sequences, of which the majority are derived from the translated genome and MAG sequence information deposited in ENA/GenBank/DDBJ databases (Bateman 2019). Over one-half of those proteins have annotations obtained from the literature by expert curators, while the remaining entries are automatically annotated using information from several databases, mainly InterPro.

The KEGG database allows estimating metabolic pathways in a metagenome. KEGG integrates information from 15 other databases by a computational database construction algorithm (Kanehisa and Goto 2000). The genomic information category, which is based on the KO (KEGG Orthologues) database, contains genomes and genes derived from different databases (RefSeq, Genbank, and NCBI Taxonomy), giving them original KEGG annotations. KEGG mapping can be performed with the KEGG Mapper tool, together with the KOALA tools (BlastKOALA and GhostKOALA), which allow for an automatic assignment of KO (KEGG orthology) identifiers used in the mapping (Kanehisa and Goto 2000; Kanehisa et al. 2016; Kanehisa and Sato 2020). PFAM is a collection of curated protein families, each represented by multiple sequence alignments generated using hidden Markov models (Finn et al. 2014). eggNOG (Evolutionary Genealogy of Genes: Non-supervised Orthologous Groups) is a database that provides orthologous gene mappings for Bacteria, Archaea, and Eukaryotes (Powell et al. 2012; Huerta-Cepas et al. 2016). Specialized databases include dbCAN for the carbohydrate-active enzyme (CAZYmes) (Yin et al. 2012); MEROPS for proteolytic enzymes (Rawlings, Barrett, and Finn 2016), their substrates, and inhibitors; and the Lipase Engineering Database (Fischer and Pleiss 2003).

2.9 Visualization of Metagenomic Data Annotation

A challenging task that researchers usually face when analyzing big data such as NGS sequencing results is representing the data to appropriately communicate the findings of the studies. Ideally, the representations should be as simple as possible, they should not represent individual entities or behaviors of the data but rather aggregated groups or tendencies and colors of different entities, all the details of the figure should be clearly distinguishable, and the resulting figures should fit in the printing area of most paper formats. The most common representations used in the field can be observed in Figure 2.4.

In metagenome analysis, stacked bar plots are the most used and abused representation of taxonomic annotation and abundance. This representation can be obtained from almost all analysis pipelines and is very useful to see major sample differences in abundance or composition. However, stacked bar plots are inadequate to see differences in minor taxa, can be hard to read (especially for color-blind people), and are not suited to represent diversity in complex communities. Several other diagrams have been used to represent diversity and abundance, each with its pros and cons.

Stack bar plots and other representations for visualizing metagenomic data omit or distort quantitative hierarchical relationships and cannot display secondary variables. Krona, Centrifuge, and other tools

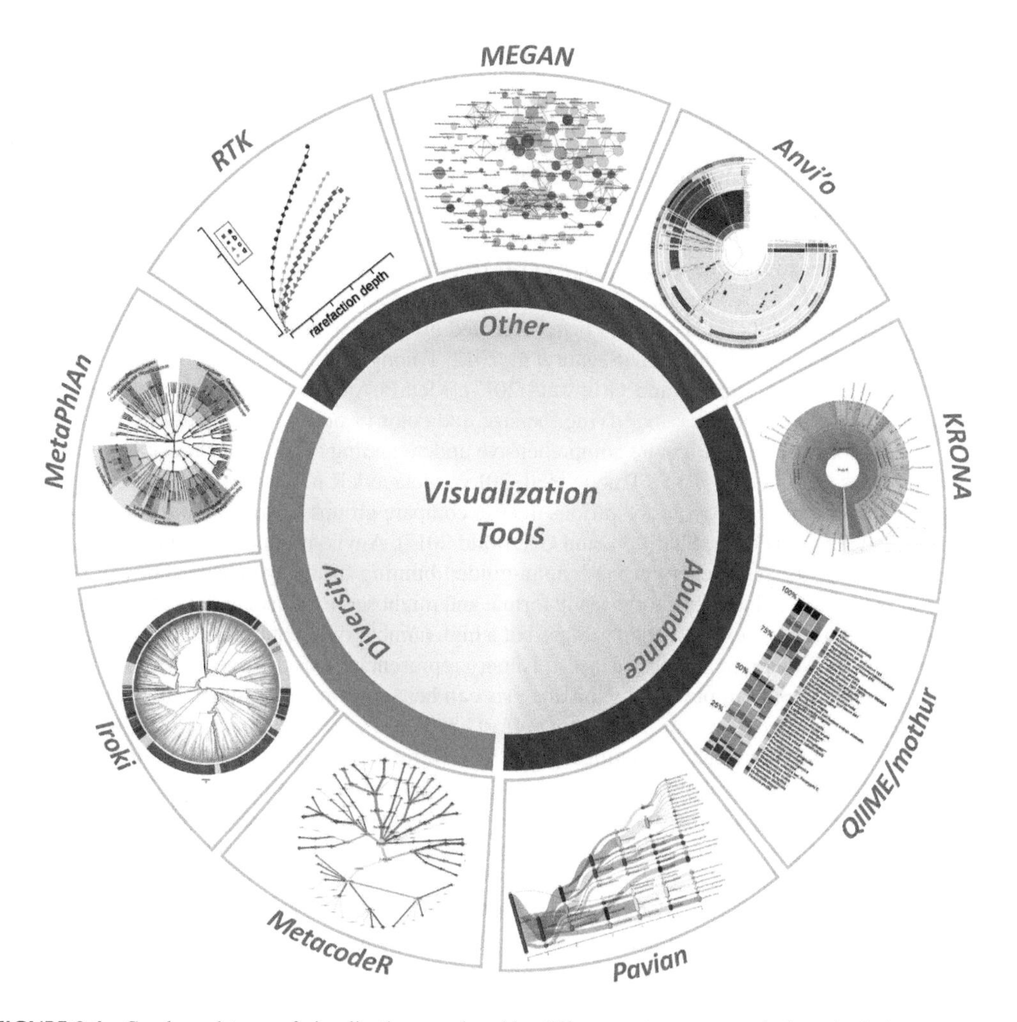

FIGURE 2.4 Graphs and types of visualizations produced by different metagenome analysis tools aiming to represent taxonomical information. Visualizations can accommodate metagenome diversity, abundance, or other relevant experimental information. Diversity can be explicitly represented as cladograms or trees in which each branch is an OTU or ASV or can be formally analyzed using rarefaction curves. Other visualizations such as stacked bar plots, pie charts, or Sankey diagrams can also incorporate abundance information. Other representations can depict the presence, co-occurrence, or differential abundance of taxa in several samples. See the main text for a detailed explanation.

enable the interactive visualization of complex metagenomic data using multi-layer pie charts, which depict the abundance of the most commons microbes of a sample. Krona distinguishes itself among others, as it has a lightweight implementation and is easily integrated into existing portals such as MG-RAST, METAREP, and Galaxy (Ondov, Bergman, and Phillippy 2011). The major disadvantages of this representation are that it can only accommodate information from one sample or experimental group (i.e., it is not suited to compare different samples), and the information that is interactively displayed (embedded in the graph) is lost when printed.

Sankey flow diagrams are equivalent to pie charts in terms of displayed information and interpretation but are more visually appealing. This representation also considers quantitative taxonomic hierarchy information, with taxonomic levels distributed in the horizontal axis. Pavian (Breitwieser and Salzberg 2020) and BioSankey (Platzer et al. 2018) are among the tools that can produce this type of diagram and, similarly, can embed information on the graph to allow the interactive representation of additional data such as sample comparisons. This representation has the same disadvantages as pie charts.

Taxonomical information has traditionally been represented using cladograms and phylogenetic trees. This simple representation is suited to display all the taxonomic entities in a sample or group at a given rank. Annotated trees and cladograms are widely used in the metagenomic field, as they can display diversity information more precisely than the previously mentioned representations. Tunable tools that allow annotation of phylogenetic trees include iTOL (Letunic and Bork 2007, 2019), EvolView (Zhang et al. 2012; Subramanian et al. 2019), and Iroki (Moore et al. 2020), among others. Most of them are not specially designed to handle the output of metagenome annotation tools. However, iTOL recently incorporated an option to read the tree files output by QIIME (Letunic and Bork 2019), while Iroki can deal with trees from QIIME, SILVAngs classifier, and other tools (Moore et al. 2020).

Phylogeny trees and cladograms can also be represented by many specialized metagenomic pipelines. That is the case of tools like MetaPhlAn (Segata et al. 2012; Truong et al. 2015), Anvi'o (Eren et al. 2015), and MetacodeR (Foster, Sharpton, and Grünwald 2017). MetaPhAn produces cladograms that can be annotated by shading the branches; modifying the size and color of nodes; and separating the different ranks of the taxonomy to achieve a more comprehensive understanding of diversity, abundance, and even sample differences (Segata et al. 2012; Truong et al. 2015). MetacodeR performs similar tree representations but also allows producing layouts of various trees to compare groups of samples and other segmentations of the study design (Foster, Sharpton, and Grünwald 2017). Anvi'o produces densely annotated and interactive trees to assist researchers in the human-guided binning of metagenomes (Eren et al. 2015). These trees can be exported in publication-ready format and might serve to display several metadata types associated with samples and experimental groups, but a high number of annotations can interfere with the readability of the final diagram and obstruct the diversity representation as the tree is shrunk.

For sample comparison, sample rarefaction analysis can help determine a sequencing depth at which all samples have the same amount of 'sequencing effort.' This is a widespread normalization strategy to avoid sequencing depth bias when estimating abundances. Rarefaction curves are also informative about the amount of ecosystem diversity that was effectively captured by the sequencing experiment and therefore are very useful during the experiment design stage. Although many tools can produce rarefaction analyses, a recently published tool called RTK (Saary et al. 2017) seems to be among the best current solutions to efficiently calculate rarefaction curves and normalize a high number of samples.

STAMP (STatistical Analysis of Metagenomic Profiles) is a graphical tool focused on sample comparison with a relevant and rigorous statistical treatment of biological effects (Parks and Beiko 2010; Parks et al. 2014). This tool provides pairwise or multiple comparisons of annotated metagenomic profiles with the added feature of reporting effect size and confidence intervals associated with an annotation feature (gene, pathway, gene ontology, or enzyme class). Effect size statistics calculated using STAMP can be complementary and sometimes pivotal in assessing the biological relevance of p-values in hypothesis testing during sample comparison.

Species co-occurrence networks can give information that is complementary to differential abundance analysis. In this case, the nodes of the network represent identified OTUs or taxa (generally species), and the connections represent a correlation coefficient that is obtained from the matrix of sample species abundance. This co-occurrence network can be obtained from several pipelines, including MEGAN (Huson et al. 2016; Bağcı et al. 2019). The direct implication of this analysis is identifying hub species that can shape or influence community structure. As discussed earlier, this could represent a valuable tool to design intervention strategies to influence soil communities, gut microbiota, water treatment biodigester microbes, and so on.

Deciding on the most appropriate graphical representation for a dataset or result is pivotal for efficient communication of experimental findings. Several pipelines can produce the same types of representation with different aesthetics and varying levels of difficulty. It is a task that should be considered with the same caution as the decision on tools for scientific computation.

2.10 Challenges and Future Perspectives

From its beginnings in the early 2000s, metagenomics has become a popular approach to evaluate the taxonomic and functional composition of microbiomes. The reduction in sequencing costs, increased computational power, and the surge of bioinformatics tools have enabled many laboratories to study microbial communities in diverse environments. As noted throughout the chapter, there are still several challenges that researchers in the field are facing.

The most pressing of subjects is arriving at a consensus on annotation pipelines. Without a unified way of processing and reporting data, much of the findings will not be reusable for validation by peers or under future hypotheses, making it harder to achieve knowledge integration. There should be a culture among bioinformaticians and biologists working with computational resources to ensure that their use of tools is guided by truly scientific reasons and not by trends or ease of use. In most cases, the highly cited tools are already evaluated independently by a benchmarking study, which should provide enough support for deciding on a pipeline. If that is not the case, several platforms (some discussed in this chapter) are available to guide researchers through performing a benchmarking study. The time spent on this task would not be fruitless, as it will reflect on the reliability of the results obtained from our methodology.

There is also a global demand from the scientific community to improve on how researchers produce, interpret, and publish data to ensure that the information complies with four criteria: It should be findable, accessible, interoperable, and reusable (Wilkinson et al. 2016). The FAIR guidelines (common jargon) aim to provide a framework for researchers, mainly those working on big data, to produce data with enough value to be used by others and enough structure to be handled by machines.

For scientists working on metagenomics, this implies that, as a general recommendation, the raw data of every experiment should be available along with the processed results. Moreover, these data should be accompanied by structured and detailed metadata that describes the experimental conditions, treatments, subject details, locations, sample processing, and any other relevant information that could be influencing the results.

Given the heterogeneity of methods, each with its own set of deficiencies, the used pipeline should also be available and preferably the used code be deposited in a reputable DOI-issuing repository so that others can access and cite it (Wilson et al. 2017). This becomes increasingly easy as researchers working in scientific computing adopt good practices when performing computational experiments. A detailed compendium of good practices for scientific computing can be found in (Wilson et al. 2017), and these should be considered as important as good pipetting practices for wet-lab experiments.

Another challenge in the field is homogenizing the nomenclature of techniques, processes, data types, and computational tasks. In the case of bioinformatics data and tasks, which are somehow less ambiguous, there is a set of ontologies (EDAM ontologies) to describe the more relevant and used operations, formats, and types (Ison et al. 2013). The proper use of these terms, although technically complex, is more normalized than the use of terms directly related to biological experiments. For example, terms such as metagenome, microbiome, metaprofiling, and metabarcoding are used interchangeably in literature to refer to the same technique. An effort to generate a common vocabulary was initially published by Marchesi and Ravel (2015). Unfortunately, some of the terms and definitions used in that text, such as metataxonomics, have not achieved widespread acceptance in the field. In the case of metataxonomics, in our opinion, the term does not properly justify its intended uses (e.g., marker-based and WGS sequencing for the sole purpose of taxonomical description of communities). On the other hand, their definition of the microbiome (i.e., covering both biotic and abiotic factors of an environment) rules out the most widespread uses of this term, which by force of habit only refer to the biotic component of an ecosystem. We encourage our readers to review the cited resources on nomenclature and actively commit to incorporating proper language in the description of scientific results. Adopting these or other future nomenclatures is a challenge that the community should face responsibly, as normalizing language will improve brevity, clarity, precision, and ultimately communication among specialists.

REFERENCES

Aßhauer, Kathrin P., Bernd Wemheuer, Rolf Daniel, and Peter Meinicke. 2015. "Tax4Fun: Predicting Functional Profiles from Metagenomic 16S RRNA Data." *Bioinformatics* 31 (17): 2882–84. https://doi.org/10.1093/bioinformatics/btv287.

Almeida, Alexandre, Alex L. Mitchell, Aleksandra Tarkowska, and Robert D. Finn. 2018. "Benchmarking Taxonomic Assignments Based on 16S RRNA Gene Profiling of the Microbiota from Commonly Sampled Environments." *GigaScience* 7 (5): 1–10. https://doi.org/10.1093/gigascience/giy054.

Amir, Amnon, Daniel McDonald, Jose A. Navas-Molina, Evguenia Kopylova, James T. Morton, Zhenjiang Zech Xu, Eric P. Kightley, et al. 2017. "Deblur Rapidly Resolves Single-Nucleotide Community Sequence Patterns." Edited by Jack A. Gilbert. *MSystems* 2 (2): 1–7. https://doi.org/10.1128/mSystems.00191-16.

Bacci, Giovanni, Alessia Bani, Marco Bazzicalupo, Maria Teresa Ceccherini, Marco Galardini, Paolo Nannipieri, Giacomo Pietramellara, and Alessio Mengoni. 2015. "Evaluation of the Performances of Ribosomal Database Project (RDP) Classifier for Taxonomic Assignment of 16S RRNA Metabarcoding Sequences Generated from Illumina-Solexa NGS." *Journal of Genomics* 3: 36–39. https://doi.org/10.7150/jgen.9204.

Badger, Jonathan H., and Gary J. Olsen. 1999. "CRITICA: Coding Region Identification Tool Invoking Comparative Analysis." *Molecular Biology and Evolution* 16 (4): 512–24. https://doi.org/10.1093/oxfordjournals.molbev.a026133.

Bağcı, Caner, Sina Beier, Anna Górska, and Daniel H. Huson. 2019. "Introduction to the Analysis of Environmental Sequences: Metagenomics with MEGAN." In *Methods in Molecular Biology*, 1910: 591–604. Humana Press Inc. https://doi.org/10.1007/978-1-4939-9074-0_19.

Balvočiute, Monika, and Daniel H. Huson. 2017. "SILVA, RDP, Greengenes, NCBI and OTT—How Do These Taxonomies Compare?" *BMC Genomics* 18 (Suppl 2): 1–8. https://doi.org/10.1186/s12864-017-3501-4.

Bateman, Alex. 2019. "UniProt: A Worldwide Hub of Protein Knowledge." *Nucleic Acids Research* 47 (D1): D506–15. https://doi.org/10.1093/nar/gky1049.

Bazinet, Adam L., and Michael P. Cummings. 2012. "A Comparative Evaluation of Sequence Classification Programs." *BMC Bioinformatics* 13 (1). https://doi.org/10.1186/1471-2105-13-92.

Bowman, Jeff S., and Hugh W. Ducklow. 2015. "Microbial Communities Can Be Described by Metabolic Structure: A General Framework and Application to a Seasonally Variable, Depth-Stratified Microbial Community from the Coastal West Antarctic Peninsula." *PLoS One* 10 (8): 1–18. https://doi.org/10.1371/journal.pone.0135868.

Bray, Nicolas L., Harold Pimentel, Páll Melsted, and Lior Pachter. 2016. "Near-Optimal Probabilistic RNA-Seq Quantification." *Nature Biotechnology* 34 (5): 525–27. https://doi.org/10.1038/nbt.3519.

Breitwieser, Florian P., Jennifer Lu, and Steven L. Salzberg. 2018. "A Review of Methods and Databases for Metagenomic Classification and Assembly." *Briefings in Bioinformatics* 20 (4): 1125–39. https://doi.org/10.1093/bib/bbx120.

Breitwieser, Florian P., and Steven L. Salzberg. 2020. "Pavian: Interactive Analysis of Metagenomics Data for Microbiome Studies and Pathogen Identification." *Bioinformatics* 36 (4): 1303–4. https://doi.org/10.1093/bioinformatics/btz715.

Buchfink, Benjamin, Chao Xie, and Daniel H. Huson. 2015. "Fast and Sensitive Protein Alignment Using DIAMOND." *Nature Methods* 12 (1): 59–60. https://doi.org/10.1038/nmeth.3176.

Callahan, Benjamin J., Paul J. McMurdie, Michael J. Rosen, Andrew W. Han, Amy Jo A. Johnson, and Susan P. Holmes. 2016. "DADA2: High-Resolution Sample Inference from Illumina Amplicon Data." *Nature Methods* 13 (7): 581–83. https://doi.org/10.1038/nmeth.3869.

Cole, J. R., B. Chai, R. J. Farris, Q. Wang, S. A. Kulam, D. M. McGarrell, G. M. Garrity, and J. M. Tiedje. 2005. "The Ribosomal Database Project (RDP-II): Sequences and Tools for High-Throughput RRNA Analysis." *Nucleic Acids Research* 33 (DATABASE ISS.): 294–6. https://doi.org/10.1093/nar/gki038.

Cole, J. R., Qiong Wang, Jordan A. Fish, Benli Chai, Donna M. McGarrell, Yanni Sun, C. Titus Brown, Andrea Porras-Alfaro, Cheryl R. Kuske, and James M. Tiedje. 2014. "Ribosomal Database Project: Data and Tools for High Throughput RRNA Analysis." *Nucleic Acids Research* 42 (D1): 633–42. https://doi.org/10.1093/nar/gkt1244.

DeSantis, T. Z., P. Hugenholtz, N. Larsen, M. Rojas, E. L. Brodie, K. Keller, T. Huber, D. Dalevi, P. Hu, and G. L. Andersen. 2006. "Greengenes, a Chimera-Checked 16S RRNA Gene Database and Workbench Compatible with ARB." *Applied and Environmental Microbiology* 72 (7): 5069–72. https://doi.org/10.1128/AEM.03006-05.

Douglas, Gavin M., Vincent J. Maffei, Jesse Zaneveld, Svetlana N. Yurgel, James R. Brown, Christopher M. Taylor, Curtis Huttenhower, and Morgan G. I. Langille. 2019. "PICRUSt2: An Improved and Extensible Approach for Metagenome Inference." *BioRxiv*. https://doi.org/10.1101/672295.

Edgar, Robert C. 2010. "Search and Clustering Orders of Magnitude Faster Than BLAST." *Bioinformatics* 26 (19): 2460–1. https://doi.org/10.1093/bioinformatics/btq461.

Edgar, Robert C. 2017. "SINAPS: Prediction of microbial traits from marker gene sequences." *BioRxiv*, no. Moran 2015. https://doi.org/10.1101/124156.

Edgar, Robert C. 2018a. "Taxonomy Annotation and Guide Tree Errors in 16S RRNA Databases." *PeerJ* 2018 (6). https://doi.org/10.7717/peerj.5030.

Edgar, Robert C. 2018b. "Updating the 97% Identity Threshold for 16S Ribosomal RNA OTUs." *Bioinformatics* 34 (14): 2371–75. https://doi.org/10.1093/bioinformatics/bty113.

Eren, A. Murat, Ozcan C. Esen, Christopher Quince, Joseph H. Vineis, Hilary G. Morrison, Mitchell L. Sogin, and Tom O. Delmont. 2015. "Anvi'o: An Advanced Analysis and Visualization Platform for 'omics Data." *PeerJ* 2015 (10): 1–29. https://doi.org/10.7717/peerj.1319.

Escobar-Zepeda, Alejandra, Elizabeth Ernestina Godoy-Lozano, Luciana Raggi, Lorenzo Segovia, Enrique Merino, Rosa María Gutiérrez-Rios, Katy Juarez, Alexei F. Licea-Navarro, Liliana Pardo-Lopez, and Alejandro Sanchez-Flores. 2018. "Analysis of Sequencing Strategies and Tools for Taxonomic Annotation: Defining Standards for Progressive Metagenomics." *Scientific Reports* 8 (1): 1–13. https://doi.org/10.1038/s41598-018-30515-5.

Faust, Karoline. 2019. "Microbial Consortium Design Benefits from Metabolic Modeling." *Trends in Biotechnology* 37 (2): 123–5. https://doi.org/10.1016/j.tibtech.2018.11.004.

Finn, Robert D., Alex Bateman, Jody Clements, Penelope Coggill, Ruth Y. Eberhardt, Sean R. Eddy, Andreas Heger, et al. 2014. "Pfam: The Protein Families Database." *Nucleic Acids Research* 42 (D1): 222–30. https://doi.org/10.1093/nar/gkt1223.

Fischer, Markus, and Jürgen Pleiss. 2003. "The Lipase Engineering Database: A Navigation and Analysis Tool for Protein Families." *Nucleic Acids Research* 31 (1): 319–21. https://doi.org/10.1093/nar/gkg015.

Fosso, Bruno, Graziano Pesole, Francesc Rosselló, and Gabrie Valiente. 2018. "Unbiased Taxonomic Annotation of Metagenomic Samples." *Journal of Computational Biology* 25 (3): 348–60. https://doi.org/10.1089/cmb.2017.0144.

Foster, Zachary S. L., Thomas J. Sharpton, and Niklaus J. Grünwald. 2017. "Metacoder: An R Package for Visualization and Manipulation of Community Taxonomic Diversity Data." *PLoS Computational Biology* 13 (2): 1–15. https://doi.org/10.1371/journal.pcbi.1005404.

Frishman, Dmitrij, Andrey Mironov, Hans Werner Mewes, and Mikhail Gelfand. 1998. "Combining Diverse Evidence for Gene Recognition in Completely Sequenced Bacterial Genomes." *Nucleic Acids Research* 26 (12): 2941–7. https://doi.org/10.1093/nar/26.12.2941.

Gardner, Paul P., Renee J. Watson, Xochitl C. Morgan, Jenny L. Draper, Robert D. Finn, Sergio E. Morales, and Matthew B. Stott. 2019. "Identifying Accurate Metagenome and Amplicon Software via a Meta-Analysis of Sequence to Taxonomy Benchmarking Studies." *PeerJ* 2019 (1): 1–19. https://doi.org/10.7717/peerj.6160.

Gardner, Paul P., James Paterson, Fatemeh Ashari-Ghomi, Sinan Umu, Stephanie McGimpsey, and Aleksandra Pawlik. 2017. "A Meta-Analysis of Bioinformatics Software Benchmarks Reveals That Publication-Bias Unduly Influences Software Accuracy." *BioRxiv*. https://doi.org/10.1101/092205.

Glöckner, Frank Oliver, Pelin Yilmaz, Christian Quast, Jan Gerken, Alan Beccati, Andreea Ciuprina, Gerrit Bruns, et al. 2017. "25 Years of Serving the Community with Ribosomal RNA Gene Reference Databases and Tools." *Journal of Biotechnology* 261 (February): 169–76. https://doi.org/10.1016/j.jbiotec.2017.06.1198.

Hillmann, Benjamin, Gabriel A. Al-Ghalith, Robin R. Shields-Cutler, Qiyun Zhu, Daryl M. Gohl, Kenneth B. Beckman, Rob Knight, and Dan Knights. 2018. "Evaluating the Information Content of Shallow Shotgun Metagenomics." Edited by John F. Rawls. *MSystems* 3 (6): e00069–18. https://doi.org/10.1128/mSystems.00069-18.

Hinchliff, Cody E., Stephen A. Smith, James F. Allman, J. Gordon Burleigh, Ruchi Chaudhary, Lyndon M. Coghill, Keith A. Crandall, et al. 2015. "Synthesis of Phylogeny and Taxonomy into a Comprehensive Tree of Life." *Proceedings of the National Academy of Sciences of the United States of America* 112 (41): 12764–9. https://doi.org/10.1073/pnas.1423041112.

Hleap, Jose S., Joanne E. Littlefair, Dirk Steinke, Paul D. N. Hebert, and Melania E. Cristescu. 2020. "Assessment of Current Taxonomic Assignment Strategies for Metabarcoding Eukaryotes." *BioRxiv*, 40. https://doi.org/10.1101/2020.07.21.214270.

Hoff, Katharina J., and Mario Stanke. 2013. "WebAUGUSTUS-a Web Service for Training AUGUSTUS and Predicting Genes in Eukaryotes." *Nucleic Acids Research* 41 (Web Server issue): 123–8. https://doi.org/10.1093/nar/gkt418.

Huerta-Cepas, Jaime, Damian Szklarczyk, Kristoffer Forslund, Helen Cook, Davide Heller, Mathias C. Walter, Thomas Rattei, et al. 2016. "EGGNOG 4.5: A Hierarchical Orthology Framework with Improved Functional Annotations for Eukaryotic, Prokaryotic and Viral Sequences." *Nucleic Acids Research* 44 (D1): D286–93. https://doi.org/10.1093/nar/gkv1248.

Huson, D. H., Sina Beier, Isabell Flade, Anna Górska, Mohamed El-Hadidi, Suparna Mitra, Hans-Joachim Joachim Ruscheweyh, and Rewati Tappu. 2016. "MEGAN Community Edition—Interactive Exploration and Analysis of Large-Scale Microbiome Sequencing Data." Edited by Timothée Poisot. *PLOS Computational Biology* 12 (6): e1004957. https://doi.org/10.1371/journal.pcbi.1004957.

Huson, D. H., Suparna Mitra, Hj Ruscheweyh, N. Weber, and S. C. Schuster. 2011. "Integrative Analysis of Environmental Sequences Using MEGAN4." *Genome Research* 21 (9): 1552–60. https://doi.org/10.1101/gr.120618.111.

Hyatt, Doug, Gwo Liang Chen, Philip F. LoCascio, Miriam L. Land, Frank W. Larimer, and Loren J. Hauser. 2010. "Prodigal: Prokaryotic Gene Recognition and Translation Initiation Site Identification." *BMC Bioinformatics* 11. https://doi.org/10.1186/1471-2105-11-119.

Hyatt, Doug, Philip F. Locascio, Loren J. Hauser, and Edward C. Uberbacher. 2012. "Gene and Translation Initiation Site Prediction in Metagenomic Sequences." *Bioinformatics* 28 (17): 2223–30. https://doi.org/10.1093/bioinformatics/bts429.

Ison, Jon, Matúš Kalaš, Inge Jonassen, Dan Bolser, Mahmut Uludag, Hamish McWilliam, James Malone, Rodrigo Lopez, Steve Pettifer, and Peter Rice. 2013. "EDAM: An Ontology of Bioinformatics Operations, Types of Data and Identifiers, Topics and Formats." *Bioinformatics* 29 (10): 1325–32. https://doi.org/10.1093/bioinformatics/btt113.

Iwai, Shoko, Thomas Weinmaier, Brian L. Schmidt, Donna G. Albertson, Neil J. Poloso, Karim Dabbagh, and Todd Z. DeSantis. 2016. "Piphillin: Improved Prediction of Metagenomic Content by Direct Inference from Human Microbiomes." *PLoS One* 11 (11): 1–18. https://doi.org/10.1371/journal.pone.0166104.

Jain, Chirag, Luis M. Rodriguez-R, Adam M. Phillippy, Konstantinos T. Konstantinidis, and Srinivas Aluru. 2018. "High Throughput ANI Analysis of 90K Prokaryotic Genomes Reveals Clear Species Boundaries." *Nature Communications* 9 (1): 1–8. https://doi.org/10.1038/s41467-018-07641-9.

Kanehisa, Minoru, and Susumu Goto. 2000. "KEGG: Kyoto Encyclopedia of Genes and Genomes." *Nucleic Acids Research.* Oxford University Press. https://doi.org/10.1093/nar/28.1.27.

Kanehisa, Minoru, and Yoko Sato. 2020. "KEGG Mapper for Inferring Cellular Functions from Protein Sequences." *Protein Science* 29 (1): 28–35. https://doi.org/10.1002/pro.3711.

Kanehisa, Minoru, Yoko Sato, Kanae Morishima, Kanehisa M. Sato, Y. Morishima, K. Minoru Kanehisa, Yoko Sato, and Kanae Morishima. 2016. "BlastKOALA and GhostKOALA: KEGG Tools for Functional Characterization of Genome and Metagenome Sequences." *Journal of Molecular Biology* 428 (4): 726–31. https://doi.org/10.1016/j.jmb.2015.11.006.

Kelley, David R., Bo Liu, Arthur L. Delcher, Mihai Pop, and Steven L. Salzberg. 2012. "Gene Prediction with Glimmer for Metagenomic Sequences Augmented by Classification and Clustering." *Nucleic Acids Research* 40 (1): 1–12. https://doi.org/10.1093/nar/gkr1067.

Khachatryan, Lusine, Rick H. de Leeuw, Margriet E. M. Kraakman, Nikos Pappas, Marije te Raa, Hailiang Mei, Peter de Knijff, and Jeroen F. J. Laros. 2020. "Taxonomic Classification and Abundance Estimation Using 16S and WGS—A Comparison Using Controlled Reference Samples." *Forensic Science International: Genetics* 46 (December 2019): 102257. https://doi.org/10.1016/j.fsigen.2020.102257.

Kiełbasa, Szymon M., Raymond Wan, Kengo Sato, Paul Horton, and Martin C. Frith. 2011. "Adaptive Seeds Tame Genomic Sequence Comparison." *Genome Research* 21 (3): 487–93. https://doi.org/10.1101/gr.113985.110.

Kim, Daehwan, Li Song, Florian P. Breitwieser, and Steven L. Salzberg. 2016. "Centrifuge: Rapid and Sensitive Classification of Metagenomic Sequences." *Genome Research* 26 (12): 1721–9. https://doi.org/10.1101/gr.210641.116.

Kõljalg, Urmas, Karl Henrik Larsson, Kessy Abarenkov, R. Henrik Nilsson, Ian J. Alexander, Ursula Eberhardt, Susanne Erland, et al. 2005. "UNITE: A Database Providing Web-Based Methods for the Molecular Identification of Ectomycorrhizal Fungi." *New Phytologist* 166 (3): 1063–68. https://doi.org/10.1111/j.1469-8137.2005.01376.x.

Langille, Morgan G. I., Jesse Zaneveld, J. Gregory Caporaso, Daniel McDonald, Dan Knights, Joshua A. Reyes, Jose C. Clemente, et al. 2013. "Predictive Functional Profiling of Microbial Communities Using 16S RRNA Marker Gene Sequences." *Nature Biotechnology* 31 (9): 814–21. https://doi.org/10.1038/nbt.2676.

Lee, Imchang, Yeong Ouk Kim, Sang Cheol Park, and Jongsik Chun. 2016. "OrthoANI: An Improved Algorithm and Software for Calculating Average Nucleotide Identity." *International Journal of Systematic and Evolutionary Microbiology* 66 (2): 1100–3. https://doi.org/10.1099/ijsem.0.000760.

Letunic, Ivica, and Peer Bork. 2007. "Interactive Tree of Life (ITOL): An Online Tool for Phylogenetic Tree Display and Annotation." *Bioinformatics* 23 (1): 127–8. https://doi.org/10.1093/bioinformatics/btl529.

Letunic, Ivica, and Peer Bork. 2019. "Interactive Tree of Life (ITOL) v4: Recent Updates and New Developments." *Nucleic Acids Research* 47 (W1): 256–9. https://doi.org/10.1093/nar/gkz239.

Levy Karin, Eli, Milot Mirdita, and Johannes Söding. 2020. "MetaEuk-Sensitive, High-Throughput Gene Discovery, and Annotation for Large-Scale Eukaryotic Metagenomics." *Microbiome* 8 (1): 1–15. https://doi.org/10.1186/s40168-020-00808-x.

Lindgreen, Stinus, Karen L. Adair, and Paul P. Gardner. 2016. "An Evaluation of the Accuracy and Speed of Metagenome Analysis Tools." *Scientific Reports* 6: 1–14. https://doi.org/10.1038/srep19233.

López-García, Adrian, Carolina Pineda-Quiroga, Raquel Atxaerandio, Adrian Pérez, Itziar Hernández, Aser García-Rodríguez, and Oscar González-Recio. 2018. "Comparison of Mothur and QIIME for the Analysis of Rumen Microbiota Composition Based on 16S RRNA Amplicon Sequences." *Frontiers in Microbiology* 9 (DEC): 1–11. https://doi.org/10.3389/fmicb.2018.03010.

Louca, S., L. W. Parfrey, and M. Doebeli. 2016. "Decoupling Function and Taxonomy in the Global Ocean Microbiome." *Science* 353 (6305): 1272–7. https://doi.org/10.1126/science.aaf4507.

Mangul, Serghei, Lana S. Martin, Brian L. Hill, Angela Ka Mei Lam, Margaret G. Distler, Alex Zelikovsky, Eleazar Eskin, and Jonathan Flint. 2019. "Systematic Benchmarking of Omics Computational Tools." *Nature Communications* 10 (1): 1–11. https://doi.org/10.1038/s41467-019-09406-4.

Marchesi, Julian R., and Jacques Ravel. 2015. "The Vocabulary of Microbiome Research: A Proposal." *Microbiome* 3 (1): 1–3. https://doi.org/10.1186/s40168-015-0094-5.

Marx, Vivien. 2020. "When Computational Pipelines Go 'Clank'." *Nature Methods* 17 (7): 659–62. https://doi.org/10.1038/s41592-020-0886-9.

McIntyre, Alexa B. R., Rachid Ounit, Ebrahim Afshinnekoo, Robert J. Prill, Elizabeth Hénaff, Noah Alexander, Samuel S. Minot, et al. 2017. "Comprehensive Benchmarking and Ensemble Approaches for Metagenomic Classifiers." *Genome Biology* 18 (1): 1–19. https://doi.org/10.1186/s13059-017-1299-7.

Menzel, Peter, Kim Lee Ng, and Anders Krogh. 2016. "Fast and Sensitive Taxonomic Classification for Metagenomics with Kaiju." *Nature Communications* 7. https://doi.org/10.1038/ncomms11257.

Moore, Ryan M., Amelia O. Harrison, Sean M. McAllister, Shawn W. Polson, and K. Eric Wommack. 2020. "Iroki: Automatic Customization and Visualization of Phylogenetic Trees." *PeerJ* 8: e8584. https://doi.org/10.7717/peerj.8584.

Nasko, Daniel J., Sergey Koren, Adam M. Phillippy, and Todd J. Treangen. 2018. "RefSeq Database Growth Influences the Accuracy of K-Mer-Based Lowest Common Ancestor Species Identification." *Genome Biology* 19 (1): 165. https://doi.org/10.1186/s13059-018-1554-6.

Ondov, Brian D., Nicholas H. Bergman, and Adam M. Phillippy. 2011. "Interactive Metagenomic Visualization in a Web Browser." *BMC Bioinformatics* 12 (September). https://doi.org/10.1186/1471-2105-12-385.

Parks, Donovan H., and Robert G. Beiko. 2010. "Identifying Biologically Relevant Differences Between Metagenomic Communities." *Bioinformatics* 26 (6): 715–21. https://doi.org/10.1093/bioinformatics/btq041.

Parks, Donovan H., Maria Chuvochina, Pierre-Alain Chaumeil, Christian Rinke, Aaron J. Mussig, and Philip Hugenholtz. 2019. "Selection of Representative Genomes for 24,706 Bacterial and Archaeal Species Clusters Provide a Complete Genome-Based Taxonomy." *BioRxiv*, 1–25. https://doi.org/10.1101/771964.

Parks, Donovan H., Maria Chuvochina, Pierre-Alain Chaumeil, Christian Rinke, Aaron J. Mussig, and Philip Hugenholtz. 2020. "A Complete Domain-to-Species Taxonomy for Bacteria and Archaea." *Nature Biotechnology* (April). https://doi.org/10.1038/s41587-020-0501-8.

Parks, Donovan H., Maria Chuvochina, David W. Waite, Christian Rinke, Adam Skarshewski, Pierre Alain Chaumeil, and Philip Hugenholtz. 2018. "A Standardized Bacterial Taxonomy Based on Genome Phylogeny Substantially Revises the Tree of Life." *Nature Biotechnology* 36 (10): 996. https://doi.org/10.1038/nbt.4229.

Parks, Donovan H., Gene W. Tyson, Philip Hugenholtz, and Robert G. Beiko. 2014. "STAMP: Statistical Analysis of Taxonomic and Functional Profiles." *Bioinformatics* 30 (21): 3123–4. https://doi.org/10.1093/bioinformatics/btu494.

Peabody, Michael A., Thea Van Rossum, Raymond Lo, and Fiona S. L. Brinkman. 2015. "Evaluation of Shotgun Metagenomics Sequence Classification Methods Using in Silico and In Vitro Simulated Communities." *BMC Bioinformatics* 16 (1). https://doi.org/10.1186/s12859-015-0788-5.

Platzer, Alexander, Julia Polzin, Klaus Rembart, Ping Penny Han, Denise Rauer, and Thomas Nussbaumer. 2018. "BioSankey: Visualization of Microbial Communities over Time." *Journal of Integrative Bioinformatics* 15 (4): 1–7. https://doi.org/10.1515/jib-2017-0063.

Powell, Sean, Damian Szklarczyk, Kalliopi Trachana, Alexander Roth, Michael Kuhn, Jean Muller, Roland Arnold, et al. 2012. "EggNOG v3.0: Orthologous Groups Covering 1133 Organisms at 41 Different Taxonomic Ranges." *Nucleic Acids Research* 40 (D1): 284–9. https://doi.org/10.1093/nar/gkr1060.

Quast, Christian, Elmar Pruesse, Pelin Yilmaz, Jan Gerken, Timmy Schweer, Pablo Yarza, Jörg Peplies, and Frank Oliver Glöckner. 2013. "The SILVA Ribosomal RNA Gene Database Project: Improved Data Processing and Web-Based Tools." *Nucleic Acids Research* 41 (D1): 590–6. https://doi.org/10.1093/nar/gks1219.

Rawlings, Neil D., Alan J. Barrett, and Robert Finn. 2016. "Twenty Years of the MEROPS Database of Proteolytic Enzymes, Their Substrates and Inhibitors." *Nucleic Acids Research* 44 (D1): D343–50. https://doi.org/10.1093/nar/gkv1118.

Rees, Jonathan A., and Karen Cranston. 2017. "Automated Assembly of a Reference Taxonomy for Phylogenetic Data Synthesis." *Biodiversity Data Journal* 5 (1). https://doi.org/10.3897/BDJ.5.e12581.

Rho, Mina, Haixu Tang, and Yuzhen Ye. 2010. "FragGeneScan: Predicting Genes in Short and Error-Prone Reads." *Nucleic Acids Research* 38 (20): e191–e191. https://doi.org/10.1093/nar/gkq747.

Rinke, Christian, Maria Chuvochina, Aaron Mussig, Pierre-Alain Chaumeil, David Waite, William Whitman, Donovan Parks, and Philip Hugenholtz. 2020. "A Rank-Normalized Archaeal Taxonomy Based on Genome Phylogeny Resolves Widespread Incomplete and Uneven Classifications." 1–24. https://doi.org/10.1101/2020.03.01.972265.

Rodriguez-R, Luis M., Santosh Gunturu, William T. Harvey, Ramon Rosselló-Mora, James M. Tiedje, James R. Cole, and Konstantinos T. Konstantinidis. 2018. "The Microbial Genomes Atlas (MiGA) Webserver: Taxonomic and Gene Diversity Analysis of Archaea and Bacteria at the Whole Genome Level." *Nucleic Acids Research* 46 (W1): W282–8. https://doi.org/10.1093/nar/gky467.

Saary, Paul, Kristoffer Forslund, Peer Bork, and Falk Hildebrand. 2017. "RTK: Efficient Rarefaction Analysis of Large Datasets." *Bioinformatics (Oxford, England)* 33 (16): 2594–5. https://doi.org/10.1093/bioinformatics/btx206.

Salzberg, Steven L., Arthur L. Deicher, Simon Kasif, and Owen White. 1998. "Microbial Gene Identification Using Interpolated Markov Models." *Nucleic Acids Research* 26 (2): 544–8. https://doi.org/10.1093/nar/26.2.544.

Schoch, Conrad L., Stacy Ciufo, Mikhail Domrachev, Carol L. Hotton, Sivakumar Kannan, Rogneda Khovanskaya, Detlef Leipe, et al. 2020. "NCBI Taxonomy: A Comprehensive Update on Curation, Resources and Tools." *Database : The Journal of Biological Databases and Curation* 2020 (2): 1–21. https://doi.org/10.1093/database/baaa062.

Sczyrba, Alexander, Peter Hofmann, Peter Belmann, David Koslicki, Stefan Janssen, Johannes Dröge, Ivan Gregor, et al. 2017. "Critical Assessment of Metagenome Interpretation—A Benchmark of Metagenomics Software." *Nature Methods* 14 (11): 1063–71. https://doi.org/10.1038/nmeth.4458.

Segata, Nicola, Levi Waldron, Annalisa Ballarini, Vagheesh Narasimhan, Olivier Jousson, and Curtis Huttenhower. 2012. "Metagenomic Microbial Community Profiling Using Unique Clade-Specific Marker Genes." *Nature Methods* 9 (8): 811–14. https://doi.org/10.1038/nmeth.2066.

Seppey, Mathieu, Mosè Manni, and Evgeny M. Zdobnov. 2020. "LEMMI: A Continuous Benchmarking Platform for Metagenomics Classifiers." *Genome Research*, 1208–16. https://doi.org/10.1101/gr.260398.119.

Siegwald, Léa, Hélène Touzet, Yves Lemoine, David Hot, Christophe Audebert, and Ségolène Caboche. 2017. "Assessment of Common and Emerging Bioinformatics Pipelines for Targeted Metagenomics." *PLoS One* 12 (1): 1–26. https://doi.org/10.1371/journal.pone.0169563.

Stanke, Mario, and Stephan Waack. 2003. "Gene Prediction with a Hidden Markov Model and a New Intron Submodel." *Bioinformatics* 19 (SUPPL.2): 215–25. https://doi.org/10.1093/bioinformatics/btg1080.

Subramanian, Balakrishnan, Shenghan Gao, Martin J. Lercher, Songnian Hu, and Wei Hua Chen. 2019. "Evolview v3: A Webserver for Visualization, Annotation, and Management of Phylogenetic Trees." *Nucleic Acids Research* 47 (W1): W270–5. https://doi.org/10.1093/nar/gkz357.

Tamames, Javier, Marta Cobo-Simón, and Fernando Puente-Sánchez. 2019. "Assessing the Performance of Different Approaches for Functional and Taxonomic Annotation of Metagenomes." *BMC Genomics* 20 (1): 1–16. https://doi.org/10.1186/s12864-019-6289-6.

Truong, Duy Tin, Eric A. Franzosa, Timothy L. Tickle, Matthias Scholz, George Weingart, Edoardo Pasolli, Adrian Tett, Curtis Huttenhower, and Nicola Segata. 2015. "MetaPhlAn2 for Enhanced Metagenomic Taxonomic Profiling." *Nature Methods* 12 (10): 902–3. https://doi.org/10.1038/nmeth.3589.

Velsko, Irina M., Laurent A. F. Frantz, Alexander Herbig, Greger Larson, and Christina Warinner. 2018. "Selection of Appropriate Metagenome Taxonomic Classifiers for Ancient Microbiome Research." *MSystems* 3 (4): 1–41. https://doi.org/10.1128/msystems.00080-18.

Wang, Qiong, George M. Garrity, James M. Tiedje, and James R. Cole. 2007. "Naïve Bayesian Classifier for Rapid Assignment of RRNA Sequences into the New Bacterial Taxonomy." *Applied and Environmental Microbiology* 73 (16): 5261–7. https://doi.org/10.1128/AEM.00062-07.

Wemheuer, Franziska, Jessica A. Taylor, Rolf Daniel, Emma Johnston, Peter Meinicke, Torsten Thomas, and Bernd Wemheuer. 2020. "Tax4Fun2: Prediction of Habitat-Specific Functional Profiles and Functional Redundancy Based on 16S RRNA Gene Sequences." *Environmental Microbiome* 15 (1): 11. https://doi.org/10.1186/s40793-020-00358-7.

Wheeler, David L., Tanya Barrett, Dennis A. Benson, Stephen H. Bryant, Kathi Canese, Vyacheslav Chetvernin, Deanna M. Church, et al. 2008. "Database Resources of the National Center for Biotechnology Information." *Nucleic Acids Research* 36 (SUPPL.1): D13–21. https://doi.org/10.1093/nar/gkm1000.

Wilkinson, Mark D., Michel Dumontier, IJsbrand Jan Aalbersberg, Gabrielle Appleton, Myles Axton, Arie Baak, Niklas Blomberg, et al. 2016. "Comment: The FAIR Guiding Principles for Scientific Data Management and Stewardship." *Scientific Data* 3: 1–9. https://doi.org/10.1038/sdata.2016.18.

Wilkinson, Toby J., Sharon A. Huws, Joan E. Edwards, Alison H. Kingston-Smith, Karen Siu-Ting, Martin Hughes, Francesco Rubino, Maximillian Friedersdorff, and Christopher J. Creevey. 2018. "CowPI: A Rumen Microbiome Focussed Version of the PICRUSt Functional Inference Software." *Frontiers in Microbiology* 9 (May): 1–10. https://doi.org/10.3389/fmicb.2018.01095.

Wilson, Greg, Jennifer Bryan, Karen Cranston, Justin Kitzes, Lex Nederbragt, and Tracy K. Teal. 2017. "Good Enough Practices in Scientific Computing." Edited by Francis Ouellette. *PLOS Computational Biology* 13 (6): e1005510. https://doi.org/10.1371/journal.pcbi.1005510.

Woloszynek, Stephen, Zhengqiao Zhao, Gregory Ditzler, Jacob R. Price, Erin R. Reichenberger, Yemin Lan, Jian Chen, et al. 2018. "Analysis Methods for Shotgun Metagenomics." In *Theoretical and Applied Aspects OfSystems Biology*, edited by F. Alves Barbosa da Silva, N. Carels, and F. Paes Silva Junior, 71–112. https://doi.org/10.1007/978-3-319-74974-7_5.

Xu, Xihui, Raphy Zarecki, Shlomit Medina, Shany Ofaim, Xiaowei Liu, Chen Chen, Shunli Hu, et al. 2019. "Modeling Microbial Communities from Atrazine Contaminated Soils Promotes the Development of Biostimulation Solutions." *ISME Journal* 13 (2): 494–508. https://doi.org/10.1038/s41396-018-0288-5.

Yang, Bo, Yong Wang, and Pei Yuan Qian. 2016. "Sensitivity and Correlation of Hypervariable Regions in 16S RRNA Genes in Phylogenetic Analysis." *BMC Bioinformatics* 17 (1): 1–8. https://doi.org/10.1186/s12859-016-0992-y.

Ye, Simon H., Katherine J. Siddle, Daniel J. Park, and Pardis C. Sabeti. 2019. "Benchmarking Metagenomics Tools for Taxonomic Classification." *Cell* 178 (4): 779–94. https://doi.org/10.1016/j.cell.2019.07.010.

Yi, Lynn, Harold Pimentel, Nicolas L. Bray, and Lior Pachter. 2018. "Gene-Level Differential Analysis at Transcript-Level Resolution." *Genome Biology* 19 (1): 1–11. https://doi.org/10.1186/s13059-018-1419-z.

Yilmaz, Pelin, Laura Wegener Parfrey, Pablo Yarza, Jan Gerken, Elmar Pruesse, Christian Quast, Timmy Schweer, Jörg Peplies, Wolfgang Ludwig, and Frank Oliver Glöckner. 2014. "The SILVA and 'All-Species Living Tree Project (LTP)' Taxonomic Frameworks." *Nucleic Acids Research* 42 (D1): 643–48. https://doi.org/10.1093/nar/gkt1209.

Yin, Yanbin, Xizeng Mao, Jincai Yang, Xin Chen, Fenglou Mao, and Ying Xu. 2012. "DbCAN: A Web Resource for Automated Carbohydrate-Active Enzyme Annotation." *Nucleic Acids Research* 40 (W1): 445–51. https://doi.org/10.1093/nar/gks479.

Yoon, Seok Hwan, Sung min Ha, Jeongmin Lim, Soonjae Kwon, and Jongsik Chun. 2017. "A Large-Scale Evaluation of Algorithms to Calculate Average Nucleotide Identity." *Antonie van Leeuwenhoek, International Journal of General and Molecular Microbiology* 110 (10): 1281–6. https://doi.org/10.1007/s10482-017-0844-4.

Zhang, Huangkai, Shenghan Gao, Martin J. Lercher, Songnian Hu, and Wei Hua Chen. 2012. "EvolView, an Online Tool for Visualizing, Annotating and Managing Phylogenetic Trees." *Nucleic Acids Research* 40 (W1): 569–72. https://doi.org/10.1093/nar/gks576.

Zhao, Yongan, Haixu Tang, and Yuzhen Ye. 2012. "RAPSearch2: A Fast and Memory-Efficient Protein Similarity Search Tool for Next-Generation Sequencing Data." *Bioinformatics* 28 (1): 125–6. https://doi.org/10.1093/bioinformatics/btr595.

Zhu, Wenhan, Alexandre Lomsadze, and Mark Borodovsky. 2010. "Ab Initio Gene Identification in Metagenomic Sequences." *Nucleic Acids Research* 38 (12): 1–15. https://doi.org/10.1093/nar/gkq275.

Section III

Metagenomics of Extreme Environments

3

Metagenomic Insights into the Microbial Communities of Desert Ecosystems

Satish Kumar*, G.C. Wakchaure, Kamlesh K. Meena, Mahesh Kumar, Ajay Kumar Singh, Jagadish Rane
ICAR-National Institute of Abiotic Stress Management, Baramati, Pune, India
Bharat Bhushan*
ICAR-Indian Institute of Maize Research, Ludhiana, Punjab, India
Ajinath Shridhar Dukare
ICAR-Central Institute of Post-Harvest Engineering and Technology, Ludhiana, India

CONTENTS

3.1 Introduction

Deserts are among the largest biomes on Earth, covering about 20% of the total global land surface (Middleton & Thomas 1997). The desert ecosystem is characterized by the limited availability of water (due to the prolonged moisture deficit), poor nutrients status of soil (low organic carbon), extreme temperatures, and high level of incident UV radiation. Desert environments are often confronted with abiotic stresses such as drought, extreme temperatures, and salinity (Eida et al. 2018). The land in deserts is often not protected by dense vegetation cover; thus, the finer material is easily blown away with wind, resulting in the frequent occurrence of sand and dust storms. Deserts are considered among the harshest terrestrial ecosystems on the Earth, posing many challenges to life forms. Hence flora, fauna, and the microbial communities of deserts have acquired adaptation to cope with the stressed conditions prevailing there. The microbial communities of desert ecosystems have a predominance of xerotolerant microbes adapted to cope with the lack of water (Potts & Webb 1994; Lebre et al. 2017). The description of deserts also encompasses polar, cold, or temperate deserts, known for receiving scant precipitation. These deserts occur at higher latitudes where arid conditions are created due to insufficient moisture in the air. Moreover, in cold deserts, the main form of precipitation is snow or fog in place of rain.

The most noted hot deserts of the world include the Sahara Desert and the Kalahari Desert in Africa, the Arabian Desert and the Syrian Desert in the Middle East, the Great Victoria Desert in Australia, the Great Basin Desert in North America, and the Atacama Desert in South America. The Antarctic desert and Arctic desert are the largest cold deserts of the world, located at the South and the North Poles of

DOI: 10.1201/9781003042570-6

the Earth surface, respectively. The Gobi Desert (Asia) is also among the noted cold deserts of the world. The cold desert receives very limited precipitation because the moisture in the air of the cold desert is often too low to result in precipitation.

Deserts are mainly categorized based on temperature and aridity, such as hot, cold, polar tundra, and polar frost deserts (Lacap-Bugler et al. 2017). They are also classified based on their geographical location in four different categories: 1) polar deserts, 2) subtropical deserts, 3) cold winter deserts, and 4) cool coastal deserts. Deserts (both hot and cold) play a great role in modeling the Earth's temperature because they reflect a major portion of the incident light compared to forest and sea (Coakley 2003). On the land surface, evapotranspiration (ET) is a principal process ensuring the exchanges of energy and water at the interface of the hydrosphere, atmosphere, and biosphere (Wang and Dickinson 2012). It is believed that for a desert ecosystem, the theoretical evapotranspiration rate is very high, but in practical terms, it may even sometimes be close to zero due to the unavailability of water (Shields 1982). Despite the extreme environmental conditions, deserts are ecologically significant, because they sustain ~6% of the human population, are a habitat to many endemic plants and animals, and store almost one-third of the total terrestrial carbon (Makhalanyane et al. 2015). Due to the great ecological and biotechnological significance attached to desert microbial communities, the current chapter aims to present updated information on desert microbial communities. We reviewed various metagenomic studies conducted on desert (both hot and cold deserts) microbial communities and summarize the salient findings of such studies. A description of the extensive taxonomic microbial diversity in desert ecosystems and the emerging exciting opportunities for future microbiology and biotechnology research are also discussed.

3.2 Life Forms of Desert Ecosystems: A Brief Description

Prima facie, deserts seem to be an inhospitable environment to living organisms; still, diverse life-forms manage to flourish under such harsh conditions. These include different forms of plants, animals, fungi, cyanobacteria, and microbes. Such living forms have evolved various molecular and physiological mechanisms to survive under desiccation and other extreme conditions. For instance, under such harsh conditions in the desert, the plants of taxonomically unrelated families of *Cactaceae* and *Agavaceae* have flourished with ease under xeric conditions. Many of the members of these two plant families have evolved a crassulacean acid metabolism (CAM), where they temporally separate the light and dark reactions of photosynthesis. CAM plants open stomata only during the night and fix carbon dioxide at night, thus avoiding water loss by temporally separating the light and dark reaction (carbon fixation) of photosynthesis. Additionally, xerotolerant microbes are also best adapted to flourish under the extreme environmental conditions of deserts (Pointing & Belnap 2012).

Now it is a widely recognized concept that the plant-associated microbiome does have a bigger role to play in the adaptation of desert plants to drought stress conditions (Marasco et al. 2012; Rolli et al. 2015; Eida et al. 2018). In the absence of dense flora and fauna in a typical desert ecosystem, microbial communities are probably the dominant drivers of ecosystem processes (Makhalanyane et al. 2015). During evolution, extreme climatic conditions have posed a selection pressure, resulting in the selected enrichment of stress-tolerant microbial taxa; thus, desert microbial communities display a low level of phylogenetic diversity and differ in composition and function compared to other water-sufficient terrestrial ecosystems (Fierer et al. 2012). Like other ecosystems, in desert ecosystems, microbes also play a fundamental role in governing the key biogeochemical cycles (Kieft & Skujinš 1991) and in deriving the food chain at different trophic levels. They play an important role as decomposers, N_2 fixers, nitrifiers, denitrifiers, phosphate solubilizers, and many other ecosystem services. Microbes associated with plants native to arid, nutrient-poor environments such as deserts can cope with a hostile environment, confer many similar benefits to host plants, and are capable of promoting plant growth and stress tolerance in crop species and also governing crucial steps in nutrient cycling in desert ecosystems.

3.3 Significance of Desert Microbial Ecology

In the last two decades, there has been mounting interest in bio-prospecting the microbial diversity of the stressed habitats as a source for novel microbes that can be potentially exploited for mitigation of multiple abiotic stressors under modern-day agriculture. Rigorous efforts made so far to develop sustainable strategies for reverse-desertification and abiotic stress-mitigation in plants have strongly endorsed the potential of microbes present in hyperarid, arid, and semi-arid ecosystems as stress alleviators. Understanding of the microbial diversity of the desert ecosystem may form the basis for deciphering the molecular basis of adaption and function in arid environments and designing microbe-mediated strategies for rehabilitation of arid lands and water resource planning in drought-prone regions.

In earlier times, culture-dependent methods were used to study the microbial communities of the deserts (Abdel-Hafez 1981; Alsohaili & Bani-Hasan 2018). Culture-based methods detect only a tiny fraction of the overall diversity present in a sample, as has been highlighted in the famous concept of the great plate count anomaly (Harwani 2013). It is proposed that more than 95% of the total microbial diversity in a natural sample cannot be captured using Petri-plate–based culture methods (Kumar et al. 2015). The advent of metagenomic techniques offered new avenues to generate information regarding the architectural dynamics and ecosystem services of the desert-microbial community, particularly associated with desert resources. This assumes importance because a detailed reference map of the structural and functional ecology of the desert microbial community would lead to the transfer of next-gen strategies for abiotic stress mitigation in agriculture through the bioactive compounds, biomolecules, and biotechnologically useful abiotic stress-tolerant gene pool. Further, metagenomic techniques can be used to delineate the interaction networks behind nutrient cycling in resource-poor desert environments. Due to the biotechnological and ecological importance attached to the microbial communities of the desert ecosystem, efforts have been made to catalog the microbial diversity of the desert ecosystem, and recent initiatives like the Atacama Database (www.atacamadb.cl.) are testimony to this fact (Contador et al. 2020). The recently described version of the Atacama Database contains 2,302 microorganisms encompassing bacteria, archaea, and eukaryotes retrieved from different environments within the desert between 1984 and 2016. Such databases serve as the baseline to understand the microbial ecology, population dynamics, seasonal behavior, and impacts of climate change on the desert ecosystem.

3.4 Metagenomic Techniques for Exploring Desert Microbial Ecology

Post-2005, we have witnessed some path-changing developments like the arrival of next-generation sequencing (NGS) platforms, making it possible to generate huge metagenomic sequence data at an affordable cost. The simplest and traditional way to study microbes is their isolation on Petri plates and their subsequent characterization. Desert microbial communities have been studied using culture-dependent approaches (Abdel-Hafez 1981; Jaouani et al. 2014; Alsohaili, & Bani-Hasan 2018; Okoro et al. 2009). Many novel *actinomycetes* have been isolated from the Atacama Desert, one of the hyperarid and most ancient deserts of the world. *Actinomycetes* of desert ecosystems are known for their biotechnological potential, as they are a source of new bioactive natural products and secondary metabolites with a range of antibiotic, anti-cancer, and anti-inflammatory properties (Bull & Asenjo 2013; Okoro et al. 2009; Goodfellow et al. 2018; Rateb et al. 2018). In one such culture-based study, Alotaibi and coworkers studied fungal and bacterial diversity in desert soil samples collected from Sabkha and desert areas in Saudi Arabia and reported 203 fungal species belonging to 33 different fungal genera, mainly represented by *Fusarium*, *Alternaria*, *Chaetomium*, *Aspergillus*, *Cochliobolus*, and *Pencillium*, whereas *Bacillus subtilis* and *Lactobacillus murinus* dominated the bacterial diversity (2020). As far as microbial load in typical cultivated soil is concerned, it has been predicted that 1 gram of garden soil contains billions of bacterial cells. Using traditional culture-based techniques, very few bacterial cells (often less than 1%) appear on Petri plates to form a bacterial colony, and hence a large microbial spectrum remains uncovered.

In recent times, culture-based approaches have also witnessed new advances in developing techniques for rapid identification of microbes using MALDI-TOF biotyping and novel culture methods (Kurli et al. 2018). Microbial identification using MALDI-TOF techniques is a simple, rapid, and low-cost approach where microbes can be identified by comparing the MS-spectra of ribosomal proteins with a database consisting of biomarker spectra of intracellular proteins primarily in the range of 2–20 kDa rRNA. MALDI-TOF MS-based protein profiling has been used to infer the identity and phylogenetic placement of bacteria isolated from desert soil (Khairnar et al. 2020). The microbial isolates of cold deserts have also been studied using MALDI-TOF MS-based bacterial identification approaches (Pandey et al. 2019). Currently, there exist few commercial offerings like Bruker-Biotyper (Bruker Daltonics, Bremen, Germany), Axima Assurance (Shimadzu, Kyoto, Japan), Vitek-MS (earlier SARAMIS AnagnosTec Germany, later acquired by bioMérieux), and Andromas (Andromas SAS, Paris, France), which operate on the same biological principle of microbial identification. However, MALDI-TOF–based bacterial identification suffers a major limitation due to the limited database, which is largely biased toward microbes of clinical or food safety relevance (Rahi et al. 2016).

The extensive efforts targeting the study of culture-based microbial diversity suffer a culture bias towards easily cultured microbial groups, resulting in under-representation of fastidious, viable but non-culturable (VBNC) microbes. In the best scenario, such approaches can capture less than 95% of the total microbial diversity. Fortunately, techniques are available to isolate DNA from environmental samples like soils, drainage water, lake water, fecal samples, rocks, and so on. The DNA isolated directly from environmental samples is called metagenome (a mix of genetic material from all the microbes present in the sample). Metagenomic DNA so extracted can be studied using various molecular ecology techniques to study microbial diversity, composition, and function. The major advantage of molecular microbial community studies is that they are not affected by the inevitable bias introduced in culture-based studies but still suffer the biases introduced due to the DNA extraction and PCR amplification procedures.

Metagenomic community structure is often inferred by amplifying and sequencing the 16S rRNA gene/gene region using environmental DNA as a template. The 16S rRNA gene is considered the best molecular phylogenetic marker and is also termed the molecular chronometer (Woese 1987). This gene has many conserved regions, along with the nine hypervariable regions (V1–V9), and also offers a sufficient size of 1,500 bp for informatics purposes. The gene is present in all the bacteria and archaea and is evolutionarily conserved species-specific signatures in the form of the DNA sequence of the hypervariable region. A large number of the studies over the decades aiming at the identification of bacteria and archaea has generated large sequence data of 16S rRNA gene, and thus the extensive reference databases Ribosomal Database Project (Cole et al. 2014), Greengenes (DeSantis et al. 2006), and Silva (Quast et al. 2012) are available. These databases serve as a reference where one can compare the 16S rRNA gene sequencing generated in a typical metagenomic study for community structure and composition analysis.

Before 2005, when the first NGS platform became a commercial offering, metagenomic approaches suffered low throughput. Various metagenomic approaches involved the study of microbial communities using culture-independent methods, like denaturing gradient gel electrophoresis (DGGE), temperature gradient gel electrophoresis (TGGE), terminal restriction length polymorphism (T-RFLP), amplified ribosomal DNA restriction analysis (ARDRA), 16S rRNA gene cloning, stable isotope probing (SIP), ribosomal intergenic spacer analysis (RISA), automated ribosomal intergenic spacer analysis (ARISA), and fluorescence in situ hybridization (FISH) (refer to Figure 3.1). The metagenomic studies conducted on desert microbial communities using the molecular techniques described in this section are summarized in Table 3.1. A detailed description of all these techniques is beyond the scope of this chapter, and readers may refer to the description presented by Paul and coworkers (2018) and Dubey and coworkers (2020).

The advent of NGS technology has expanded the scale and scope of metagenomic studies, resulting in a million-fold increase in the throughput. Over the last decade, a continuous drop in the cost of DNA sequencing has enabled more and more laboratories to undertake projects on metagenomics, and a huge amount of metagenome sequence data has already been generated. Depending on the sequencing strategy, NGS-based sequencing data can be generated for a targeted gene like the 16S rRNA gene or functional marker genes, an approach called amplicon sequencing or targeted sequencing. Contrary to this, if the sequencing data from a metagenome are generated without targeting any specific gene region,

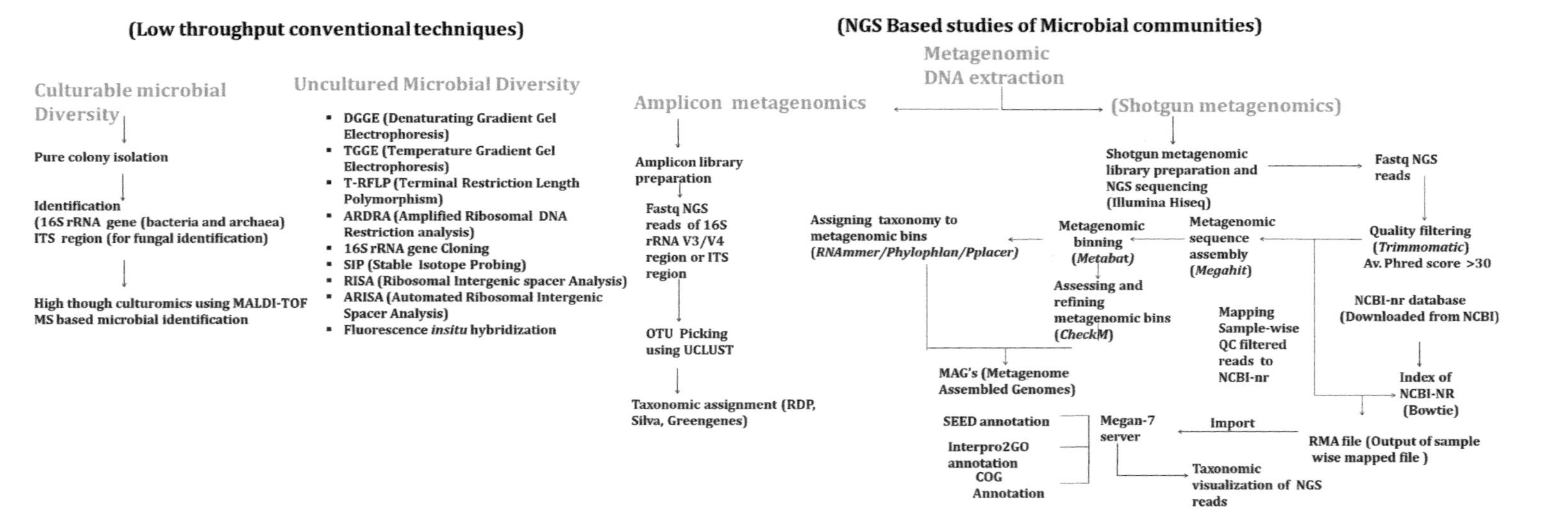

FIGURE 3.1 Metagenomic techniques employed to explore the microbial communities are outlined under two categories as (1) low throughput conventional techniques and (2) NGS-based high throughput techniques. A brief outline workflow used to analyze the amplicon and shotgun sequence NGS datasets is also depicted.

the approach is called shotgun metagenome sequencing. A 16S rRNA gene-based amplicon sequencing approach can be used to generate the information of the community structure and composition, but no functional profiling can be done using this approach. However, some bioinformatics tools like PICRUSt (Douglas et al. 2018) and Tax4Fun (Abhauer et al. 2015) are designed to indirectly predict microbiome function based on the 16S rRNA gene data alone. However, such approaches are indirect predictions and hence suffer poor reliability. Such 16S rRNA gene sequence-based imputed metagenomics and functional prediction tools like Tax4Fun have been successfully employed to infer the metabolic potential of microbial communities of the date palm (*Phoenix dactylifera*) of the Sahara Desert (Mosqueira et al. 2019). Similarly, PICRUSt has also been employed to predict the functional potential from amplicon sequence data of 16S rRNA gene sequences of the rhizobacterial community of *Cistanthe longiscapa* (*Montiaceae*), an endemic annual herb of the Atacama Desert (Astorga-Eló et al. 2020).

Shotgun metagenomics is one of the straightforward approaches to study the functional and metabolic potential of uncultivated bacteria in environmental samples. Shotgun metagenomic techniques have been applied effectively to generate information about the microbial communities of the desert ecosystem (Vikram et al. 2016). Unlike 16S rRNA gene-based metagenomic studies, shotgun metagenomics involves the random sequencing of metagenomic DNA aimed to sequence all the genes present in environmental samples. The shotgun metagenomic reads so generated can also be used for metagenomic assembly. Assembly-based metagenomics of sufficiently sequenced (depth and coverage) shotgun metagenomic sequence data can be used for metagenomic genome reconstruction. In this approach, the quality-filtered metagenome-derived short NGS reads are subjected to *de novo* metagenomic sequence assembly using the assembly tool MEGAHIT (Li et al. 2015) or metaSPAdes (Nurk et al. 2017). The resulting contigs from the metagenomic sequence assembly are binned into metagenomic bins based on the differential coverage and the tetra-nucleotide frequencies of contigs using bioinformatics tools like metaBAT2 (Kang et al. 2019). The genomic bins so obtained can be refined using tools like CheckM (Parks et al. 2015). Subsequently, the individual bins can be refined to generate draft genomes, also called metagenome assembled genomes (MAGs). Such reconstructed genomes or MAGs from shotgun metagenomic sequence data can subsequently be annotated to gain an understanding of the physiology of uncultured microbial members.

In one such shotgun metagenomic study by Crits-Christoph and co-workers, the authors generated metagenomic sequence data on hypersaline endolithic communities of halites from the Atacama Desert using shotgun metagenomics (2016). The authors extracted metagenomic DNA from five halite nodules and generated 9.6 Gb of high-quality paired-end, metagenomic shotgun sequences taxonomically belonging to *Archaea* (71%), *Bacteria* (27%), and lesser *Eukarya* (1%). Using metagenomic assembly, they could construct the 1.1-Mbp genome of the nanohaloarchaeon (termed SG9) with 1,292 CDS and 46.4% GC content. Nanohaloarchea is an uncultured group under the domain archaea, phylogenetically closer to members of the family Halobacteriaceae, but no cultured representative of this group has been isolated so far. Nanohaloarchaea is a very important microbial group whose existence and biology have been decoded through metagenomics. Taxonomically, Nanohaloarchaea is one of the few known classes of the phylum Euryarchaeota. Other major classes of phylum Euryarchaeota include Halobacteria, Methanobacteria, Methanopyri, Methanomicrobia, Methanococci, Archaeoglobi, and Thermococci. The members of Nanohaloarchaea largely inhabit highly saline environments, and their MAGs have been recovered from diverse habitats like hypersaline lakes, solar salterns, halite nodules, the hyperarid Atacama Desert, and soda lakes (Narasingarao et al. 2012; Martinez-Garcia et al. 2014; Crits-Christoph et al. 2016; Finstad et al. 2017; Vavourakis et al. 2016). The MAGs of Nanohaloarchea were first reported in the water of hypersaline lakes (only in 0.1 μm filtered fraction), namely Tyrrell, located in Victoria, Australia. The MAGs of the two Nanohaloarchaea (*Ca. Nanosalina* sp. J07AB43 and *Ca. Nanosalinarum* sp. J07AB56) reported by Narasingarao et al. from the Tyrrell Lake metagenome were found to be substantially different in terms of a smaller genome size of 1.2 Mb and fewer coding sequence (CDS) (1,407 and 1,677 for the two MAGs) compared to the previously sequenced extreme halophiles in the archaeal domain (Narasingarao et al. 2012). On one side, as a general observation, all the halophilic archaea, as well as the halophilic bacterium *Salinibacter ruber*, have G+C content of 60% or even greater; the GC content of the two MAGs *Ca. Nanosalina* sp. J07AB43 and *Ca. Nanosalinarum* sp. J07AB56 was 43% and 56%, respectively. The estimated total genome size of Nanohaloarchaea was also found to be very small (1.2 Mb) compared to other known archaeal genomes, which range from 2.7 to 5.4 Mbp. However, the presence of all the archaeal housekeeping genes in the reported MAGs was evidence that the recovered MAGs were nearly complete. The comparison of the genome-derived proteome of two Nanohaloarchaeal members with other archaeal and extremely halophilic bacterial genomes revealed highly unusual amino acid compositions supporting the 'salt-in' strategy for osmotic balance. The proteome was found to be over-represented by amino acids with negatively charged side chains (aspartic and glutamic acid) and under-represented by residues with bulky hydrophobic side chains (tryptophan, phenylalanine, and isoleucine).

Crits-Christoph and co-workers also predicted salt adaptation strategies in MAGs of Nanohaloarchaea based on the pI (isoelectric pH of genome-encoded proteins). Out of the total predicted proteins, 79% of predicted proteins could not be assigned to a known function, showing the unique physiology of the nanohaloarchaeon SG9 (2016). However, genome annotation revealed the presence of the three potassium uptake systems, Trk, Ktr, and HKT, in Nanohaloarchea SG9. The presence of Trk, Ktr, and HKT uptake systems derives potassium ion selective influx into the cytoplasm of Nanohaloarchrae SG9 to survive under extreme salt conditions. A detailed description of the MAGs of Nanohaloarchea reported so far describing their genome size, GC%, number of CDS, source metagenome, and salt adaptation strategy has been presented in a nice review by Kumar and coworkers (2020).

Metagenomic techniques employed to explore microbial communities are outlined under two categories, low-throughput conventional techniques and NGS-based high-throughput techniques. A brief outline workflow used to analyze amplicon and shotgun sequence NGS datasets is also discussed. The application of a few of these metagenomic techniques for the study of desert microbial communities is presented in Table 3.1

3.5 Desert Vegetation and Associated Microbiome

One-third of the total land surface area on the earth is chronically challenged with the availability of sufficient water (Peel et al. 2007). The flora of such arid and semi-arid habitats is simultaneously subjected to multiple abiotic stresses, including chronic drought and extreme temperature fluctuations. The prevailing conditions of low water availability, high temperatures, and salinity influence the plant diversity and distribution in desert ecosystems. However, the members of Cactaceae, together with agaves and many other xerophytes, manage to survive in such extreme climatic conditions. Cacti, in particular, manage to survive long periods of drought while maintaining a positive tissue water status (Edwards & Donoghue 2006). These plants have succulent bodies that are used for water storage, crassulacean acid metabolism for photosynthesis, and many other morphological adaptations.

The encounter with drought stress brings out changes in the soil microbial community, which also affect the root microbiome of the plants. Drought stress–mediated changes in the soil microbiome alter the available pool of bacteria from which plants recruit endophytic communities growing in such soil (Naylor and Coleman-Derr 2018). In the recent past, many studies have been conducted to understand key factors like the influence of host genotype, the influence of the soil physicochemical properties, the seasonal variations, the agricultural management, and the biogeography of the plant species on the plant microbiome (Caruso et al. 2011; Lundberg et al. 2012; Desgarennes et al. 2014; Coleman-Derr et al. 2016). In an amplicon sequencing-based study of the desert microbiome, Eida et al. used Illumina's MiSeq based amplicon sequencing of the V3–V4 region of the 16S rRNA gene to reveal the microbial community structure in the soil, rhizosphere, and endosphere of four desert plants, *Ribulus terrestris*, *Zygophyllum simplex*, *Panicum turgidum*, and *Euphorbia granulate*, and reported the dominance of seven bacterial phyla, with Actinobacteria and Proteobacteria being the most abundant phyla in the endosphere, rhizosphere, and bulk soil (2018).

A sound understanding of the interplay of factors like drought and how exactly it affects root-associated bacterial communities is an essential step in developing strategies to combat drought stress–mediated agricultural losses. Microbes isolated from desert plants have been shown to exhibit rock-weathering activity, and they have also been shown to aid in growth promotion in cacti developing on rocky cliffs and large rocks (Puente et al. 2004; Puente 2009). The advent of metagenomics and the development of

TABLE 3.1
Metagenomic Studies of Desert Microbial Communities, along with a Description of the Molecular Approach Used and the Salient Research Findings

Study Site	Metagenomic Approaches Employed	Salient Findings	Type of Sample Analyzed	Reference
Atacama Desert, South America	Illumina MiSeq sequencing of 16S rRNA gene and ITS2 regions	Rhizosphere microbiota differed in composition compared to surrounding bulk soil Diversity and richness of fungal OTUs were negatively correlated with aridity, whereas diversity, composition, and structure of the bacterial community were not influenced by aridity	Rhizospheric soil DNA of *Cistanthe longiscapa*, a native annual plant of the Atacama desert, and the bulk soil metagenome was sequenced	Araya et al. 2020
Namib Desert, Namibia	T-RFLP fingerprinting based on 16S rRNA gene (different primers used for bacteria and archaea) and ITS gene	There were noted soil niche–specific bacterial, fungal, and archaeal communities across samples. However, the abiotic drivers of community structure were domain specific	Surface soil samples from nine edaphic environments (niches) were collected in the central Namib Desert, Namibia, including gravel plain soil sample, dune soils, interdune soils, riverbed soils, salt pan	Johnson et al. 2017
Lejía Lake soil in Atacama Desert	Illumina MiSeq sequencing of the V1–V3 region of the 16S rRNA gene (2 × 300 bp)	The soil samples from Lejía Lake were dominated by *Bacteroidetes*, *Proteobacteria*, and *Firmicutes*, with 29.2%, 28.2%, and 28.1% of the relative abundance, respectively, with *Halomonas* as a dominating genus	Three soil samples from Lejía lakeshore were sequenced	Mandakovic et al. 2018
Negev Desert in Israel	Phospholipid fatty acid (PLFA) and DGGE	DGGE-based microbial profiling suggested differences in community structure of shrub-covered and inter-shrub soil of both sites	Soil samples from desert soil covered with shrubs like *Zygophyllum dumosum* and *Hammada scopariain* were studied The semi-arid soil covered with shrubs *Noaea mucronata* and *Thymelaea hirsutainshru* and the winter shrub soil	Ben-David et al. 2011

Guanajuato and Jalisco states in Mexico and Philip L. Boyd Deep Canyon Desert Reserve in California	Illumina amplicon sequencing of ITS2 and 16S regions	It was observed that the plant compartment had a stronger influence on prokaryotic community structure, whereas the fungal communities were mainly influenced by biogeography Cultivation of *A. tequilana* negatively influenced prokaryotic diversity	Samples were taken from the soil, root zone, rhizosphere, phyllosphere, root endosphere, and leaf endosphere of three agave species	Coleman-Derr et al. 2016
Namib Desert	Illumina MiSeq sequencing of soil metagenome using bacterial/ archaeal 515f and 806r primers	Microbial communities were dominated by *Bacteroidetes* (29% of all sequences), *Proteobacteria* (23%, mostly *Alphaproteobacteria* [17%] and *Betaproteobacteria* [3%]), and *Actinobacteria* (22%), followed by *Firmicutes* (4%), *Acidobacteria* (4%), *Chloroflexi* (4%), and *Verrucomicrobia* (3%)	128 soil samples collected from gravel plains in the central Namib Desert were included in the study	Armstrong et al. 2016
Namib Desert	Metagenome shotgun sequencing using Illumina's Hiseq-2000 paired-end technology	Bacterial phylotypes (93%) were predominant, whereas archaea (0.43%) and fungi (5.6%) were present a minor portion of microbial hypolith communities, along with many double-stranded viruses as the part of the microbial community. Cyanobacteria carried the major photosynthetic activity and hence the primary production	50 hypolith samples were collected from the Namib Desert	Vikram et al. 2016
Yungay region of the Atacama Desert	16S rRNA gene cloning	*Actinobacteria*, *Proteobacteria*, *Firmicutes*, and TM7 division bacteria were detected. Phylum *Actinobacteria* dominated samples, as 94% of total cloned sequences belonged to *Actinobacteria*	Surface soil samples and corresponding subsurface soil samples were collected and studied using PFLA and 16S rRNA gene cloning	Connon et al. 2007

the experimental framework to study microbial communities in a high-throughput manner have opened ways to explore the role of xerophyte-associated microbiomes as an important facet in drought stress research. It is now a well-recognized fact that all plants and animals are 'holobionts' (host + all associated microbes) that form tight associations with microorganisms (Vandenkoornhuyse et al. 2015). These connections between the host and all microbial symbionts form a biological unit on which evolutionary processes act. Reports on the colonization of bacterial, archaeal, and fungal community members suggest that the adaptation of plant agaves and cacti to arid environments could be mediated not only by their inherent characteristics (CAM metabolism, arid-adapted morphology) but also by their symbiotic microorganisms (Citlali et al. 2018). It is also believed that native plants like cacti and agaves, which have evolved in deserts under the selection pressure of arid conditions, have only selected those microbes that helped to impart tolerance to arid conditions during evolution (Eida et al. 2018). However, there is still meager information available about the structural, compositional, and functional aspects of the microbial communities associated with xerophytes (Fonseca-Garcia et al. 2016). Recently, a few encouraging findings suggest the study of the plant microbiome as an important facet in abiotic stress research. The plant-associated microbes of the xerophytes may play an important role in conferring the ability of xerophytes to thrive and produce large amounts of biomass in arid conditions. Now it has been shown that the plant-associated microbes are transmitted vertically from generation to generation, suggesting they are important breeding criteria for the selection of genotypes (Gopal & Gupta 2016; Wei and Jousset 2017; Mitter et al. 2017). Mitter et al. (2017) successfully showed the vertical transmission of the endophyte *Paraburkholderia phytofirmans* PsJN introduced in the flowers of wheat and few other monocots and its subsequent inclusion in progeny seed microbiomes. The study of the metagenome of associated desert plants has unfolded many details of the stress tolerance strategy of the associated microbes. In one such study, Finkel and coworkers employed a shotgun metagenomic approach for understanding the salt tolerance mechanism of the uncultured microbes native to the leaves of a salt-secreting desert tree, *Tamarix aphylla*, popularly known as salt cedar (Finkel et al. 2016). The leaves of *Tamarix aphylla* form salt crystals, and leaf microbial communities are largely dominated by microbial members of the Halomonadaceae family (Finkel et al. 2011). Using shotgun metagenomics, authors generated 2.3×10^8 overlapping paired-end reads from Tamarix leaf samples and constructed 17 MAGs, of which 6 were more than 80% complete. The resulted 17 genomic bins varied in size from 200 kb to 3.5 Mb, with genomic completeness ranging from 47% to 99%. Out of the 17 metagenomic bins, 3 bins, EB6, EB8, and SB1, belonged to Oceanospirillales that require high salt concentration (Garrity et al. 2015). In a habitat such as the Tamarix phyllosphere, with a high concentration of Na^+ ions and a low concentration of H^+ ions, it was expected that Na^+/H^+ antiport would be the prevalent mechanism of salt tolerance. The genome annotation of the metagenomic bins revealed the presence of multiple copies of the Na^+/H^+ antiporter genes in all the metagenomics bins except those that originated from insect endosymbiont microbes (Kumar et al. 2020). However, the isoelectric pH (pI) prediction of the annotated MAGs and the amino acid usage pattern in metagenome genome-encoded proteins can give a fair view of the salt stress adaptation strategy of such uncultured microbes.

3.6 Microbiomes of Cold Deserts

Low temperature is quite prevalent on the Earth, mainly at polar regions, glaciers, tundra, and permafrost alpine habitats. About 98% of Antarctica is covered with ice and hence termed a polar desert. Alpine ecosystems also have low temperatures and represent many biotypes, such as soils, bare rocks, permafrost, glaciers, and snow. In broader terms, 70% of the earth is covered with seawater, of which two-thirds has a temperature of around 2°C (Gilichinsky & Wagener 1995). Apart from the Antarctic desert, Gobi Desert (Eastern Asia), Patagonian Desert (South America), Great Basin (United States), Karakum Desert (Turkmenistan), Colorado Plateau (Turkmenistan), Kyzylkum Desert (Central Asia), Taklamakan Desert (China), and Columbia Basin (United States) are the noted cold deserts of the world. Initially, it was believed that microbes are metabolically inactive at subzero temperatures;

however, it is now shown that microbial activity can occur at temperatures as low as –39°C (Panikov et al. 2006), although the microbial communities of temperate climates like the tundra northeast of Siberia do contain very low cell densities due to the freezing temperatures, poor nutrient status, low water availability, and high incident radiation (Smith et al. 2006). Despite the environmental extremes, the cold desert mineral soils of the Antarctic Dry Valley are studied using culture-based approaches, and the bacterial isolates of genera *Pseudomonas*, *Flavobacterium*, *Arthrobacter*, *Brevibacterium*, *Cellulomonas*, and *Corynebacterium*, *Bacillus*, *Micrococcus*, *Nocardia*, and *Streptomyces* have been isolated (Cameron et al. 1972; Friedmann 1993). The isolation of the microbial isolates of genera *Bacillus*, Pseudomonas species, *Lysinibacillus*, *Microbacterium*, *Paenarthrobacter*, *Alcaligenes*, *Serratia*, *Carnobacterium*, *Rhodococcus*, and *Stenotrophomonas* have also been reported from the Pindari Glacier region and cold desert of the Indian Himalayan region (Pandey et al. 2019). Viable microbial cells have also been isolated from permafrost samples taken throughout the cold regions of Earth (Khlebnikova et al. 1990). For instance, a few bacterial isolates like *Micrococcus antarcticus*, capable of growing even at 0°C, have been isolated from the Chinese Great Wall station in Antarctica (Liu et al. 2000). Based on their T_{min}, T_{opt}, and T_{max}, the microbes are classified as psychrotrophic or psychrophilic microorganisms, the latter being restricted to permanently cold habitats like permafrost (Morita 1975). Culture-based analysis of alpine cold habitats reported the isolation of microbial isolates by producing a large amount of biotechnologically important compounds like catalase, lipase, urease, protease, amylase, and many organic acids (Marasco et al. 2012). It is also reported that some strains of *Streptomyces* isolated from permafrost in Spitsbergen could produce an exceptionally high amount of trehalose and glycerol, enabling this strain to survive in cold and dry conditions (Ivanova et al. 2008).

With the arrival of metagenomics, the microbial communities of the cold deserts have also been studied in a culture-independent manner (Lee et al. 2012; Van Horn et al. 2013; Garrido-Benavent et al. 2020). Studies employing metagenomic approaches have shown that the Antarctic desert soils contain much higher levels of microbial diversity than previously thought (Cowan et al. 2014). Studies on microbial mats of the McMurdo Dry Valleys have shown surprisingly high rates of carbon (CO_2) and dinitrogen (N_2) fixation, comparable to rates in warmer temperate or tropical environments, during a brief period of the austral summer when the temperature rises above the freezing point, resulting in the availability of liquid water for the photosynthetic activity mainly exhibited by Cyanobacterial mats (Sohm et al. 2020). Metagenomic approaches have been applied to study the different biotypes of cold deserts that include permafrost soils (Yergeau et al. 2010; Mackelprang et al. 2011; Ji et al. 2020), glacier fore fields (Brankatschk et al. 2011; Franzetti et al. 2020), and cold desert endoliths (Friedmann 1982; Coleine et al. 2020). The Tibetan plateau represents the largest area under alpine permafrost where studies on microbial communities are extensively conducted. A wide spectrum of microbial isolates from Chinese permafrost has been retrieved, which were comparable with the microbial community of permafrost in the Arctic (Hu et al. 2015). The microbial communities of permafrost samples are reported to be dominated by *Actinobacteria*, due to their ability to form resting cysts, which are essentially the resting stage (Soina et al. 2004).

3.7 Conclusions and Future Research Needs

Various culture-based and uncultured metagenomic studies have shown the unique microbial diversity in the hot and cold deserts of the globe and operational elemental cycles even in very low-temperature environments like permafrost. The microbial diversity of the desert ecosystem offers a unique opportunity to research how to exploit this hidden microbial treasure for its biotechnological potential. The gene pool of arid and hyperarid desert microbial communities and cold deserts can be exploited to gain a deeper understanding of the stress tolerance mechanism of the native microbes, which are good candidates for transgenic research for imparting similar attributes to major crops. Moreover, such microbes are a great source of stress-resistant microbial enzymes greatly suited to industrial processes where extremes of temperature are used, requiring thermo-stable enzymes.

REFERENCES

Abdel-Hafez, S. I. I. 1981. "Halophilic fungi of desert soils in Saudi Arabia." *Mycopathologia*, 75(2), 75–80.

Abhauer, K. P., Wemheuer, B., Daniel, R., & Meinicke, P. 2015. "Tax4Fun: Predicting functional profiles from metagenomic 16S rRNA data." *Bioinformatics*, 31(17), 2882–2884.

Alotaibi, M. O., Sonbol, H. S., Alwakeel, S. S., Suliman, R. S., Fodah, R. A., Jaffal, A. S. A., et al. 2020. "Microbial diversity of some Sabkha and desert sites in Saudi Arabia." *Saudi Journal of Biological Sciences*, 27(10), 2778–2789.

Alsohaili, S. A., & Bani-Hasan, B. M. 2018. "Morphological and molecular identification of fungi isolated from different environmental sources in the northern eastern desert of Jordan." *Jordan Journal of Biological Sciences*, 11(3).

Araya, J. P., González, M., Cardinale, M., Schnell, S., & Stoll, A. 2020. "Microbiome dynamics associated with the Atacama flowering desert." *Frontiers in Microbiology*, 10, 3160.

Armstrong, A., Valverde, A., Ramond, J. B., Makhalanyane, T. P., Jansson, J. K., Hopkins, D. W., et al. 2016. "Temporal dynamics of hot desert microbial communities reveal structural and functional responses to water input." *Scientific Reports*, 6, 34434.

Astorga-Eló, M., Zhang, Q., Larama, G., Stoll, A., Sadowsky, M. J., & Jorquera, M. A. 2020. "Composition, predicted functions and co-occurrence networks of rhizobacterial communities impacting flowering desert events in the Atacama Desert, Chile." *Frontiers in Microbiology*, 11, 571.

Ben-David, E. A., Zaady, E., Sher, Y., & Nejidat, A. 2011. "Assessment of the spatial distribution of soil microbial communities in patchy arid and semi-arid landscapes of the Negev Desert using combined PLFA and DGGE analyses." *FEMS Microbiology Ecology*, 76(3), 492–503.

Brankatschk, R., Towe, S., Kleineidam, K., Schloter, M., & Zeyer, J. 2011. "Abundances and potential activities of nitrogen cycling microbial communities along a chronosequence of a glacier forefield." *The ISME Journal*, 5(6), 1025–1037.

Bull, A. T., & Asenjo, J. A. 2013. "Microbiology of hyper-arid environments: Recent insights from the Atacama Desert, Chile." *Antonie Van Leeuwenhoek*, 103(6), 1173–1179.

Cameron, R. E., RE, C., FA, M., & RM, J. 1972. "Bacterial species in soil and air of the Antarctic continent." *Antarctic Journal*, 7, 187–198.

Caruso, T., Chan, Y., Lacap, D. C., Lau, M. C. Y., McKay, C. P., & Pointing, S. B. 2011. "Stochastic and deterministic processes interact in the assembly of desert microbial communities on a global scale." *The ISME Journal*, 5, 1406–1413. doi: 10.1038/ismej.2011.21.

Citlali, F. G., Desgarennes, D., Flores-Núñez, V. M., & Partida-Martínez, L. P. 2018. "The microbiome of desert CAM plants: Lessons from amplicon sequencing and metagenomics." In *Metagenomics* (pp. 231–254). Academic Press, United Kingdom.

Coakley, J. A. 2003. "Reflectance and albedo, surface." *Encyclopedia of the Atmosphere*, 1914–1923.

Cole, J. R., Wang, Q., Fish, J. A., Chai, B., McGarrell, D. M., Sun, Y., et al. 2014. "Ribosomal database project: Data and tools for high throughput rRNA analysis." *Nucleic Acids Research*, 42, D633–D642.

Coleine, C., Albanese, D., Onofri, S., Tringe, S. G., Pennacchio, C., Donati, C., et al. 2020. "Metagenomes in the borderline ecosystems of the Antarctic cryptoendolithic communities." *Microbiology Resource Announcements*, 9(10).

Coleman-Derr, D., Desgarennes, D., Fonseca-Garcia, C., Gross, S., Clingenpeel, S., Woyke, T., et al. 2016. "Plant compartment and biogeography affect microbiome composition in cultivated and native agave species." *New Phytologist*, 209, 798–811. doi: 10.1111/nph.13697.

Connon, S. A., Lester, E. D., Shafaat, H. S., Obenhuber, D. C., & Ponce, A. 2007. Bacterial diversity in hyper-arid Atacama Desert soils." *Journal of Geophysical Research: Biogeosciences*, 112(G4).

Contador, C. A., Veas-Castillo, L., Tapia, E., Antipán, M., Miranda, N., Ruiz-Tagle, B., et al. 2020. "Atacama database: A platform of the microbiome of the Atacama Desert." *Antonie van Leeuwenhoek*, 113(2), 185–195.

Cowan, D. A., Makhalanyane, T. P., Dennis, P. G., & Hopkins, D. W. 2014. "Microbial ecology and biogeochemistry of continental Antarctic soils." *Frontiers in Microbiology*, 5, 154.

Crits-Christoph, A., Gelsinger, D. R., Ma, B., Wierzchos, J., Ravel, J., Davila, A., et al. 2016. "Functional interactions of archaea, bacteria and viruses in a hypersaline endolithic community." *Environmental Microbiology*, 18, 2064–2077.

DeSantis, T. Z., Hugenholtz, P., Larsen, N., Rojas, M., Brodie, E. L., Keller, K., & Andersen, G. L. 2006. "Greengenes, a chimera-checked 16S rRNA gene database and workbench compatible with ARB." *Applied and Environmental Microbiology*, 72(7), 5069–5072.

Desgarennes, D., Garrido, E., Torres-Gomez, M. J., Peña-Cabriales, J. J., & Partida-Martinez, L. P. 2014. "Diazotrophic potential among bacterial communities associated with wild and cultivated *Agave* species." *FEMS Microbiology Ecology*, 90, 844–857. doi: 10.1111/1574-6941.12438.

Douglas, G. M., Beiko, R. G., & Langille, M. G. 2018. "Predicting the functional potential of the microbiome from marker genes using PICRUSt." In *Microbiome analysis* (pp. 169–177). Humana Press, New York.

Dubey, R. K., Tripathi, V., Prabha, R., Chaurasia, R., Singh, D. P., Rao, C. S., et al. 2020. "Methods for exploring soil microbial diversity." In *Unravelling the soil microbiome* (pp. 23–32). Springer, Cham.

Edwards, E. J., & Donoghue, M. J. 2006. "Pereskia and the origin of the cactus life-form." *The American Naturalist*, 167(6), 777–793.

Eida, A. A., Ziegler, M., Lafi, F. F., Michell, C. T., Voolstra, C. R., Hirt, H., & Saad, M. M. 2018. "Desert plant-bacteria reveal host influence and beneficial plant growth properties." *PLoS One*, 13(12), e0208223.

Fierer, N., Leff, J. W., Adams, B. J., Nielsen, U. N., Bates, S. T., Lauber, C. L., et al. 2012. "Cross-biome metagenomic analyses of soil microbial communities and their functional attributes." *PNAS, USA*, 109(52), 21390–21395.

Finkel, O. M., Burch, A. Y., Lindow, S. E., Post, A. F., & Belkin, S. 2011. "Geographical location determines the population structure in phyllosphere microbial communities of a salt-excreting desert tree." *Applied and Environmental Microbiology*, 77(21), 7647–7655.

Finkel, O. M., Delmont, T. O., Post, A. F., & Belkin, S. 2016. "Metagenomic signatures of bacterial adaptation to life in the phyllosphere of a salt-secreting desert tree." *Applied and Environmental Microbiology*, 82(9), 2854–2861.

Finstad, K. M., Probst, A. J., Thomas, B. C., Andersen, G. L., Demergasso, C., Echeverría, A., et al. 2017. "Microbial community structure and the persistence of cyanobacterial populations in salt crusts of the hyperarid Atacama Desert from genome-resolved metagenomics." *Frontiers in Microbiology*, 8, 1435.

Fonseca-García, C., Coleman-Derr, D., Garrido, E., Visel, A., Tringe, S. G., & Partida-Martínez, L. P. 2016. "The cacti microbiome: Interplay between habitat-filtering and host-specificity." *Frontiers in Microbiology*, 7, 150.

Franzetti, A., Pittino, F., Gandolfi, I., Azzoni, R. S., Diolaiuti, G., Smiraglia, C., & Ambrosini, R. 2020. "Early ecological succession patterns of bacterial, fungal and plant communities along a chronosequence in a recently deglaciated area of the Italian Alps." *FEMS Microbiology Ecology*, 96(10), fiaa165.

Friedmann, E. I. 1982. "Endolithic microorganisms in the Antarctic cold desert." *Science*, 215(4536), 1045–1053.

Friedmann, E. I. 1993. *Antarctic microbiology* (p. 634). Wiley-Liss, New York.

Garrido-Benavent, I., Pérez-Ortega, S., Durán, J., Ascaso, C., Pointing, S. B., Rodríguez-Cielos, et al. 2020. "Differential colonization and succession of microbial communities in rock and soil substrates on a maritime Antarctic glacier forefield." *Frontiers in Microbiology*, 11, 126.

Garrity, G. M., Bell, J. A., & Liburn, T. 2015. "*Oceanospirillales* ord. nov." In W. B. Whitman, ed., *Bergey's Manual of Systematics of Archaea and Bacteria* (p. 1). doi: 10.1002/9781118960608.obm0010.

Gilichinsky, D., & Wagener, S. 1995. "Microbial life in permafrost: A historical review." *Permafrost and Periglacial Processes*, 6, 243–250.

Goodfellow, M., Nouioui, I., Sanderson, R., Xie, F., & Bull, A. T. 2018. "Rare taxa and dark microbial matter: Novel bioactive actinobacteria abound in Atacama Desert soils." *Antonie van Leeuwenhoek*, 111(8), 1315–1332.

Gopal, M., & Gupta, A. 2016. "Microbiome selection could spur next-generation plant breeding strategies." *Frontiers in Microbiology*, 7, 1971.

Harwani, D. 2013. "The great plate count anomaly and the unculturable bacteria." *Microbiology*, 2(9), 350–351.

Hu, W., Zhang, Q., Tian, T., Cheng, G., An, L., & Feng, H. 2015. "The microbial diversity, distribution, and ecology of permafrost in China: A review." *Extremophiles*, 19(4), 693–705.

Ivanova, V., Lyutskanova, D., Stoilova-Disheva, M., Kolarova, M., Aleksieva, K., Raykovska, V., et al. 2008. "Isolation and identification of α, α-Trehalose and glycerol from an Arctic psychrotolerant *Streptomyces* sp. SB9 and their possible role in the strain's survival." *Preparative Biochemistry and Biotechnology*, 39(1), 46–56.

Jaouani, A., Neifar, M., Prigione, V., Ayari, A., Sbissi, I., Ben Amor, S., et al. 2014. "Diversity and enzymatic profiling of halotolerant micromycetes from Sebkha El Melah, a Saharan salt flat in southern Tunisia." *BioMed Research International* 439197. http://dx.doi.org/10.1155/2014/439197.

Ji, M., Kong, W., Liang, C., Zhou, T., Jia, H., & Dong, X. 2020 "Permafrost thawing exhibits a greater influence on bacterial richness and community structure than permafrost age in Arctic permafrost soils." *The Cryosphere*, 14(11), 3907–3916.

Johnson, R. M., Ramond, J. B., Gunnigle, E., Seely, M., & Cowan, D. A. 2017. "Namib Desert edaphic bacterial, fungal and archaeal communities assemble through deterministic processes but are influenced by different abiotic parameters." *Extremophiles*, 21(2), 381–392.

Kang, D. D., Li, F., Kirton, E., Thomas, A., Egan, R., An, H., & Wang, Z. 2019. "MetaBAT 2: An adaptive binning algorithm for robust and efficient genome reconstruction from metagenome assemblies." *PeerJ*, 7, e7359.

Khairnar, M., Hagir, A., Narayan, A., Jain, K., Madamwar, D., Shouche, Y. S., & Rahi, P. 2020. "*Rhizobium desertarenae* sp. nov., isolated from the saline desert soil from the Rann of Kachchh, India." *bioRxiv*.

Khlebnikova, G. M., Gilichinskii, D. A., Fedorov-Davydov, D. G., & Vorobeva, E. A. 1990. "Quantitative evaluation of microorganisms in permafrost deposits and buried soils." *Microbiology (New York)*, 59(1), 106–112.

Kieft, T., & Skujinš, J. 1991. "Soil microbiology in reclamation of arid and semiarid lands." In J. Skujins, ed., *Semiarid lands and deserts: Soil resource and reclamation* (pp. 209–256). Marcel Dekker, New York.

Kumar, S., Krishnani, K. K., Bhushan, B., & Brahmane, M. P. 2015. "Metagenomics: Retrospect and prospects in high throughput age." *Biotechnology Research International*, 21735.

Kumar, S., Paul, D., Bhushan, B., Wakchaure, G. C., Meena, K. K., & Shouche, Y. 2020. "Traversing the 'omic' landscape of microbial halotolerance for key molecular processes and new insights." *Critical Reviews in Microbiology*, 46(6), 631–653.

Kurli, R., Chaudhari, D., Pansare, A. N., Khairnar, M., Shouche, Y. S., & Rahi, P. 2018. "Cultivable microbial diversity associated with cellular phones." *Frontiers in Microbiology*, 9, 1229.

Lacap-Bugler, D. C., Lee, K. K., Archer, S., Gillman, L. N., Lau, M. C., Leuzinger, S., et al. 2017. "Global diversity of desert hypolithic cyanobacteria." *Frontiers in Microbiology*, 8, 867.

Lebre, P. H., De Maayer, P., & Cowan, D. A. 2017. "Xerotolerant bacteria: Surviving through a dry spell." *Nature Reviews Microbiology*, 15(5), 285–296.

Lee, C. K., Barbier, B. A., Bottos, E. M., McDonald, I. R., & Cary, S. C. 2012. "The inter-valley soil comparative survey: The ecology of Dry Valley edaphic microbial communities." *The ISME Journal*, 6(5), 1046–1057.

Li, D., Liu, C. M., Luo, R., Sadakane, K., & Lam, T. W. 2015. "MEGAHIT: An ultra-fast single-node solution for large and complex metagenomics assembly via succinct de Bruijn graph." *Bioinformatics*, 31(10), 1674–1676.

Liu, H., Xu, Y., Ma, Y., & Zhou, P. 2000. "Characterization of *Micrococcus antarcticus* sp. nov., a psychrophilic bacterium from Antarctica." *International Journal of Systematic and Evolutionary Microbiology*, 50, 715–719.

Lundberg, D. S., Lebeis, S. L., Herrera Paredes, S., Yourstone, S., Gehring, J., Malfatti, S., et al. 2012. "Defining the core *Arabidopsis thaliana* root microbiome." *Nature*, 488, 86–90. doi: 10.1038/nature11237.

Mackelprang, R., Waldrop, M. P., DeAngelis, K. M., David, M. M., Chavarria, K. L., & Blazewicz, S. 2011. "Metagenomic analysis of a permafrost microbial community reveals a rapid response to thaw." *Nature*, 480(7377), 368–371.

Makhalanyane, T. P., Valverde, A., Gunnigle, E., Frossard, A., Ramond, J. B., & Cowan, D. A. 2015. "Microbial ecology of hot desert edaphic systems." *FEMS Microbiology Reviews*, 39(2), 203–221.

Mandakovic, D., Maldonado, J., Pulgar, R., Cabrera, P., Gaete, A., Urtuvia, V., & González, M. 2018. "Microbiome analysis and bacterial isolation from Lejía Lake soil in Atacama Desert." *Extremophiles*, 22(4), 665–673.

Marasco, R., Rolli, E., Ettoumi, B., Vigani, G., Mapelli, F., Borin, S., et al. 2012 "A drought resistance-promoting microbiome is selected by root system under desert farming." *PLoS One*, 7(10), e48479.

Martinez-Garcia, M., Santos, F., Moreno-Paz, M., Parro, V., & Anton, J. 2014. "Unveiling viral–host interactions within the 'microbial dark matter'." *Nature Communications*, 5, 4542.

Middleton, N., & Thomas, D. 1997. *World atlas of desertification*. 2nd ed. Arnold, Hodder Headline Group, London.

Mitter, B., Pfaffenbichler, N., Flavell, R., Compant, S., Antonielli, L., Petric, A., et al. 2017. "A new approach to modify plant microbiomes and traits by introducing beneficial bacteria at flowering into progeny seeds." *Frontiers in Microbiology*, 8, 11.

Morita, R. Y. 1975. "Psychrophilic bacteria." *Bacteriological Reviews*, 39(2), 144.

Mosqueira, M. J., Marasco, R., Fusi, M., Michoud, G., Merlino, G., Cherif, A., & Daffonchio, D. 2019. "Consistent bacterial selection by date palm root system across heterogeneous desert oasis agroecosystems." *Scientific Reports*, 9(1), 1–12.

Narasingarao, P., Podell, S., Ugalde, J. A., Brochier-Armanet, C., Emerson, J. B., Brocks, J. J., et al. 2012. "*De novo* metagenomic assembly reveals abundant novel major lineage of Archaea in hypersaline microbial communities." *The ISME Journal*, 6, 81.

Naylor, D., & Coleman-Derr, D. 2018. "Drought stress and root-associated bacterial communities." *Frontiers in Plant Science*, 8, 2223.

Nurk, S., Meleshko, D., Korobeynikov, A., & Pevzner, P. A. 2017. "metaSPAdes: A new versatile metagenomic assembler." *Genome Research*, 27(5), 824–834.

Okoro, C. K., Brown, R., Jones, A. L., Andrews, B. A., Asenjo, J. A., Goodfellow, M., & Bull, A. T. 2009. "Diversity of culturable actinomycetes in hyper-arid soils of the Atacama Desert, Chile." *Antonie Van Leeuwenhoek*, 95(2), 121–133.

Pandey, A., Jain, R., Sharma, A., Dhakar, K., Kaira, G. S., Rahi, P., et al. 2019. "16S rRNA gene sequencing and MALDI-TOF mass spectrometry-based comparative assessment and bioprospection of psychrotolerant bacteria isolated from high altitudes under mountain ecosystem." *SN Applied Sciences*, 1(3), 278.

Panikov, N. S., Flanagan, P. W., Oechel, W. C., Mastepanov, M. A., & Christensen, T. R. 2006. "Microbial activity in soils frozen to below –39 C." *Soil Biology and Biochemistry*, 38(4), 785–794.

Parks, D. H., Imelfort, M., Skennerton, C. T., Hugenholtz, P., & Tyson, G. W. 2015. "CheckM: Assessing the quality of microbial genomes recovered from isolates, single cells, and metagenomes." *Genome Research*, 25(7), 1043–1055.

Paul, D., Kumar, S., Mishra, M., Parab, S., Banskar, S., & Shouche, Y. S. 2018. "Molecular genomic techniques for identification of soil microbial community structure and dynamics." In *Advances in soil microbiology: Recent trends and future prospects* (pp. 9–33). Springer, Singapore.

Peel, M. C., Finlayson, B. L., & McMahon, T. A. 2007. "Updated world map of the Köppen-Geiger climate classification." *Hydrology and Earth System Sciences*, 11(5), 1633–1644.

Pointing, S. B., & Belnap, J. 2012. "Microbial colonization and controls in dryland systems." *Nature Reviews Microbiology*, 10(8), 551–562.

Potts, M., & Webb, S. J. 1994. "Desiccation tolerance of prokaryotes." *Microbiology Reviews*, 58, 755–805.

Puente, M. E., Li, C. Y., & Bashan, Y. 2004. "Microbial populations and activities in the rhizoplane of rock-weathering desert plants. I. Root colonization and weathering of igneous rocks." *Plant Biology*, 6(5), 629–642.

Puente, M. E., Li, C. Y., & Bashan, Y. 2009. "Rock-degrading endophytic bacteria in cacti." *Environmental and Experimental Botany*, 66, 389–401. doi: 10.1016/j.envexpbot.2009.04.010.

Quast, C., Pruesse, E., Yilmaz, P., Gerken, J., Schweer, T., Yarza, P., et al. 2012. "The SILVA ribosomal RNA gene database project: Improved data processing and web-based tools." *Nucleic Acids Research*, 41(D1), D590–D596.

Rahi, P., Prakash, O., & Shouche, Y. S. 2016. "Matrix-assisted laser desorption/ionization time-of-flight mass-spectrometry (MALDI-TOF MS) based microbial identifications: Challenges and scopes for microbial ecologists." *Frontiers in Microbiology*, 7, 1359.

Rateb, M. E., Ebel, R., & Jaspars, M. 2018. "Natural product diversity of actinobacteria in the Atacama Desert." *Antonie van Leeuwenhoek*, 111(8), 1467–1477.

Rolli, E., Marasco, R., Vigani, G., Ettoumi, B., Mapelli, F., Deangelis, M. L., et al. 2015. "Improved plant resistance to drought is promoted by the root-associated microbiome as a water stress-dependent trait." *Environmental Microbiology*, 17(2), 316–331.

Shields, S. J. 1982. "Evapotranspiration in a desert environment." MSc Thesis, University of Arizona.

Smith, J. J., Tow, L. A., Stafford, W., Cary, C., & Cowan, D. A. 2006. "Bacterial diversity in three different Antarctic cold desert mineral soils." *Microbial Ecology*, 51(4), 413–421.

Sohm, J., Niederberger, T., Parker, A., Tirindelli, J., Gunderson, T., Cary, S., et al. 2020. "Microbial mats of the McMurdo Dry Valleys, Antarctica: Oases of biological activity in a very cold desert." *Frontiers in Microbiology*, 11.

Soina, V. S., Mulyukin, A. L., Demkina, E. V., Vorobyova, E. A., & El-Registan, G. I. 2004. "The structure of resting bacterial populations in soil and subsoil permafrost." *Astrobiology*, 4(3), 345–358.

Van Horn, D. J., Van Horn, M. L., Barrett, J. E., Gooseff, M. N., Altrichter, A. E., Geyer, K. M., et al. 2013. "Factors controlling soil microbial biomass and bacterial diversity and community composition in a cold desert ecosystem: Role of geographic scale." *PLoS One*, 8(6), e66103.

Vandenkoornhuyse, P., Quaiser, A., Duhamel, M., Le Van, A., & Dufresne, A. 2015. "The importance of the microbiome of the plant holobiont." *New Phytologist*, 206, 1196–1206. doi: 10.1111/nph.13312.

Vavourakis, C. D., Ghai, R., Rodriguez-Valera, F., Sorokin, D. Y., Tringe, S. G., Hugenholtz, P., & Muyzer, G. 2016. "Metagenomic insights into the uncultured diversity and physiology of microbes in four hypersaline soda lake brines." *Frontiers in Microbiology*, 7, 211.

Vikram, S., Guerrero, L. D., Makhalanyane, T. P., Le, P. T., Seely, M., & Cowan, D. A. 2016. "Metagenomic analysis provides insights into functional capacity in a hyperarid desert soil niche community." *Environmental Microbiology*, 18(6), 1875–1888.

Wang, K., & Dickinson, R. E. 2012. "A review of global terrestrial evapotranspiration: Observation, modeling, climatology, and climatic variability." *Reviews of Geophysics*, 50(2).

Wei, Z., & Jousset, A. 2017. "Plant breeding goes microbial." *Trends in Plant Science*, 22(7), 555–558.

Woese, C. R. 1987. "Bacterial evolution." *Microbiological Reviews*, 51(2), 221.

Yergeau, E., Hogues, H., Whyte, L. G., & Greer, C. W. 2010. "The functional potential of high Arctic permafrost revealed by metagenomic sequencing, qPCR and microarray analyses." *The ISME Journal*, 4(9), 1206–1214.

4

Metagenomic Approaches for Exploration of Halophilic Prokaryotes in Coastal Areas

Jamseel Moopantakath, Madangchanok Imchen, Athira CH, Ranjith Kumavath*
Department of Genomic Science, Central University of Kerala, Tejaswini Hills, Periya, Kasaragod, Kerala, India

CONTENTS

4.1 Introduction

The ocean and its coastline cover up to 70% of the Earth's surface (Austin et al. 2019). The coastline constantly experiences natural abiotic and anthropological actions, including thermal expansion, storms, alteration of wind, acidification, sand mining, and so on (Foti et al. 2020). There is also a strong distinction in various biotic and abiotic features among the Atlantic, Arctic, Indian, Pacific, and Southern Oceans (Bollmann 2010; Moopantakath et al. 2021). The physicochemical properties of marine biosystems also vary based on the geographical location, sea, and surrounding water bodies (Delcroix et al. 2005; Lee et al. 2006; Pradhan and Suleiman 2009). Such differences enrich unique microbial communities. Coastal ecosystems are interconnected with lakes, rivers, mangroves, brackish water, and hypersaline saltpan (Moopantakath et al. 2020). These biosystems play crucial roles in the biogeochemical cycle and ecological sustainability. Furthermore, the sediments of the coastal ecosystem, including the shorelines between the land and sea, harbor a rich deposit of minerals and nutrients. Marine ecosystems are therefore enriched with a diverse set of species under plants, animals, fishes, crustaceans, and microbial domains. Coastal ecosystems are also faced with rapid changes in pH, temperature, and salinity (Proum et al. 2018). These frequent modulations in the environmental conditions enrich unique extremophilic microbes. Marine microorganisms are a prolific source of metabolites with pharmaceutical,

DOI: 10.1201/9781003042570-7

industrial, agricultural, and biotechnological applications (Prasad and Murugadas 2018). Metabolites of marine sources such as terpenoids, phenolic compounds, and other metabolites have shown high biological activities (Abad et al. 2011; Parthasarathy et al. 2020). The actinobacterial species *Streptomyces* sp. are hyperproducers of a secondary metabolite that has shown potential scope in antimicrobial and anticancer drug discovery (Al-Dhabi et al. 2020). Furthermore, other marine microbes have also been prolific sources of bioactive compounds. For instance, cellulase and xylanase enzymes were isolated from *Trichoderma pleuroticola* in the coastal area (Korkmaz et al. 2017). *Nocardiopsis* sp. DW-4, isolated from the Indian coastal ecosystem, was found to exhibit amylase activity (Kalasava et al. 2020). Biomolecules such as Borrelidin-C and -D derived from halophilic *Nocardiopsis* spp. have antimicrobial activity against *Salmonella enterica* (Kim et al. 2017). The crude extract of salt-tolerant fungi such as *Aspergillus flavus* and *Aspergillus gracilis* exhibits antibacterial and antioxidant activity (Ali et al. 2014). The marine halophilic organism also plays a crucial role in the sulfur, carbon, and nitrogen cycle (Kamimura et al. 2003; Könneke et al. 2005; Murata et al. 2017). However, advances in urbanization and industrialization have altered the indigenous flora and fauna due to the discharge of solid and liquid waste from household, agricultural, and industrial sources (Lokhande et al. 2011; Rodríguez et al. 2018). Therefore, it is essential to preserve the environment and its microflora for the functionality and sustenance of the ecosystems.

The diverse extreme halophiles in coastal biosystems serve as an excellent reservoir to study microbial diversity and exploitation of its genetic resources (Moopantakath et al. 2020). However, halophiles and their metabolites are less explored in these biosystems. Given the importance of halophiles in ecological as well as industrial processes, elucidation of halophilic diversity is paramount. The taxonomic and functional characteristics of marine microbes have been studied in the past decades through culturable methods. However, due to the high abundance of yet-uncultured microbes, culturable methods could miss out on the actual microbial diversity. Metagenomics analysis involves the direct extraction of DNA from the environmental sample, followed by sequencing with next-generation platforms. It can provide information on the total microbial diversity and relative abundance. Expression-based metagenomic methods that clone random or specific genes from environmental DNA can be used to screen biomolecules with various industrial and pharmaceutical applications (Ghosh et al. 2019).

4.2 Adaptation Mechanisms of Halophiles in Saline Environments

Extremophiles such as halophiles, psychrophiles, thermophiles, and alkaliphiles are organisms that can survive extreme environments. Halophiles are a type of extremophile that can survive in high saline conditions. They are classified as halotolerant, slightly halophilic, moderate halophile, and extreme halophiles (Verma et al. 2020). Extreme halophilic organisms such as haloarchaea can survive in extreme salt conditions such as salterns, saline lakes, and sea. Haloarcheal species are also detected in less saline conditions, albeit in low abundance (Moopantakath et al. 2020). Haloarchaea requires moderate to high salt concentration to maintain its membrane integrity. In a non-saline solution, the cells undergo rapid lysis. Halophiles maintain an osmotic balance between the cell and the environment. They adapt to high salinity via two mechanisms, high-salt-in or low-salt-in. The high-salt-in mechanism maintains extracellular and intracellular ion gradients. It involves the accumulation of K^+ and Cl– and exclusion of Na^+ ions. The K^+ efflux and K^+ uptake channel, Tok1 and Trk1/Trk2, respectively, were identified in *Hortaea werneckii* (Oren 2008; Gunde-Cimerman et al. 2018). Extreme haloarchaea maintains high-salt-in by maintaining intracellular and extracellular electrochemical gradients. Subsequently, the ATP requirement is balanced through ADP and inorganic phosphate with the help of ATP synthase, which leads to an influx of H^+ ions (Cheng et al. 2016). The Na^+/H^+ antiporter maintains both osmotic regulation and pH in the cell. An active chloride pump, such as light-independent symport and light-dependent retinal protein, also maintains osmotic regulation inside the host cell. The second major mechanism is low-salt-in, which implements diverse organic compatible solutes such as sugars, amino acids, betaines, ectonies, and polyols (Figure 4.1). Zwitterionic are molecules with an equal number of functional positive and negative charges. They stabilize proteins, biochemical pathways, and balance enzymatic conditions in higher salt conditions. In *Chromohalobacter* spp., solutes such as glycine, betaine, and ectoine are involved in

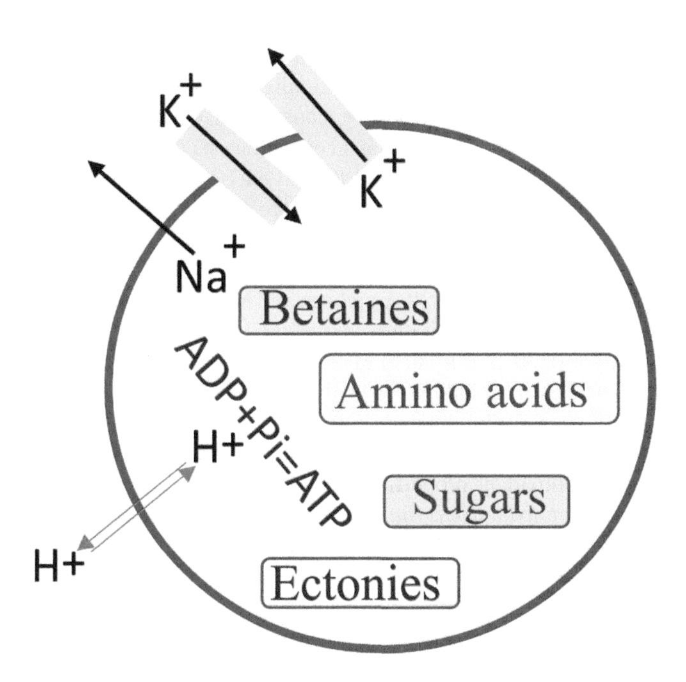

FIGURE 4.1 Halotolerant microbes maintains osmotic balance through the high-salt-in or low-salt-in mechanism. The compatible solutes such as sugars, amino acids, betaines, ectonies, polyols, and Na^+ /H^+ antiporter maintains both osmotic regulation and pH in the cell through influx /efflux of ions and zwitterionic interactions.

zwitterionic interactions. In the case of *Halorhodospira* species, a special uncharged organic trehalose osmolyte maintains osmotic regulation inside the cell. Similarly, anionic organic osmoregulators such as sulfotrehalose and β-glutamate also play a crucial role in osmotic regulation in halophilic bacteria (Verma et al. 2020).

Recent studies have highlighted the importance of several genes that were overexpressed in increased salinity. Transcriptional regulatory genes such as *rrnAC2519*, *cdc6A*, *gch3*, and *flaC* were found to express in *Haloarcula* species upon increased salinity (Thombre et al. 2016). On the other hand, *Haloferax* species expressed transcription regulators, chaperonin proteins, and ribosomal protein S7 in saline conditions (Bidle et al. 2008). The carotenoid compound bacterioruberin also plays an important role in maintaining membrane fluidity and osmotic balance in *Haloarcula marismortui* and *H. mediterranei* (Camacho et al. 2014; D'Souza et al. 1997).

4.3 Industrial Applications of Halophilic Secondary Metabolites

Halophiles are a well-known source of metabolites with biological applications. Secondary metabolites from halophiles are of special interest due to their stability in harsh saline conditions. Pigments from halophiles have gained considerable interest due to their potential role in anticancer, antimicrobial, and several other applications (El-Naggar et al. 2017; Kumari et al. 2020; Subramanian and Jayaraman 2020). Haloarcheal species are also known to synthesize the red-pigmented compound bacterioruberin, which has anticancer and antioxidant properties (Zalazar et al. 2019). Pigmented β-carotene produced by *Dunaliella* spp. has been proposed as a food additive and in the pharmaceutical industry (Priyadarshani and Rath 2012; Mobin and Alam 2017). In addition to pigments, halophiles are also a source of several other enzymes. For instance, *Haloferax sulfurifontis* GUMFAZ2, isolated from Goa, could synthesize xylanase-free cellulase at 5 M NaCl (Malik et al. 2019). Likewise, *Halomonas boliviensis* synthesizes Polyhydroxyalkanoates (PHBs) suitable for the plastic industry (Quillaguamán et al. 2010). Other potential secondary metabolites of interest are hydroxyectoine from *Marinococcus* M52, which serves as a cryopreservative agent (Schiraldi et al. 2006), and biotransformation of hydrocarbons by Haloarchaeal strain EH4 (Bonete et al. 2015). Recent studies have highlighted that the halophilic *Streptomyces* spp. were enriched in oil spills, which can degrade oil (Ugochukwu and Okorie 2020). *Halomonas* genus can degrade synthetic dyes such as Azo dye at high NaCl concentrations (Tian et al. 2019).

Extreme halophiles have several advantages in industrial mass production due to less chance of contamination and cost-effectiveness in up- and downstream processes, and they also serve as a promising host for biotechnological applications (Rodrigo-Baños et al. 2015). Industrially significant enzymes from halophiles have been cloned in fast-growing bacteria for the hyperproduction of desired enzymes (Gadda and McAllister-Wilkins 2003). In addition to haloarchaea and halobacteria, a rich diversity of halophilic fungus are present in the coastal ecosystem with potential applications in pharmaceuticals, industry, and bioremediation (Ravindran et al. 2012; Butinar et al. 2005; Bano et al. 2018).

The synthesis of nanoparticles from halophiles is also an active area of research. Zinc nanoparticles synthesized by *Alkalibacillus* sp. W7 exhibited antimicrobial activity against *Candida albicans* and *Escherichia coli* (MH Al-Kordy et al. 2020). Similarly, silver nanoparticles synthesized in *Streptomyces violaceus* MM72 exhibited antioxidant properties and antimicrobial activity against *Escherichia coli, Pseudomonas aeruginosa, Staphylococcus aureus*, and *Bacillus subtilis* (Sivasankar et al. 2018). Selenium nanoparticles synthesized by *Halococcus salifodinae* BK18 exhibited anticancer activity against HeLa cell line (Srivastava et al. 2014).

4.4 Methods for Screening Halophilic Microbes and Their Diversity

Delineating species diversity and relative abundance in natural settings is essential to better understand their potential roles in biogeochemical cycles and bioprospection. Microbial diversity studies have mostly been performed through a traditional culture-dependent technique that requires culturing of the microbes in artificial media. Culture-independent PCR-based techniques include terminal restriction fragment length polymorphism (T-RFLP) and denaturing gradient gel electrophoresis (DGGE). However, with the advent of next-generation sequencing (NGS) technologies, metagenomics has become the prime choice among culture-independent approaches.

4.4.1 Culture-Dependent Screening

Culture-dependent methods require optimum nutrients and physical parameters for the growth of halophiles. Various types of media are available for optimum growth, such as Sehgal and Gibbons, Mullakhanbai and Larsen, Tomlinson and Hochstein, and Caton modified media (Table 4.1). The main composition of the media includes NaCl, KCl, $MgSO_4$, yeast extract, and tryptone/peptone. Salinity gradients are maintained from 1M to 6M NaCl based on the optimum growth characteristics and production of the desired metabolite (Schneegurt 2012). Once pure cultures are obtained, the conventional method of microbial characterization has to be performed for the identification of the isolates, such as Sanger sequencing of the full-length 16s rRNA gene.

4.4.2 Metagenomic Analysis of Halophilic Prokaryotes

Metagenomics is the study of microbial community/diversity directly from the environmental DNA, bypassing the culture step. Several discoveries and advances have been made in halophilic studies by implementing metagenomics (Table 4.2). Through this approach, it is possible to detect yet-uncultured microbial species (Roh et al. 2006). The extraction of environmental DNA is the primary, yet crucial, step. Although environmental DNA isolation can be performed through the standard sodium dodecyl sulphate lysis protocol, the downstream process for PCR amplification can be tedious due to the co-extraction of PCR inhibitors such as humic acids, bilirubin, bile salts, urobilinogen, and polysaccharides, which are most commonly present in sediments and saline environments (Yankson and Steck 2009; Miao et al. 2014; Kashi 2016).

TABLE 4.1

List of Media Used for Isolation and Culture of Halophiles

Media	Composition (g/L)	Temperature	Isolated organism	Reference
Sehgal and Gibbons	• NaCl–250 • KCl–2 • $MgSO_4 \cdot 7H_2O$–20 • $FeCl_2$–0.023 • $Na_3C_6H_5O_7$–3 • Casamino acids–7.5 • Yeast extract–10	37°C	*Haloferax larsenii* RG3D.1	Kanekar et al. 2017; Sehgal and Gibbons 1960
Mullakhanbhai and Larsen	• NaCl–125 • K_2SO_4–5 • $MgCl_2 \cdot 6H_2O$–50 • $CaCl_2 \cdot 6H_2O$–0.12 • Yeast extract–5 • Tryptone/Peptone–5	37°C	*Halobacterium volcanii*	Mullakhanbai and Larsen 1975
Caton	• NaCl–220 • KCl–5 • KNO_3–1 • $MgSO_4 \cdot 7H_2O$–10 • $CaCl_2 \cdot 6H_2O$–0.2 • $Na_3C_6H_5O_7$–3 • Yeast extract–1 • Tryptone/Peptone–5	37°C	*Halomonas, Idiomarina, Salinivibrio, Bacteroidetes, Bacillus, Salibacillus, Oceanobacillus, Halobacillus.*	Caton et al. 2004
Forsyth and Kushner	• NaCl-29–174 • $MgSO_4 \cdot 7H_2O$–0.1 • $(NH_4)_2SO_4$–2 • K_2HPO_4–3.12 • KH_2PO_4–0.28 • NH_4Cl–2 • Glucose–10	37°C	Vibrio costicola	Forsyth et al. 1970
ZoBell marine media	• Peptone–5 • Yeast extract–1 • $C_6H_5FeO_7$–0.1 • NaCl–19.45 • $MgCl_2$–8.8 • Na_2SO_4–3.24 • $CaCl_2$–1.8 • KCl–0.550 • $NaHCO_3$–0.160 • KBr–0.080 • $SrCl_2$–0.034 • Na_2SiO_3–0.004 • NH_4NO_3–0.0016 • Na_2HPO_4–0.008 • NaF–0.0024	22°C-30° 37 °C	*Halomonas glaciei* *Bacillus aquimaris*	Shivanand et al. 2009 Ramya et al. 2020; Reddy et al. 2003; Lyman and Richard 1940; ZoBell 1941
Chemically defined medium (CDM)	• NaCl–12.5 • $MgCl_2.6H_2O$–5 • K_2S04–0.5 • $CaCl_2$–0.02 • NH_4Cl–0.5 mL of 1M • Glycerol–0.5 mL of 10% • Sodium succinate–4.5 mL of 10% • K_2HP0_4–0.2 mL 0.5M • $C_{12}H_{18}C_{12}N_4OS$–0.008 • Biotin–0.0001	37°C	*Haloferax volcanii*	Kauri et al. 1990

TABLE 4.2

Some of the Recent Advances in Novel Discoveries and Findings Regarding Halophilic Diversity and Metabolism Using Metagenomic Approaches

Sr. No.	Biosystem	Major findings	References
1	Deep sea cold seep	Novel halophilic *Sulfurovum* spp.	Sun et al. 2020
2	Mangrove, Brazil	Novel lipase LipG7	Araujo et al. 2020
3	Atlantis II brine	A unique thioredoxin protein	AbdelWahed et al. 2020
4	Hypersaline lake	Novel esterase hAGEst	Tutuncu et al. 2019
5	Hypersaline environment	Importance of Fructokinase in the metabolism of sucrose and fructose in hypersaline environment	Williams et al. 2019
6	Deep brine pool	Novel thioredoxin reductase (ATII-TrxR)	Badiea et al. 2019
7	Deep-sea, Central Pacific Ocean	Two novel esterases, DMWf18-543 and DMWf18-558	Huo et al. 2018
8	Saltmarshes	Draft genomes of Bacteroidetes, Balneolaeota and Halobacteria	Vera-Gargallo and Ventosa, 2018
9	Sea	Draft genome of novel *Candidatus Scalindua rubra*	Speth et al. 2017
10	Terrestrial subsurface, complex sediment and aquifer	Two novel phyla involved in carbon and/or hydrogen cycling	Castelle et al. 2015

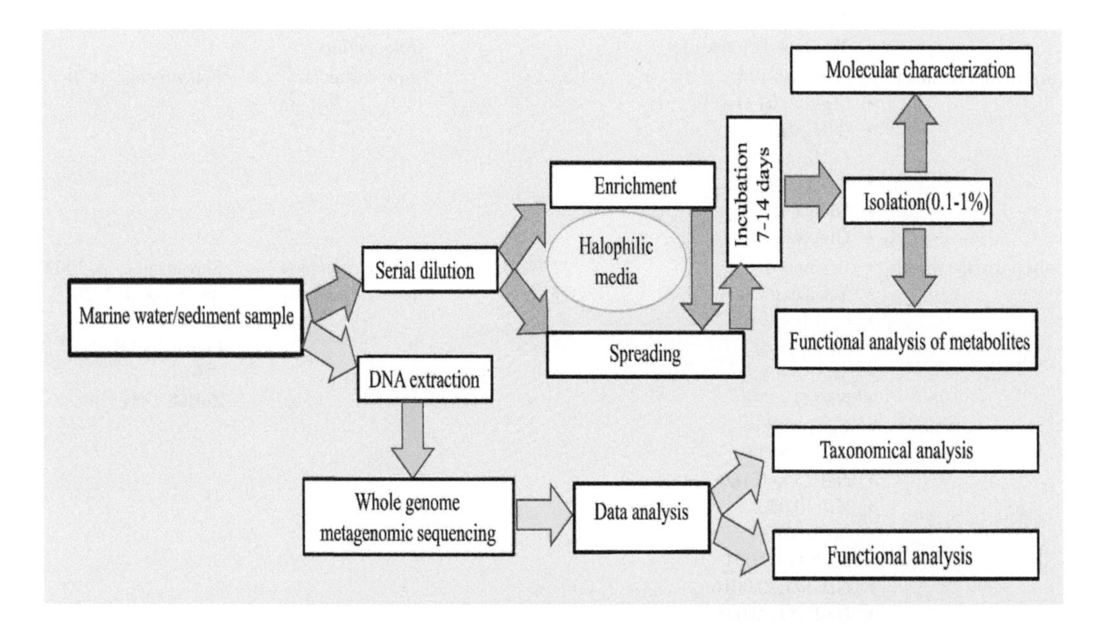

FIGURE 4.2 A schematic representation of the culture-dependent and independent approaches to study halophilic microbial communities. In culture-dependent approach, environmental samples are initially enriched in selective halophilic broth or directly spreaded on halophilic selective plates. The resultant colonies on the plates are subsequently processed for molecular identification such as 16s rRNA gene sequencing or biochemical analysis. In culture-independent whole metagenome approach, the environmental DNA is directly sequenced and annotated for taxonomic and functional analysis.

4.4.2.1 Amplicon-Based Metagenomics

Amplicon-based metagenomics, more specifically metabarcoding or metataxonomics, involves the amplifying the 16s rRNA gene hypervariable regions of prokaryotes using universal (Kim and Park 2014) or phylum-specific primers in a high-throughput NGS platform (Baker et al. 2003; Casanueva et al. 2008) (Figure 4.2). The robustness of metagenomic amplicon sequencing for taxonomical analysis of microbial communities has been implemented in several studies. An in-depth analysis of haloarchaea in the Arabian Sea coast of India revealed the dominance of haloarchaea such as *Haloferax*, *Haladaptatus*, *Haloterrigena*, and *Natrialba* (Moopantakath et al. 2020). In another recent study, sequencing of nearly full-length 16s rRNA gene identified a putative novel halophilic *Sulfurovum* spp. from the South China Sea (Sun et al. 2020).

4.4.2.2 Whole Metagenome Sequencing

The rapid emergence of highly sophisticated NGS instruments has significantly reduced the cost of whole metagenome sequencing (WGS). In the whole metagenome approach, complete metagenomic DNA is extracted from the sample and sheared into fragments, followed by high-throughput sequencing using next-generation sequencing instruments.

The inference of taxonomic and functional information can be further extended to genome-resolved metagenomic (GRM) analysis from whole metagenome data. This provides an interesting approach to derived whole genomes known as metagenome-assembled genomes (MAGs) of uncultivable microbes (Chen et al. 2020). GRM studies of archaea in terrestrial subsurfaces, complex sediments, and aquifers led to the finding of novel species under the Methanosarcinales order and two novel phyla involved in carbon and/or hydrogen cycling (Castelle et al. 2015). Similarly, MAGs constructed from metagenome datasets from the Red Sea identified a novel species, *Candidatus Scalindua rubra*, that performs osmo-adaptation with compatible solutes rather than a salt-in strategy, which is common in marine anammox bacteria (Speth et al. 2017).

Recent advances in whole genome data analysis for halophiles include the application of machine learning for gene prediction. Hidden Markov model (HMM) profiles generated for the analysis of halo-archaeal fructokinase and ketohexokinase genes identified that the metabolism of sucrose and fructose in a hypersaline environment could be majorly carried out by fructokinase rather than ketohexokinase, as evident from the higher abundance of the haloarchaeal fructokinase genes (Williams et al. 2019).

Whole metagenome analysis provides in-depth clues of the total genetic diversity and abundance, along with taxonomical information. This provides valuable information on the potential production and diversity of secondary metabolites. For instance, shotgun metagenomic analysis of Canadian seawater identified the dominance of genes encoding for pigments, carbohydrates, vitamins, and taxonomic predominance by Nitrosopumilus, Flavobacteriales, and Oceanospirillaceae (Yergeau et al. 2017). Nitrogen-fixing microbes Planctomycetes and Proteobacteria, along with the *nifH* gene, dominated in the surface of various ocean metagenomes (Delmont et al. 2018). In contrast, enrichment of arsenite oxidase genes was observed in Diamante Lake despite the abundance of halophile *Halorubrum* and *Natronomonas*, indicating arsenic contamination (Rascovan et al. 2016). Such studies show the importance of shotgun metagenomic analysis in marine and other saline environments. Despite such a wide application and the advantages of the whole metagenome, most microbial community studies have utilized the 16s rRNA gene metataxonomic approach, because the cost per sample for the whole metagenome is several folds higher for the latter.

4.4.2.3 Expression-Based Functional Metagenomics

Expression-based functional metagenomics is the analysis of random metagenomic DNA expressed in a suitable host for the bioprospection of bioactive molecules. It primarily involves the extraction of metagenomic DNA from samples followed by restriction digestion and ligation into a suitable vector (plasmid, cosmid, or phage) for the expression of the inserts. The construct is transformed into a suitable host such as *Escherichia coli* DH5α and screened for the desired compound with special selective or differential media such as halophilic media (Lewin et al. 2017). Using such a technique, saline-tolerant DNA/RNA helicases, endonuclease III, cell surface glycoproteins, pyrophosphate-energized proton pumps, and hypothetical proteins, halotolerant esterases, have been discovered (De Santi et al. 2016; Cheng et al. 2016; Mirete et al. 2015). In another approach, an *E. coli* KNabc strain deficient of Na^+/H^+ antiporter was used as the host to screen for novel Na^+/H^+ antiporter (Xiang et al. 2010). The technique relies on the complementation of the Na^+/H^+ antiporter from the insert with the antiporter-deficient host. The authors discovered a putative novel antiporter, *m-nha*, from an ancient brine well in China, with a maximum similarly of 92% to *Nha*H in *Halobacillus dabanensis* D-8T. The clone with the *m-nha* insert exhibited increased resistance to Na^+ and alkaline conditions (Xiang et al. 2010).

The main advantage of expression-based functional metagenomics over NGS-based sequencing is the potential to discover completely novel genes, since expression-based functional metagenomics does not rely on prior knowledge of the sequence. On the contrary, the technique requires a large quantity of DNA input, skills, more time, and space for executing the work (Lam et al. 2015). Nonetheless, the strength of the technique has been exploited for the bioprospection of several novel enzymes from various environmental samples. For instance, in addition to the saline natural environment, functional metagenomics has also been implemented to screen microbial genes conferring resistance to Na^+ in the human gut microbiome. This has led to the discovery of membrane protein-encoding gene *brpA* under β-carotene monooxygenase that enhances salt tolerance and renders red/orange pigmentation to the cell membrane and pellet (Culligan et al. 2014). Similarly, a multidrug transporter permease protein-coding gene was discovered that confers tolerance to salinity (Gupta et al. 2020). Despite the advantages and strength of functional metagenomics, its implementation in saline tolerance studies has received relatively less attention compared to commercially important enzymes such as proteases, lipases, and cellulase (Kumar et al. 2020; Prayogo et al. 2020).

4.5.1 Diversity of Halophiles in Saltpan Biosystems

Saltpans are hypersaline aquatic environments, highly enriched with minerals, with about threefold higher salinity than the surrounding seawater (Bell 2012; Fernández et al. 2014; Naghoni et al. 2017). Extreme environments are a potential source of novel microbes and metabolites of industrial and pharmaceutical interest (Malik et al. 2020). The nutrients and microflora in the saltpan are also influenced by season and geographical location (Rathore et al. 2017; Mora-Ruiz et al. 2018; Kaviarasan et al. 2019). Several haloarchaeal species under the *Haloferax, Haloarcula*, and *Natrinema* genera have been successfully isolated from the saltpan through culture-dependent techniques (Table 4.3) (Karthikeyan et al. 2013; Thombre et al. 2016). However, due to the difficulties faced in media formulation, prolonged incubation, and differential enrichment, surveillance of haloarcheal communities is more appropriate through metagenomics (Vogel et al. 2009). Halophilic microbial communities in coastal ecosystems including saltpan have been explored through metagenomics in several locations. Metagenomic studies have revealed that the *Halorubrum* genus is predominant in the Australian saltpan (Dry Creek, Bajool, and Lara) (Oh et al. 2010). Similarly, *Halorubrum* was among the dominant genera in the saltpan from the Mediterranean and Atlantic regions. In addition, *Haloquadratum* and *Natronomonas* were also predominant in the Mediterranean and Atlantic saltpan. It is interesting to note that *Haloquadratum* is yet-uncultured halophiles. Furthermore, functional analysis confirmed the presence of sulfate adenylyltransferase, phosphoadenylylsulfate reductase, and sulfite reductase, indicating the possible roles of halophiles in the sulfur cycle in such extreme conditions (Fernández et al. 2014). The authors also noted that the photosynthetic genes, *psbA* and *psbD*, were absent, while rhodopsin genes were detected in the saltern metagenome. Genes pertaining to solutes that play a crucial role in osmotic regulation through a 'salt-out' strategy such as glutamate synthase, betaine transporters, and glycerol kinase also dominated in the saltpan (Fernández et al. 2014). Metagenome-assembled genomes obtained from the Archaeal metagenome datasets of hypersaline lakes in Victoria, Australia, revealed the uniqueness of amino acid composition and metabolism in *Nanohaloarchaea* (Narasingarao et al. 2012). The taxonomic composition of soda pans was predominated by genus *Actinomycetales*, while the functional genes were enriched with functions associated with respiratory metabolism rather than fermentation metabolism (Szabó et al. 2017).

4.5.2 Metagenomic Diversity of Halophiles in Sea and Coastline

The salinity of seawater varies in different geographical locations. For instance, the Red Sea and the Dead Sea have comparatively higher salinity (~35%) than Arabian seas (>3.6%). The Red Sea coastline is also characterized by oxygen-deficient warm brine pools and high UV radiation exposure (Gurvich 2006;

TABLE 4.3

List of Dominant Family/Genus/Species in Various Marine Biosystems and Locations

S. No	Sampling location	Biosystem	Halotolerant organism	References
1	Arabian seacoast, India	Mangrove, Lake, Island, Seashore, Estuary, Salt pan, River	*Haladaptatus, Haloterrigena, Natrialba, Natronorubrum, Halostagnicola, Natronococcus, Haloferax, Halogranum*	Moopantakath et al. 2020
2	Karak Salt Mine, Pakistan	Salt Mine,	*Haloarcula, Halorubrum, Halorhabdus, Natronomonas,*	Cycil et al. 2020
3	Mediterranean Sea coast of Camargue	Seacoast	*Halomonas, Salinimicrobium, Fodinibius, Pontibacter, Nitriliruptor, Marinobacter, Sphingomonas*	Osman et al. 2019
4	Goa, India	Mangrove	*Halomonadaceae, Halomonas*	Haldar and Nazareth 2018
5	Pannonian steppe	Saltpan	*Bacteroidetes Proteobacteria Altererythrobacter, Loktanella, Salinarimona, Seohaeicola. Actinomycetales*	Szabó et al. 2017
6	Red sea	Sea	*Nitrosopumilus*	Behzad et al. 2016
7	Sundarbans-India	Mangrove	*Halosarcina, Halolamina, Haloferax, Halorientalis, Halogranum*	Bhattacharyya et al. 2015
8	Gulf of Cambay sediments	Seacoast	*Firmicutes, Proteobacterial, Bacteroidetes, Acidobacteria Chloroflexi, Actinobacteria Planctomycetes*	Keshri et al. 2015
9	Mediterranean and Atlantic	Saltpan	*Halorubrum, Haloquadratum, Natronomonas*	Fernández et al. 2014
10	Dead sea	Sea	*Halorhabdus, Natronomonas Burkholderia, Chryseobacterium, Salinibacter Nanohaloarchaea, Haloquadratum, Halosarcina, Haloplanus, Halomarina*	Rhodes et al. 2012
11	Sichang island, Thailand	Seacoast	*Halothermothrix orenii*	Somboonna et al. 2012
12	Dry Creek, Bajool, Lara, Australia	Saltpan	*Halorubrum*	Oh et al. 2010

Rossbach et al. 2020). Such differences in biochemical properties lead to differential enrichment of taxonomical groups. *Nitrosopumilus* spp. dominates in the Red Sea, with a high abundance of functional genes related to DNA repair and osmolyte C1 oxidation (Behzad et al. 2016). The Dead Sea, the lowest continental point, is enriched with *Halorhabdus* and *Natronomonas* spp. (Jacob et al. 2017). However, there have been reports of extreme changes in the bacterial microbial community of the Dead Sea, from 18% to <0.5% from June to September 1992, indicating that seasonal changes could have a major influence on the microbial community. Functional genes such as *cbbL*, *nifH*, *amoA*, and *apsA* were most abundant, indicating an active metabolism of ammonia and nitrogen fixation in the intertidal region. Seashore areas in Sichang Island, Thailand, were enriched with *Halothermothrix orenii*, which are reported to survive polyextremophilic conditions (Cayol et al. 1994; Somboonna et al. 2012). A new subfamily of salt-tolerant lipase and esterase, LipG, LipEH166, estKT4, estKT7, and estKT9, were discovered from tidal flats (Kim et al. 2009; Lee et al. 2006; Jeon et al. 2012) (Table 4.3).

4.6 Conclusions and Future Prospects

Halophiles are a salt-loving microorganism that exhibits unique biochemical characteristics and metabolites of various pharmaceutical and industrial interests. Saltpan, intertidal mudflats, mangroves, and the related biosystem are the major natural reservoirs of halophiles in coastal areas. Traditional culture techniques have provided substantial information on the halophilic community. However, culture-independent methods are necessary to obtain a true snapshot of the microbial community. Halophilic microbial communities in marine biosystems are unique to each biosystem and exhibit spatiotemporal differences. Such uniqueness of extreme halophiles could be harnessed for pharmaceutical and industrial applications. The advantages of halophilic cultures can be attributed to their minimal microbial contamination due to the high-saline medium, cost-effective upstream and downstream process, and retention of activity in extreme conditions. Furthermore, pigments such as lycopene and bacterioruberin from halophilic organisms are active against communicable and non-communicable diseases and are excellent food colorants.

4.7 Acknowledgments

The authors are grateful for EEQ/2018/001085 for partial financial assistance, JM thanks the ICMR-SRF fellowship, Govt. of India, and the research facilities supported by the Central University of Kerala, IM thank the fellowship by DBT-RA Program in Biotechnology & Life Sciences, Govt. of India, and the research facilities supported by the Central University of Kerala.

REFERENCES

Abad, M. J., L. M. Bedoya, and P. Bermejo. "Marine compounds and their antimicrobial activities." *Science Against Microbial Pathogens: Communicating Current Research and Technological Advances* 51 (2011): 1293–1306.

AbdelWahed, Mohamed G., Elham A. Badiea, Amged Ouf, and Ahmed A. Sayed. "Molecular and functional characterization of unique thermo-halophilic thioredoxin from the metagenome of an exotic environment." *International Journal of Biological Macromolecules* 153 (2020): 767–778.

Al-Dhabi, Naif Abdullah, Galal Ali Esmail, Abdul-Kareem Mohammed Ghilan, Mariadhas Valan Arasu, Veeramuthu Duraipandiyan, and Karuppiah Ponmurugan. "Chemical constituents of *Streptomyces* sp. strain Al-Dhabi-97 isolated from the marine region of Saudi Arabia with antibacterial and anticancer properties." *Journal of Infection and Public Health* 13, no. 2 (2020): 235–243.

Ali, Imran, Napa Siwarungson, Hunsa Punnapayak, Pongtharin Lotrakul, Sehanat Prasongsuk, Wichanee Bankeeree, and Sudip K. Rakshit. "Screening of potential biotechnological applications from obligate halophilic fungi, isolated from a man-made solar saltern located in Phetchaburi province, Thailand." *Pakistan Journal of Botany* 46 (2014): 983–988.

Araujo, Francisco J., Denise C. Hissa, Gabrielly O. Silva, André S. L. M. Antunes, Vanessa L. R. Nogueira, Luciana R. B. Gonçalves, and Vânia M. M. Melo. "A novel bacterial carboxylesterase identified in a metagenome derived-clone from Brazilian mangrove sediments." *Molecular Biology Reports* 47 (2020): 3919–3928.

Austin, Alexandra, Alessandra Chieff, Julianne Depardieu, Max Gratton, Thomas Holdaway, Wei Tian Lee, Mary Libera, et al. "Novel concepts for offshore ocean farming." In *SNAME Maritime Convention.* Tacoma, Washington. The Society of Naval Architects and Marine Engineers, 2019.

Badiea, Elham A., Ahmed A. Sayed, Mohamad Maged, Walid M. Fouad, Mahmoud M. Said, and Amr Y. Esmat. "A novel thermostable and halophilic thioredoxin reductase from the Red Sea Atlantis II hot brine pool." *PLoS One* 14, no. 5 (2019): e0217565.

Baker, G. C., Jacques J. Smith, and Donald A. Cowan. "Review and re-analysis of domain-specific 16S primers." *Journal of Microbiological Methods* 55, no. 3 (2003): 541–555.

Bano, Amna, Javaid Hussain, Ali Akbar, Khalid Mehmood, Muhammad Anwar, Muhammad Sharif Hasni, Sami Ullah, Sumbal Sajid, and Imran Ali. "Biosorption of heavy metals by obligate halophilic fungi." *Chemosphere* 199 (2018): 218–222.

Behzad, Hayedeh, Martin Augusto Ibarra, Katsuhiko Mineta, and Takashi Gojobori. "Metagenomic studies of the Red Sea." *Gene* 576, no. 2 (2016): 717–723.

Bell, Elanor, ed. *Life at Extremes: Environments, Organisms, and Strategies for Survival.* Australian Marine Mammal Centre, Australia: Vol. 1. Cabi (2012).

Bhattacharyya, Anish, Niladri Shekhar Majumder, Pijush Basak, Shayantan Mukherji, Debojyoti Roy, Sudip Nag, Anwesha Haldar, et al. "Diversity and distribution of Archaea in the mangrove sediment of Sundarbans." *Archaea* 2015 (2015).

Bidle, Kelly A., P. Aaron Kirkland, Jennifer L. Nannen, and Julie A. Maupin-Furlow. "Proteomic analysis of *Haloferax volcanii* reveals salinity-mediated regulation of the stress response protein PspA." *Microbiology (Reading, England)* 154, no. Pt 5 (2008): 1436.

Bollmann, Moritz. *World Ocean Review: Living with the Oceans.* (2010). Hamburg, Germany, Maribus GmbH, 2010.

Bonete, María José, Vanesa Bautista, Julia Esclapez, M. García-Bonete, Carmen Pire, Mónica Camacho, Javier Torregrosa-Crespo, and R. Martínez-Espinosa. "New uses of haloarchaeal species in bioremediation processes." *Advances in Bioremediation of Wastewater and Polluted Soil. 1st ed. InTech* (2015): 23–49.

Butinar, Lorena, Polona Zalar, Jens C. Frisvad, and Nina Gunde-Cimerman. "The genus Eurotium—Members of indigenous fungal community in hypersaline waters of salterns." *FEMS Microbiology Ecology* 51, no. 2 (2005): 155–166.

Camacho-Córdova, David Isidoro, Rosa María Camacho-Ruíz, Jesús Antonio Córdova-López, and Jesús Cervantes-Martínez. "Estimation of bacterioruberin by Raman spectroscopy during the growth of halophilic archaeon *Haloarcula marismortui*." *Applied Optics* 53, no. 31 (2014): 7470–7475.

Casanueva, Ana, Ncebakazi Galada, Gillian C. Baker, William D. Grant, Shaun Heaphy, Brian Jones, Ma Yanhe, Antonio Ventosa, Jenny Blamey, and Don A. Cowan. "Nanoarchaeal 16S rRNA gene sequences are widely dispersed in hyperthermophilic and mesophilic halophilic environments." *Extremophiles* 12, no. 5 (2008): 651–656.

Castelle, Cindy J., Kelly C. Wrighton, Brian C. Thomas, Laura A. Hug, Christopher T. Brown, Michael J. Wilkins, Kyle R. Frischkorn et al. "Genomic expansion of domain archaea highlights roles for organisms from new phyla in anaerobic carbon cycling." *Current Biology* 25, no. 6 (2015): 690–701.

Caton, Todd M., Lisa R. Witte, H. D. Ngyuen, Julie A. Buchheim, Mark A. Buchheim, and Mark A. Schneegurt. "Halotolerant aerobic heterotrophic bacteria from the Great Salt Plains of Oklahoma." *Microbial Ecology* 48, no. 4 (2004): 449–462.

Cayol, J. L., Bernard Ollivier, B. K. C. Patel, G. Prensier, J. Guezennec, and J. L. Garcia. "Isolation and characterization of *Halothermothrix orenii* gen. nov., sp. nov., a halophilic, thermophilic, fermentative, strictly anaerobic bacterium." *International Journal of Systematic and Evolutionary Microbiology* 44, no. 3 (1994): 534–540.

Chen, Lin-Xing, Karthik Anantharaman, Alon Shaiber, A. Murat Eren, and Jillian F. Banfield. "Accurate and complete genomes from metagenomes." *Genome Research* 30, no. 3 (2020): 315–333.

Cheng, Bin, Yiwei Meng, Yanbing Cui, Chunfang Li, Fei Tao, Huijia Yin, Chunyu Yang, and Ping Xu. "Alkaline response of a halotolerant alkaliphilic *Halomonas* strain and functional diversity of its Na^+ $(K^+)/H^+$ antiporters." *Journal of Biological Chemistry* 291, no. 50 (2016): 26056–26065.

Culligan, Eamonn P., Roy D. Sleator, Julian R. Marchesi, and Colin Hill. "Metagenomic identification of a novel salt tolerance gene from the human gut microbiome which encodes a membrane protein with homology to a brp/blh-family β-carotene 15, 15′-monooxygenase." *PLoS One* 9, no. 7 (2014): e103318.

Cycil, Leena Mavis, Shiladitya DasSarma, Wolf Pecher, Ryan McDonald, Maria AbdulSalam, and Fariha Hasan. "Metagenomic insights into the diversity of halophilic microorganisms indigenous to the Karak Salt Mine, Pakistan." *Frontiers in Microbiology* 11 (2020): 1567.

Delcroix, Thierry, Michael J. McPhaden, Alain Dessier, and Yves Gouriou. "Time and space scales for sea surface salinity in the tropical oceans." *Deep Sea Research Part I: Oceanographic Research Papers* 52, no. 5 (2005): 787–813.

Delmont, Tom O., Christopher Quince, Alon Shaiber, Özcan C. Esen, Sonny T. M. Lee, Michael S. Rappé, Sandra L. McLellan, Sebastian Lücker, and A. Murat Eren. "Nitrogen-fixing populations of Planctomycetes and Proteobacteria are abundant in surface ocean metagenomes." *Nature Microbiology* 3, no. 7 (2018): 804–813.

De Santi, Concetta, Bjørn Altermark, Marcin Miroslaw Pierechod, Luca Ambrosino, Donatella de Pascale, and Nils-Peder Willassen. "Characterization of a cold-active and salt tolerant esterase identified by functional screening of Arctic metagenomic libraries." *BMC Biochemistry* 17, no. 1 (2016): 1–13.

D'Souza, Sandra E., Wijaya Altekar, and S. F. D'Souza. "Adaptive response of *Haloferax mediterranei* to low concentrations of NaCl (< 20%) in the growth medium." *Archives of Microbiology* 168, no. 1 (1997): 68–71.

El-Naggar, Noura El-Ahmady, and Sara M. El-Ewasy. "Bioproduction, characterization, anticancer and antioxidant activities of extracellular melanin pigment produced by newly isolated microbial cell factories *Streptomyces glaucescens* NEAE-H." *Scientific Reports* 7, no. 1 (2017): 1–19.

Fernández, Ana B., Blanca Vera-Gargallo, Cristina Sánchez-Porro, Rohit Ghai, R. Thane Papke, Francisco Rodriguez-Valera, and Antonio Ventosa. "Comparison of prokaryotic community structure from Mediterranean and Atlantic saltern concentrator ponds by a metagenomic approach." *Frontiers in Microbiology* 5 (2014): 196.

Forsyth, M. P., and D. J. Kushner. "Nutrition and distribution of salt response in populations of moderately halophilic bacteria." *Canadian Journal of Microbiology* 16, no. 4 (1970): 253–261.

Foti, Enrico, Rosaria Ester Musumeci, and Martina Stagnitti. "Coastal defence techniques and climate change: a review." *Rendiconti Lincei. Scienze Fisiche e Naturali* 31, no. 1 (2020): 123–138.

Gadda, Giovanni, and Elien Elizabeth McAllister-Wilkins. "Cloning, expression, and purification of choline dehydrogenase from the moderate halophile *Halomonas elongata*." *Applied and Environmental Microbiology* 69, no. 4 (2003): 2126–2132.

Ghosh, Arpita, Aditya Mehta, and Asif M. Khan. "Metagenomic analysis and its applications." *Encyclopedia of Bioinformatics and Computational Biology* 3 (2019): 184–193.

Gunde-Cimerman, Nina, Ana Plemenitaš, and Aharon Oren. "Strategies of adaptation of microorganisms of the three domains of life to high salt concentrations." *FEMS Microbiology Reviews* 42, no. 3 (2018): 353–375.

Gupta, Sonika, Parul Sharma, Kamal Dev, and Anuradha Sourirajan. "Isolation of gene conferring salt tolerance from halophilic bacteria of Lunsu, Himachal Pradesh, India." *Journal of Genetic Engineering and Biotechnology* 18, no. 1 (2020): 1–7.

Gurvich, Evgeny G. *Metalliferous Sediments of the World Ocean: Fundamental Theory of Deep-Sea Hydrothermal Sedimentation*. Springer, Berlin, 2006.

Haldar, Shyamalina, and Sarita W. Nazareth. "Taxonomic diversity of bacteria from mangrove sediments of Goa: metagenomic and functional analysis." *3 Biotech* 8, no. 10 (2018): 1–10.

Huo, Ying-Yi, Shu-Ling Jian, Hong Cheng, Zhen Rong, Heng-Lin Cui, and Xue-Wei Xu. "Two novel deep-sea sediment metagenome-derived esterases: residue 199 is the determinant of substrate specificity and preference." *Microbial Cell Factories* 17, no. 1 (2018): 1–12.

Jacob, Jacob H., Emad I. Hussein, Muhamad Ali K. Shakhatreh, and Christopher T. Cornelison. "Microbial community analysis of the hypersaline water of the Dead Sea using high-throughput amplicon sequencing." *MicrobiologyOpen* 6, no. 5 (2017): e00500.

Jeon, Jeong Ho, Hyun Sook Lee, Jun Tae Kim, Sang-Jin Kim, Sang Ho Choi, Sung Gyun Kang, and Jung-Hyun Lee. "Identification of a new subfamily of salt-tolerant esterases from a metagenomic library of tidal flat sediment." *Applied Microbiology and Biotechnology* 93, no. 2 (2012): 623–631.

Kalasava, Ashish B., Kruti G. Dangar, and Satya P. Singh. "Optimization of amylase production from *Nocardiopsis* sp." DW-4 Isolated from Dwarka, Coastal Region of Gujarat, April 6, 2020.

Kamimura, Kazuo, Emi Higashino, Souichi Moriya, and Tsuyoshi Sugio. "Marine acidophilic sulfur-oxidizing bacterium requiring salts for the oxidation of reduced inorganic sulfur compounds." *Extremophiles* 7, no. 2 (2003): 95–99.

Kanekar, Pradnya P., Snehal O. Kulkarni, Prashant K. Dhakephalkar, Kantimati G. Kulkarni, and Neha Saxena. "Isolation of a halophilic, bacteriorhodopsin-producing Archaeon, *Haloferax larsenii* RG3D. 1 from the rocky beach of Malvan, west coast of India." *Geomicrobiology Journal* 34, no. 3 (2017): 242–248.

Karthikeyan, P., Sarita G. Bhat, and M. Chandrasekaran. "Halocin SH10 production by an extreme haloarchaeon *Natrinema* sp. BTSH10 isolated from salt pans of South India." *Saudi Journal of Biological Sciences* 20, no. 2 (2013): 205–212.

Kashi, Fereshteh Jookar. "An improved procedure of the metagenomic DNA extraction from saline soil, sediment and salt." *International Letters of Natural Sciences* 60 (2016).

Kauri, Tiiu, Rebecca Wallace, and Donn J. Kushner. "Nutrition of the halophilic archaebacterium, *Haloferax volcanii*." *Systematic and Applied Microbiology* 13, no. 1 (1990): 14–18.

Kaviarasan, Thanamegam, Hans Uwe Dahms, Murugaiah Santhosh Gokul, Santhaseelan Henciya, Krishnan Muthukumar, Shiva Shankar, and Rathinam Arthur James. "Seasonal species variation of sediment organic carbon stocks in salt marshes of Tuticorin area, Southern India." *Wetlands* 39, no. 3 (2019): 483–494.

Keshri, Jitendra, Basit Yousuf, Avinash Mishra, and Bhavanath Jha. "The abundance of functional genes, cbbL, nifH, amoA and apsA, and bacterial community structure of intertidal soil from Arabian Sea." *Microbiological Research* 175 (2015): 57–66.

Kim, Eun-Young, Ki-Hoon Oh, Mi-Hwa Lee, Chul-Hyung Kang, Tae-Kwang Oh, and Jung-Hoon Yoon. "Novel cold-adapted alkaline lipase from an intertidal flat metagenome and proposal for a new family of bacterial lipases." *Applied and Environmental Microbiology* 75, no. 1 (2009): 257–260.

Kim, Jungwoo, Daniel Shin, Seong-Hwan Kim, Wanki Park, Yoonho Shin, Won Kyung Kim, Sang Kook Lee, Ki-Bong Oh, Jongheon Shin, and Dong-Chan Oh. "Borrelidins C–E: new Antibacterial macrolides from a saltern-derived halophilic *Nocardiopsis* sp." *Marine Drugs* 15, no. 6 (2017): 166.

Kim, Min-Soo, and Eun-Jin Park. "Bacterial communities of traditional salted and fermented seafoods from Jeju Island of Korea using 16S rRNA gene clone library analysis." *Journal of Food Science* 79, no. 5 (2014): M927–M934.

Könneke, Martin, Anne E. Bernhard, R. José, Christopher B. Walker, John B. Waterbury, and David A. Stahl. "Isolation of an autotrophic ammonia-oxidizing marine archaeon." *Nature* 437, no. 7058 (2005): 543–546.

Korkmaz, Melih N., Sennur C. Ozdemir, and Ataç Uzel. "Xylanase production from marine derived *Trichoderma pleuroticola* 08ÇK001 strain isolated from Mediterranean coastal sediments." *Journal of Basic Microbiology* 57, no. 10 (2017): 839–851.

Kumar, Satish, Dhiraj Paul, Bharat Bhushan, G. C. Wakchaure, Kamlesh K. Meena, and Yogesh Shouche. "Traversing the 'omic' landscape of microbial halotolerance for key molecular processes and new insights." *Critical Reviews in Microbiology* 46, no. 6 (2020): 631–653.

Kumari, K. S., Shivakrishna, P., Al-Attar, A. M. and Ganduri, V. R. "Antibacterial and cytotoxicity activities of bioactive compounds from Micrococcus species OUS9 isolated from sea water." *Journal of King Saud University-Science* 32, no. 6 (2020): 2818–2825.

Lam, Kathy N., Jiujun Cheng, Katja Engel, Josh D. Neufeld, and Trevor C. Charles. "Current and future resources for functional metagenomics." *Frontiers in Microbiology* 6 (2015): 1196.

Lee, Mi-Hwa, Choong-Hwan Lee, Tae-Kwang Oh, Jae Kwang Song, and Jung-Hoon Yoon. "Isolation and characterization of a novel lipase from a metagenomic library of tidal flat sediments: evidence for a new family of bacterial lipases." *Applied and Environmental Microbiology* 72, no. 11 (2006): 7406–7409.

Lewin, Anna, Rahmi Lale, and Alexander Wentzel. "Expression platforms for functional metagenomics: emerging technology options beyond *Escherichia coli*." In *Functional Metagenomics: Tools and Applications*, pp. 13–44. Springer, Cham, 2017.

Lokhande, Ram S., Pravin U. Singare, and Deepali S. Pimple. "Study on physico-chemical parameters of waste water effluents from Taloja industrial area of Mumbai, India." *International Journal of Ecosystem* 1, no. 1 (2011): 1–9.

Lyman, John, and Richard H. Fleming. "Composition of sea water." *The Journal of Marine Research* 3, no. 2 (1940): 134–146.

Malik, Alisha D., and Irene J. Furtado. "*Haloferax sulfurifontis* GUMFAZ2 producing xylanase-free cellulase retrieved from Haliclona sp. inhabiting rocky shore of Anjuna, Goa-India." *Journal of basic microbiology* 59, no. 7 (2019): 692–700.

Malik, Kamla, Nisha Kumari, Sushil Ahlawat, Upendra Kumar, and Meena Sindhu. "Extremophile microorganisms and their industrial applications." In *Microbial Diversity, Interventions and Scope*, pp. 137–156. Springer, Singapore, 2020.

MH Al-Kordy, Hend, Soraya A. Sabry, and Mona E. M. Mabrouk. "Photocatalytic and antimicrobial activity of zinc oxide nanoparticles synthesized by halophilic *Alkalibacillus* sp. w7 isolated from a salt lake." *Egyptian Journal of Aquatic Biology and Fisheries* 24, no. 4 (2020): 43–56.

Miao, Tianjin, Song Gao, Shengwei Jiang, Guoshi Kan, Pengju Liu, Xianming Wu, Yingfeng An, and Shuo Yao. "A method suitable for DNA extraction from humus-rich soil." *Biotechnology Letters* 36, no. 11 (2014): 2223–2228.

Mirete, Salvador, Merit R. Mora-Ruiz, María Lamprecht-Grandío, Carolina G. de Figueras, Ramon Rosselló-Móra, and José E. González-Pastor. "Salt resistance genes revealed by functional metagenomics from brines and moderate-salinity rhizosphere within a hypersaline environment." *Frontiers in Microbiology* 6 (2015): 1121.

Mobin, Saleh, and Firoz Alam. "Some promising microalgal species for commercial applications: a review." *Energy Procedia* 110 (2017): 510–517.

Moopantakath, Jamseel, Madangchanok Imchen, Busi Siddhardha, and Ranjith Kumavath. "16s rRNA metagenomic analysis reveals predominance of Crtl and CruF genes in Arabian Sea coast of India." *Science of the Total Environment* 743 (2020): 140699.

Moopantakath, Jamseel, Madangchanok Imchen, Ranjith Kumavath, and Rosa M. Martínez-Espinosa. "Ubiquitousness of *Haloferax* and carotenoid producing genes in Arabian sea coastal biosystems of India." *Marine Drugs* 19 (2021): 442.

Mora-Ruiz, M. del R., Ana Cifuentes, Francisca Font-Verdera, C. Pérez-Fernández, Maria Eugenia Farias, B. González, Alejandro Orfila, and Ramón Rosselló-Móra. "Biogeographical patterns of bacterial and archaeal communities from distant hypersaline environments." Systematic and Applied Microbiology 41, no. 2 (2018): 139–150.

Mullakhanbhai, Moiz F., and Helge Larsen. "*Halobacterium volcanii* spec. nov., a Dead Sea halobacterium with a moderate salt requirement." *Archives of Microbiology* 104, no. 1 (1975): 207–214.

Murata, Kazuyoshi, Qinfen Zhang, Jesús Gerardo Galaz-Montoya, Caroline Fu, Maureen L. Coleman, Marcia S. Osburne, Michael F. Schmid, Matthew B. Sullivan, Sallie W. Chisholm, and Wah Chiu. "Visualizing adsorption of cyanophage P-SSP7 onto marine Prochlorococcus." *Scientific Reports* 7, no. 1 (2017): 1–10.

Naghoni, Ali, Giti Emtiazi, Mohammad Ali Amoozegar, Mariana Silvia Cretoiu, Lucas J. Stal, Zahra Etemadifar, Seyed Abolhassan Shahzadeh Fazeli, and Henk Bolhuis. "Microbial diversity in the hypersaline Lake Meyghan, Iran." *Scientific Reports* 7, no. 1 (2017): 1–13.

Narasingarao, Priya, Sheila Podell, Juan A. Ugalde, Céline Brochier-Armanet, Joanne B. Emerson, Jochen J. Brocks, Karla B. Heidelberg, Jillian F. Banfield, and Eric E. Allen. "De novo metagenomic assembly reveals abundant novel major lineage of Archaea in hypersaline microbial communities." *The ISME Journal* 6, no. 1 (2012): 81–93.

Oh, Dickson, Kate Porter, Brendan Russ, David Burns, and Mike Dyall-Smith. "Diversity of Haloquadratum and other haloarchaea in three, geographically distant, Australian saltern crystallizer ponds." *Extremophiles* 14, no. 2 (2010): 161–169.

Oren, Aharon. "Microbial life at high salt concentrations: phylogenetic and metabolic diversity." *Saline Systems* 4, no. 1 (2008): 1–13.

Osman, Jorge R., Christophe Regeard, Catherine Badel, Gustavo Fernandes, and Michael S. DuBow. "Variation of bacterial biodiversity from saline soils and estuary sediments present near the Mediterranean Sea coast of Camargue (France)." *Antonie Van Leeuwenhoek* 112, no. 3 (2019): 351–365.

Parthasarathy, Ramalingam, Manjegowda Chandrika, HC Yashavantha Rao, Subban Kamalraj, Chelliah Jayabaskaran, and Arivalagan Pugazhendhi. "Molecular profiling of marine endophytic fungi from green algae: assessment of antibacterial and anticancer activities." *Process Biochemistry* 96 (2020): 11–20.

Pradhan, Biswajeet, and Zuraimi Suleiman. "Landcover mapping and spectral analysis using multi-sensor satellite data fusion techniques: case study in Tioman Island, Malaysia." *Journal of Geomatics* 3, no. 2 (2009): 71–78.

Prasad, M. M., and V. Murugadas. *Marine Microorganisms: Potential Resources of Biomolecules of Human Healthcare Importance.* ICAR–Central Institute of Fisheries Technology, Cochin, 2018.

Prayogo, Fitra Adi, Anto Budiharjo, Hermin Pancasakti Kusumaningrum, Wijanarka Wijanarka, Agung Suprihadi, and Nurhayati Nurhayati. "Metagenomic applications in exploration and development of novel enzymes from nature: A review." *Journal of Genetic Engineering and Biotechnology* 18, no. 1 (2020): 1–10.

Priyadarshani, Indira, and Biswajit Rath. "Commercial and industrial applications of micro algae—A review." *Journal of Algal Biomass Utilization* 3, no. 4 (2012): 89–100.

Proum, Sorya, Jose H. Santos, Lee Hoon Lim, and David J. Marshall. "Tidal and seasonal variation in carbonate chemistry, pH and salinity for a mineral-acidified tropical estuarine system." *Regional Studies in Marine Science* 17 (2018): 17–27.

Quillaguamán, Jorge, Héctor Guzmán, Doan Van-Thuoc, and Rajni Hatti-Kaul. "Synthesis and production of polyhydroxyalkanoates by halophiles: Current potential and future prospects." *Applied Microbiology and Biotechnology* 85, no. 6 (2010): 1687–1696.

Ramya, P., D. Sangeetha, E. S. Anooj, and Lekshmi Gangadhar. "Isolation, identification and screening of exopolysaccharides from marine bacteria." *Annals of Tropical Medicine and Health* 23 (2020): 23–940.

Rascovan, Nicolás, Javier Maldonado, Martín P. Vazquez, and María Eugenia Farías. "Metagenomic study of red biofilms from Diamante Lake reveals ancient arsenic bioenergetics in haloarchaea." *The ISME Journal* 10, no. 2 (2016): 299–309.

Rathore, Aditya P., Doongar R. Chaudhary, and Bhavanath Jha. "Seasonal patterns of microbial community structure and enzyme activities in coastal saline soils of perennial halophytes." *Land Degradation & Development* 28, no. 5 (2017): 1779–1790.

Ravindran, Chinnarajan, Govindaswamy R. Varatharajan, Raju Rajasabapathy, S. Vijayakanth, Alagu Harish Kumar, and Ram M. Meena. "A role for antioxidants in acclimation of marine derived pathogenic fungus (NIOCC 1) to salt stress." *Microbial Pathogenesis* 53, no. 3–4 (2012): 168–179.

Reddy, G., P. Raghavan, N. Sarita, J. Prakash, Narayana Nagesh, Daniel Delille, and Shivaji Shivaji. "*Halomonas glaciei* sp. nov. isolated from fast ice of Adelie Land, Antarctica." *Extremophiles* 7, no. 1 (2003): 55–61.

Rhodes, Matthew E., Aharon Oren, and Christopher H. House. "Dynamics and persistence of Dead Sea microbial populations as shown by high-throughput sequencing of rRNA." *Applied and Environmental Microbiology* 78, no. 7 (2012): 2489–2492.

Rodrigo-Baños, Montserrat, Inés Garbayo, Carlos Vílchez, María José Bonete, and Rosa María Martínez-Espinosa. "Carotenoids from Haloarchaea and their potential in biotechnology." *Marine Drugs* 13, no. 9 (2015): 5508–5532.

Rodríguez, Juanjo, Christine M. J. Gallampois, Sari Timonen, Agneta Andersson, Hanna Sinkko, Peter Haglund, Åsa M. M. Berglund, et al. "Effects of organic pollutants on bacterial communities under future climate change scenarios." *Frontiers in Microbiology* 9 (2018): 2926.

Roh, Changhyun, Francois Villatte, Byung-Gee Kim, and Rolf D. Schmid. "Comparative study of methods for extraction and purification of environmental DNA from soil and sludge samples." *Applied Biochemistry and Biotechnology* 134, no. 2 (2006): 97–112.

Rossbach, Susann, Sebastian Overmans, Altynay Kaidarova, Jürgen Kosel, Susana Agusti, and Carlos M. Duarte. "Giant clams in shallow reefs: UV-resistance mechanisms of Tridacninae in the Red Sea." *Coral Reefs* 39, no. 5 (2020): 1345–1360.

Schiraldi, Chiara, Carmelina Maresca, Angela Catapano, Erwin A. Galinski, and Mario De Rosa. "High-yield cultivation of Marinococcus M52 for production and recovery of hydroxyectoine." *Research in Microbiology* 157, no. 7 (2006): 693–699.

Schneegurt, Mark A. "Media and conditions for the growth of halophilic and halotolerant bacteria and archaea." In *Advances in Understanding the Biology of Halophilic Microorganisms*, pp. 35–58. Springer, Dordrecht, 2012.

Sehgal, S. N., and N. E. Gibbons. "Effect of some metal ions on the growth of *Halobacterium cutirubrum*." *Canadian Journal of Microbiology* 6, no. 2 (1960): 165–169.

Shivanand, Pooja, and Gurunathan Jayaraman. "Production of extracellular protease from halotolerant bacterium, *Bacillus aquimaris* strain VITP4 isolated from Kumta coast." *Process Biochemistry* 44, no. 10 (2009): 1088–1094.

Sivasankar, Palaniappan, Palaniappan Seedevi, Subramaniam Poongodi, Murugesan Sivakumar, Tamilselvi Murugan, Loganathan Sivakumar, Kannan Sivakumar, and Thangavel Balasubramanian. "Characterization, antimicrobial and antioxidant property of exopolysaccharide mediated silver nanoparticles synthesized by *Streptomyces violaceus* MM72." *Carbohydrate Polymers* 181 (2018): 752–759.

Somboonna, Naraporn, Anunchai Assawamakin, Alisa Wilantho, Sithichoke Tangphatsornruang, and Sissades Tongsima. "Metagenomic profiles of free-living archaea, bacteria and small eukaryotes in coastal areas of Sichang Island, Thailand." *BMC Genomics* 13, no. 7 (2012): 1–19. BioMed Central.

Speth, Daan R., Ilias Lagkouvardos, Yong Wang, Pei-Yuan Qian, Bas E. Dutilh, and Mike S. M. Jetten. "Draft genome of *Scalindua rubra*, obtained from the interface above the discovery deep brine in the Red Sea, sheds light on potential salt adaptation strategies in anammox bacteria." *Microbial Ecology* 74, no. 1 (2017): 1–5.

Srivastava, Pallavee, Judith M. Braganca, and Meenal Kowshik. "In vivo synthesis of selenium nanoparticles by *Halococcus salifodinae* BK18 and their anti-proliferative properties against HeLa cell line." *Biotechnology Progress* 30, no. 6 (2014): 1480–1487.

Subramanian, Prathiba, and Jayaraman Gurunathan. "Differential production of pigments by halophilic bacteria under the effect of salt and evaluation of their antioxidant activity." *Applied Biochemistry and Biotechnology* 190, no. 2 (2020): 391–409.

Sun, Qing-Lei, Jian Zhang, Min-Xiao Wang, Lei Cao, Zeng-Feng Du, Yuan-Yuan Sun, Shi-Qi Liu, Chao-Lun Li, and Li Sun. "High-throughput sequencing reveals a potentially novel Sulfurovum species dominating the microbial communities of the seawater—Sediment interface of a deep-sea cold seep in South China Sea." *Microorganisms* 8, no. 5 (2020): 687.

Szabó, Attila, Kristóf Korponai, Csaba Kerepesi, Boglárka Somogyi, Lajos Vörös, Dániel Bartha, Károly Márialigeti, and Tamás Felföldi. "Soda pans of the Pannonian steppe harbor unique bacterial communities adapted to multiple extreme conditions." *Extremophiles* 21, no. 3 (2017): 639–649.

Thombre, Rebecca S., Vinaya D. Shinde, Radhika S. Oke, Sunil Kumar Dhar, and Yogesh S. Shouche. "Biology and survival of extremely halophilic archaeon *Haloarcula marismortui* RR12 isolated from Mumbai salterns, India in response to salinity stress." *Scientific Reports* 6, no. 1 (2016): 1–10.

Tian, Fang, Guang Guo, Can Zhang, Feng Yang, Zhixin Hu, Chong Liu, and Shi-wei Wang. "Isolation, cloning and characterization of an azoreductase and the effect of salinity on its expression in a halophilic bacterium." *International Journal of Biological Macromolecules* 123 (2019): 1062–1069.

Tutuncu, Havva Esra, Nurgul Balci, Melek Tuter, and Nevin Gul Karaguler. "Recombinant production and characterization of a novel esterase from a hypersaline lake, Acıgöl, by metagenomic approach." *Extremophiles* 23, no. 5 (2019): 507–520.

Ugochukwu, Ekenwosu Joseph, and Peter Ugochukwu Okorie. "Isolation and characterization of streptomycetes with potential to decompose organic compounds during bioremediation of arable soil." *Sustinere: Journal of Environment and Sustainability* 4, no. 1 (2020): 16–23.

Vera-Gargallo, Blanca, and Antonio Ventosa. "Metagenomic insights into the phylogenetic and metabolic diversity of the prokaryotic community dwelling in hypersaline soils from the Odiel saltmarshes (SW Spain)." *Genes* 9, no. 3 (2018): 152.

Verma, Ashish, Sachin Kumar, and Preeti Mehta. "Physiological and genomic perspective of halophiles among different salt concentrations." In *Physiological and Biotechnological Aspects of Extremophiles*, pp. 137–151. Academic Press, London, United Kingdom, Cambridge, United States, 2020.

Vogel, Timothy M., Penny R. Hirsch, Pascal Simonet, Janet K. Jansson, James M. Tiedje, Jan Dirk Van Elsas, Renaud Nalin, Laurent Philippot, and Mark J. Bailey. "Advantages of the metagenomic approach for soil exploration: Reply from Vogel et al." *Nature Reviews Microbiology* 7, no. 10 (2009): 756–757.

Williams, Timothy J., Michelle A. Allen, Yan Liao, Mark J. Raftery, and Ricardo Cavicchioli. "Sucrose metabolism in haloarchaea: Reassessment using genomics, proteomics, and metagenomics." *Applied and Environmental Microbiology* 85, no. 6 (2019).

Xiang, Wenliang, Jie Zhang, Lin Li, Huazhong Liang, Hai Luo, Jian Zhao, Zhirong Yang, and Qun Sun. "Screening a novel Na^+/H^+ antiporter gene from a metagenomic library of halophiles colonizing in the Dagong Ancient Brine Well in China." *FEMS Microbiology Letters* 306, no. 1 (2010): 22–29.

Yankson, Kweku K., and Todd R. Steck. "Strategy for extracting DNA from clay soil and detecting a specific target sequence via selective enrichment and real-time (quantitative) PCR amplification." *Applied and Environmental Microbiology* 75, no. 18 (2009): 6017–6021.

Yergeau, Etienne, Christine Michel, Julien Tremblay, Andrea Niemi, Thomas L. King, Joanne Wyglinski, Kenneth Lee, and Charles W. Greer. "Metagenomic survey of the taxonomic and functional microbial communities of seawater and sea ice from the Canadian Arctic." *Scientific Reports* 7, no. 1 (2017): 1–10.

Zalazar, Lucia, Pablo Pagola, María Victoria Miró, Maria Sandra Churio, Micaela Cerletti, Celeste Martínez, María Iniesta-Cuerda, Ana J. Soler, Andreina Cesari, and R. De Castro. "Bacterioruberin extracts from a genetically modified hyperpigmented *Haloferax volcanii* strain: Antioxidant activity and bioactive properties on sperm cells." *Journal of Applied Microbiology* 126, no. 3 (2019): 796–810.

ZoBell, Claude E. "Studies on marine bacteria. I. The cultural requirements of heterotrophic aerobes." *The Journal of Marine Research* 4 (1941): 41–75.

5

Metagenome Assembly for Gut Microbial Functional Diversity Associated with Xenobiotic Degradation

Ashok Kumar Sharma*
University of Minnesota, Twin Cities
Vinay Shankar Dubey
Yoga and Psychotherapy Association of India Sagar, India

CONTENTS

5.1 Introduction

Pharmacokinetics (absorption, distribution, metabolism, and elimination) decides the fate of any xenobiotic entering the human body that passes through various organs before exiting (Aimone and de Lannoy 2014). The oral route is the most preferred route for any xenobiotic to enter the human body. The majority of absorption occurs in the intestines, from where the xenobiotic is transported to the liver, either through systemic circulation or via portal vein, where it may be bio-transformed/metabolized (Clarke et al. 2019). The resulting metabolites, along with untransformed xenobiotics, are secreted back to the intestines through bile or via enterohepatic circulation. Even parenterally/topically administered drugs and their resulting metabolites can pass to the intestines through biliary secretion. Therefore, most xenobiotics entering the body spend a significant time in the intestines, where dense microbial communities reside, known as gut microbiomes. These play an important role in deciding the overall bioavailability of xenobiotic molecules (Koppel, Maini Rekdal, and Balskus 2017; Spanogiannopoulos et al. 2016).

From the last decade, multiple evidence has suggested a prominent role of the gut microbiome in the transformation of xenobiotics such as dietary chemicals, therapeutic drugs, and environmental toxins as well as host-derived molecules (Wang et al. 2011; Rowland et al. 2018). Higher microbial diversity in the gut offers a huge reservoir of yet-unexplored metabolic enzymes capable of carrying out these transformations, which further impacts the overall pharmacokinetic and pharmacodynamic properties of a given molecule (Goldman, Peppercorn, and Goldin 1974; Sousa et al. 2008). Additionally, variability in the gut microbiome due to genetic and environmental factors may also contribute to individual specific xenobiotic responses. Huge efforts have been made to provide detailed mechanistic insights on the role of microbial enzymes in xenobiotic metabolism (Saitta et al. 2014; Koppel, Maini Rekdal, and Balskus 2017; Ulmer et al. 2014).

However, our understanding of 1) species-specific metabolism, 2) unknown microbial metabolic enzymes, 3) microbial and host contribution, and 4) the ultimate fate of xenobiotic molecules is still

DOI: 10.1201/9781003042570-8

limited due to the complexity of microbial communities. Methods such as the *ex vivo* culturing model of the gut microbiome (Chankhamjon et al. 2019), colonization of specific microbial species in germ-free mice (Zimmermann et al. 2019a) and population-specific variations in the response toward a specific xenobiotic molecule have been used extensively in the past and made significant advancements to the field. Still, we need more comprehensive experimental and/or computational screening methods to elucidate the species-specific contribution to xenobiotic metabolism along with identification of metabolic products and their absorption, distribution, metabolism, and elimination (ADMET) properties.

5.2 Gut Microbial Mechanisms Involved in the Xenobiotic Metabolism

Apart from major metabolism in the liver and other tissues, microorganisms play a significant role in biotransformation of xenobiotics through a different mechanism. Gut microbial communities can alter the therapeutic response by either direct or indirect involvement. In the direct involvement process, microbes directly alter the dynamics/bioactivity of xenobiotics by converting them into metabolic products. This alteration of xenobiotics can lead to the activation, inhibition or toxicity of a given xenobiotic (Table 5.1). The main biotransformation processes include reduction, hydrolysis, oxidation, deacylation, demethylation, *o*-dealkylation, dehydroxylation, decarboxylation, deamination, and acetylation (Wilson and Nicholson 2017). The other very important way by which the microbiome regulates xenobiotic metabolism is indirect involvement. These processes are even more complex, and our understanding is still limited. Microbiome may regulate these processes in several ways. 1) Via altering the host metabolic enzymes/transporters; for example, phenobarbitals are more efficiently metabolized by Cytochrome P450 in germ-free than colonized mice (*Bacillus megaterium*) due to increased expression of Cytochrome P450 in germ-free vs. colonized mice (Björkholm et al. 2009). 2) Via overlapping host response to xenobiotics and some metabolites of microbial origin. This may be the result of competition among xenobiotics and microbial metabolites for similar host enzymes/proteins that are involved in various vital functions such as absorption, transportation, detoxification or elimination of xenobiotics, which may result in altered xenobiotic concentration in blood plasma and consequently reduced efficacy or increased toxicity of a given xenobiotic, respectively. This affects the impact of host–microbial interaction on increased toxicity and half-life of xenobiotics, for example, P-cresol produced by microorganisms and acetaminophen binds to the host protein known as human cytosolic sulfotransferase 1A1 (SULT1A1). This competition among acetaminophen and P-cresol is responsible for the altered host abilities of acetaminophen detoxification because the toxic metabolite N-acetyl-p-benzoquinone imine (NAPQI) accumulates in the host

TABLE 5.1

Example of Altered Bioactivities Resulting from Xenobiotic–Microbiome Interactions

Activation	Inhibition	Toxicity
1. Prontosil: NA (Gingell, Bridges, and Williams 1971) 2. Levodopa: *Enterococcus faecalis* (van Kessel et al. 2019) 3. Sulfasalazine: NA (Cooke 1969)	1. Digoxin: *Eggerthella lenta* (Haiser et al. 2013) 2. Methotrexate: NA (Sayers, MacGregor, and Carding 2018) 3. Bromazepam, Clonazepam: NA (Elmer and Remmel 1984) 4. Tacrolimus: *Faecalibacterium prausnitzii* (Guo et al. 2019)	1. Irinotecan: *Escherichia coli (Wallace* et al. *2010)* 2. 5-Fluorouracil: NA (Scott et al. 2017) 3. NSAIDs (diclofenac, indomethacin, and ketoprofen): (Saitta et al. 2014) 4. Melamine: *Klebsiella terrigena* (Zheng et al. 2013) 5. Carboplatin: *Prevotella copri* (Yu et al. 2019) 6. Sorivudine: *Bacteroides* (Nakayama et al. 1997) 7. Nitrazepam: *Clostridium leptum* (Rafii et al. 1997) 8. L-carnitine—NA (Koeth et al. 2013) 9. Cycasin: NA (Spatz et al. 1967)

NSAID: nonsteroidal anti-inflammatory drugs; NA: not available.

(Clayton et al. 2009). 3) Drug efficacy may also be influenced by host–microbial interactions. For example, simvastatin treatment response is positively associated with microbiota-produced secondary bile acids such as glycolithocholic acid, taurolithocholic acid, and lithocholic acid. Although, exact mechanism related to these associations remains poorly understood, one possible hypothesis suggests the competition between primary bile acids and statis for similar intestinal transporters. Therefore, metabolism of bile acids by gut microbes may regulate effective statin therapy (Kaddurah-Daouk et al. 2011). Apart from this direct and indirect involvement, there are several other population-wide studies that showed the potential role of microbial communities in xenobiotic metabolism, but the mechanism involved in this process is yet unknown. For example, 1) the vaginal microbiome contributes to the variable response of the anti-retroviral drug tenofovir and leads to the identification of microbes such as *Lactobacillus* and *Gardnerella vaginalis* and their associations with drug efficacy (Klatt et al. 2017), and 2) another study reported a positive association between the relative abundance of gut bacterium *Faecalibacterium prausnitzii* and tacrolimus dosing in kidney transplant recipients (Lee et al. 2015).

5.3 Ongoing Efforts to Elucidate Xenobiotic Metabolism by Gut Microbial Communities

In the last few years, several computational and functional screening methods have been developed for the identification of microbial-mediated xenobiotic degradation. 1) DrugBug, which predicts most similar substrates (based on structural similarities) to the given xenobiotic molecule using a combination of machine learning and chemoinformatics approaches (Sharma et al. 2017). Identified substrates can be linked to their respective microbial enzymes in the preconstructed database. 2) MicrobeFDT, which links xenobiotic compounds to microbial enzymes and their known toxicities using a similar approach of structural similarity (Guthrie, Wolfson, and Kelly 2019). 3) A combination of experimental and computational strategy was used to unravel the contributions of microbes and hosts to drug metabolism (Zimmermann et al. 2019a). 4) Another advancement in the field was made through elucidating the metabolic abilities of 76 selected bacterial species to metabolize a total of 271 orally administered drugs. The authors used a combination of high-throughput genomic analyses along with mass spectrometry to further detect the microbiome genes involved in drug metabolism (Zimmermann et al. 2019b). 5) microbiome-derived metabolism (MDM) screening to characterize drug microbiome interactions by systematic mapping of the human gut microbiome's ability to biotransform small molecules (Chankhamjon et al. 2019). Overall, the combination of computational and experimental approaches provides an important framework to elucidate the contribution of the gut microbiome in xenobiotic metabolism. However, the vast majority of xenobiotic-degrading enzymes encoded by microbial communities are yet to be explored and can be explored using direct sequencing of genetic material using next-generation sequencing (NGS) technologies.

5.4 Identification of Xenobiotic-Degrading Enzymes in Metagenomic Datasets

Next-generation sequencing technologies have revolutionized our understanding of microbial communities via providing higher sequencing depth. As a result, a more refined view of microbial communities can be achieved at significantly reduced costs. These advancements, specifically higher sequencing depth, helped researchers quantify microbial features directly from genetic content. NGS techniques are immensely useful to understand interindividual variations in the xenobiotic response via microbial community profiling using 16S rRNA, metagenomic, and transcriptomic sequencing along with targeted and/ or untargeted metabolomics approaches. These approaches are generally used to identify the microbial species and their genes that are significantly discriminating among responders vs. non-responders. These approaches, specifically metagenomics, can be utilized for identification of various novel xenobiotic-degrading enzymes (XDEs).

TABLE 5.2
List of Most Commonly Used Metagenomic Assemblers

Assembler	Approach	Reference
IDBA-UD	de Bruijin graph	(Peng et al. 2012)
Megahit	de Bruijin graph	(Li et al. 2015)
MetaVelvet	de Bruijin graph	(Namiki et al. 2012)
MetaVelvet-SL	de Bruijin graph	(Afiahayati, Sato, and Sakakibara 2014)
Ray Meta	de Bruijin graph	(Boisvert et al. 2012)
SOAPdenovo2	de Bruijin graph	(Luo et al. 2012)
metaSPAdes	de Bruijin graph	(Nurk et al. 2017)
Omega	Overlap graph based	(Haider et al. 2014)
Genovo	Overlap graph based	(Laserson, Jojic, and Koller 2011)

5.4.1 Metagenomic Assembly and Evaluation

In metagenomic analysis, the most crucial step is to assemble the overlapping reads into continuous fragments and generate contigs and scaffolds (Figure 5.1). This process is generally known as metagenomic assembly, and it allows an even more detailed view of the genetic content of microbial communities. Larger assembled fragments also warrant more accurate detection of protein encoding regions compared to partial gene fragments obtained from unassembled reads. Therefore, the assembly process may serve as an important determinant in sensitivity and specificity of identified metabolic enzymes for subsequent downstream analysis. Most metagenomic assemblers developed so far either use a) an overlap graph-based (OGB) consensus assembly approach or b) a de Bruijn graph (dBG)–based approach (Table 5.2). the former approach is used to find the overlap between reads to construct consensus sequences, whereas the latter approach is based on the fragmentation and re-arrangement of *k*-mers generated by a sliding window size of *k* across reads.

There are several challenges that may occur during the assembly of metagenomic sequences, such as 1) lower microbial abundance and unknown diversity—incomplete genomic fragments can be generated due to lower abundance of multiple unknown microbial species in a community. 2) Closely related species in the community may be responsible for introducing overlapping *k*-mers and can lead to chimeric contigs and failure to capture species-specific contribution to the downstream genes. For a metagenomic assembly approach, it is important to conserve as much of the less abundant species sequence as possible. Therefore, selection of appropriate metagenomic assembly approaches that generate a high-quality assembly is an important step to be considered before the gene prediction step.

Performance evaluation of metagenomic assembly methods can be done using various standard statistical measures such as number of contigs/scaffolds, length, and genomic coverage. The most commonly used parameter is N50 size, defined as the length of contigs/scaffolds for which more than 50% of the assembled sequences are equal to or greater than this length (Mäkinen, Salmela, and Ylinen 2012). A higher N50 value is generally considered to indicate a good assembly that facilitates full-length recovery of gene sequences. Methods such as MetaQuast (alignment based) and DeepMAsED (deep learning based) are specifically useful to assess the quality of metagenomic assemblies (Mineeva et al. 2020; Mikheenko, Saveliev, and Gurevich 2016).

5.4.2 Gene Prediction and Annotation

The use of larger assembled contigs (generally >1,000 bp) generated for the prediction of protein coding genes (Step 4 in Figure 5.1) is the next step in metagenomic analysis. There are several algorithms, such as MetaGeneAnnotator (Noguchi, Taniguchi, and Itoh 2008), Orphelia (Hoff et al. 2009), Prodigal (Hyatt et al. 2010), Glimmer-MG (Kelley and Delcher 2011), FragGeneScan (Rho, Tang, and Ye 2010), and MetaGeneMark (Zhu, Lomsadze, and Borodovsky 2010), that use GC content, mono-codon and dicodon frequencies, and start and stop codons for gene prediction in metagenomic datasets. Annotation

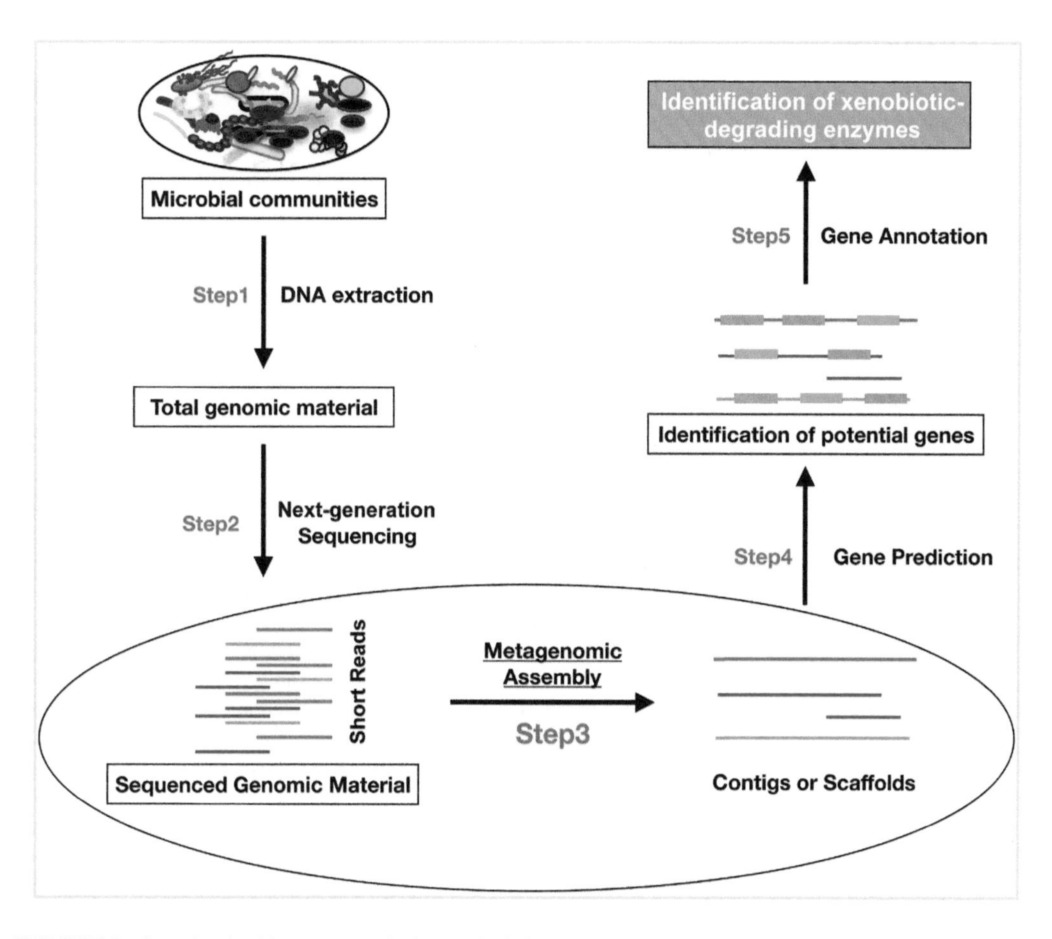

FIGURE 5.1 Steps involved in metagenomic data analysis from next-generation sequencing to xenobiotic identification.

of predicted genes to find the functional potential of the microbial community is the final step in metagenomic analysis. Depending upon the function of interest, these genes can be annotated via mapping against reference databases using various homology (similarity)-based or machine learning–based approaches. To predict XDEs, all proteins coding genes can be mapped using BLAST (Mount 2007), BLAT (Kent 2002), Diamond (Buchfink, Xie, and Huson 2015), or HMMER (Finn, Clements, and Eddy 2011) against the xenobiotic-degrading enzymes downloaded from databases such as DrugBank (Wishart et al. 2018), KEGG (Kanehisa and Goto 2000), and UniProt (Acids Research 2017). These similarity-based approaches will help users quantify the total XDEs in the given metagenomic community.

Another approach is to search for metagenomic datasets for specific enzyme types that show variable responses toward a particular xenobiotic molecule. There are many enzymes and enzyme families known to have an impact on xenobiotics, for example, beta-glucosidases, beta-glucuronidases, sulfatases, azoreductases, nitroreductases, proteases, glycosidases, and transferases (Abdelsalam et al. 2020). Differential abundances of these enzymes/enzyme families may lead to interindividual variations in drug response. These enzyme families can be quantified in metagenomic sequences using a 'chemically guided functional profiling' approach (Levin et al. 2017). This approach can be used to determine the abundance and distribution of various enzyme family members in microbial communities and provides new insights on yet-unexplored microbial functions. A very first step is to generate a sequence similarity network to categorize large numbers of enzymes belonging to the enzyme family of interest into subgroups and then use the ShortBRED program to detect unique short peptide markers. Furthermore, these markers can be quantified in raw metagenomic datasets to identify the differential abundance patterns across datasets.

REFERENCES

Abdelsalam, Nehal Adel, Ahmed Tarek Ramadan, Marwa Tarek ElRakaiby, and Ramy Karam Aziz. 2020. "Toxicomicrobiomics: The Human Microbiome vs. Pharmaceutical, Dietary, and Environmental Xenobiotics." *Frontiers in Pharmacology* 11 (April): 390.

Acids research, Nucleic, and 2017. 2017. "UniProt: The Universal Protein Knowledgebase." *Nucleic Acids Research* 45 (D1): D158–69.

Afiahayati, Kengo Sato, and Yasubumi Sakakibara. 2014. "MetaVelvet-SL: An Extension of the Velvet Assembler to a De Novo Metagenomic Assembler Utilizing Supervised Learning." *DNA Research: An International Journal for Rapid Publication of Reports on Genes and Genomes* 22 (1): 69–77.

Aimone, Lisa D., and Inés A. M. de Lannoy. 2014. "Overview of Pharmacokinetics." *Current Protocols in Pharmacology/Editorial Board, S.J. Enna. . .* [et al.] 66 (September): 7.1.1–31.

Björkholm, Britta, Chek Mei Bok, Annelie Lundin, Joseph Rafter, Martin Lloyd Hibberd, and Sven Pettersson. 2009. "Intestinal Microbiota Regulate Xenobiotic Metabolism in the Liver." *PLoS One* 4 (9): e6958.

Boisvert, Sébastien, Frédéric Raymond, Elénie Godzaridis, François Laviolette, and Jacques Corbeil. 2012. "Ray Meta: Scalable De Novo Metagenome Assembly and Profiling." *Genome Biology* 13 (12): R122.

Buchfink, Benjamin, Chao Xie, and Daniel H. Huson. 2015. "Fast and Sensitive Protein Alignment Using DIAMOND." *Nature Methods* 12 (1): 59–60.

Chankhamjon, Pranatchareeya, Bahar Javdan, Jaime Lopez, Raphaella Hull, Seema Chatterjee, and Mohamed S. Donia. 2019. "Systematic Mapping of Drug Metabolism by the Human Gut Microbiome." *BioRxiv*. https://oar.princeton.edu/jspui/handle/88435/pr1qz10.

Clarke, G., K. V. Sandhu, B. T. Griffin, T. G. Dinan, J. F. Cryan, and N. P. Hyland. 2019. "Gut reactions: Breaking down xenobiotic–microbiome interactions." *Pharmacological Reviews* 71 (2): 198–224.

Clayton, T. Andrew, David Baker, John C. Lindon, Jeremy R. Everett, and Jeremy K. Nicholson. 2009. "Pharmacometabonomic Identification of a Significant Host-Microbiome Metabolic Interaction Affecting Human Drug Metabolism." *Proceedings of the National Academy of Sciences of the United States of America* 106 (34): 14728–33.

Cooke, E. M. 1969. "Faecal Flora of Patients with Ulcerative Colitis during Treatment with Salicylazosulphapyridine." *Gut* 10 (7): 565–8.

Elmer, G. W., and R. P. Remmel. 1984. "Role of the Intestinal Microflora in Clonazepam Metabolism in the Rat." *Xenobiotica; the Fate of Foreign Compounds in Biological Systems* 14 (11): 829–40.

Finn, Robert D., Jody Clements, and Sean R. Eddy. 2011. "HMMER Web Server: Interactive Sequence Similarity Searching." *Nucleic Acids Research* 39 (Web Server issue): W29–37.

Gingell, R., J. W. Bridges, and R. T. Williams. 1971. "The Role of the Gut Flora in the Metabolism of Prontosil and Neoprontosil in the Rat." *Xenobiotica; the Fate of Foreign Compounds in Biological Systems* 1 (2): 143–56.

Goldman, P., M. A. Peppercorn, and B. R. Goldin. 1974. "Metabolism of Drugs by Microorganisms in the Intestine." *The American Journal of Clinical Nutrition* 27 (11): 1348–55.

Guo, Yukuang, Camila Manoel Crnkovic, Kyoung-Jae Won, Xiaotong Yang, John Richard Lee, Jimmy Orjala, Hyunwoo Lee, and Hyunyoung Jeong. 2019. "Commensal Gut Bacteria Convert the Immunosuppressant Tacrolimus to Less Potent Metabolites." *Drug Metabolism and Disposition: The Biological Fate of Chemicals* 47 (3): 194–202.

Guthrie, Leah, Sarah Wolfson, and Libusha Kelly. 2019. "The Human Gut Chemical Landscape Predicts Microbe-Mediated Biotransformation of Foods and Drugs." *eLife* 8 (June). https://doi.org/10.7554/eLife.42866.

Haider, Bahlul, Tae-Hyuk Ahn, Brian Bushnell, Juanjuan Chai, Alex Copeland, and Chongle Pan. 2014. "Omega: An Overlap-Graph De Novo Assembler for Metagenomics." *Bioinformatics* 30 (19): 2717–22.

Haiser, Henry J., David B. Gootenberg, Kelly Chatman, Gopal Sirasani, Emily P. Balskus, and Peter J. Turnbaugh. 2013. "Predicting and Manipulating Cardiac Drug Inactivation by the Human Gut Bacterium *Eggerthella lenta*." *Science* 341 (6143): 295–8.

Hoff, Katharina J., Thomas Lingner, Peter Meinicke, and Maike Tech. 2009. "Orphelia: Predicting Genes in Metagenomic Sequencing Reads." *Nucleic Acids Research* 37 (Web Server issue): W101–5.

Hyatt, Doug, Gwo-Liang Chen, Philip F. Locascio, Miriam L. Land, Frank W. Larimer, and Loren J. Hauser. 2010. "Prodigal: Prokaryotic Gene Recognition and Translation Initiation Site Identification." *BMC Bioinformatics* 11 (March): 119.

Kaddurah-Daouk, Rima, Rebecca A. Baillie, Hongjie Zhu, Zhao-Bang Zeng, Michelle M. Wiest, Uyen Thao Nguyen, Katie Wojnoonski, Steven M. Watkins, Miles Trupp, and Ronald M. Krauss. 2011. "Enteric Microbiome Metabolites Correlate with Response to Simvastatin Treatment." *PLoS One* 6 (10): e25482.

Kanehisa, M., and S. Goto. 2000. "KEGG: Kyoto Encyclopedia of Genes and Genomes." *Nucleic Acids Research* 28 (1): 27–30.

Kelley, David R., and Art L. Delcher. 2011. "Glimmer-MG Release Notes Version 0.1." www.cbcb.umd.edu/software/glimmer-mg/downloads/glimmer-mg_manual.pdf.

Kent, W. James. 2002. "BLAT—The BLAST-Like Alignment Tool." *Genome Research* 12 (4): 656–64.

Kessel, Sebastiaan P. van, Alexandra K. Frye, Ahmed O. El-Gendy, Maria Castejon, Ali Keshavarzian, Gertjan van Dijk, and Sahar El Aidy. 2019. "Gut Bacterial Tyrosine Decarboxylases Restrict Levels of Levodopa in the Treatment of Parkinson's Disease." *Nature Communications* 10 (1): 1–11.

Klatt, Nichole R., Ryan Cheu, Kenzie Birse, Alexander S. Zevin, Michelle Perner, Laura Noël-Romas, Anneke Grobler, et al. 2017. "Vaginal Bacteria Modify HIV Tenofovir Microbicide Efficacy in African Women." *Science* 356 (6341): 938–45.

Koeth, Robert A., Zeneng Wang, Bruce S. Levison, Jennifer A. Buffa, Elin Org, Brendan T. Sheehy, Earl B. Britt, et al. 2013. "Intestinal Microbiota Metabolism of L-Carnitine, a Nutrient in Red Meat, Promotes Atherosclerosis." *Nature Medicine* 19 (5): 576–85.

Koppel, Nitzan, Vayu Maini Rekdal, and Emily P. Balskus. 2017. "Chemical Transformation of Xenobiotics by the Human Gut Microbiota." *Science* 356 (6344). https://doi.org/10.1126/science.aag2770.

Laserson, Jonathan, Vladimir Jojic, and Daphne Koller. 2011. "Genovo: De Novo Assembly for Metagenomes." *Journal of Computational Biology: A Journal of Computational Molecular Cell Biology* 18 (3): 429–43.

Lee, John R., Thangamani Muthukumar, Darshana Dadhania, Ying Taur, Robert R. Jenq, Nora C. Toussaint, Lilan Ling, Eric Pamer, and Manikkam Suthanthiran. 2015. "Gut Microbiota and Tacrolimus Dosing in Kidney Transplantation." *PLoS One* 10 (3): e0122399.

Levin, B. J., Y. Y. Huang, S. C. Peck, Y. Wei, A. Martínez-Del Campo, J. A. Marks, E. A. Franzosa, C. Huttenhower, and E. P. Balskus. 2017. "A Prominent Glycyl Radical Enzyme in Human Gut Microbiomes Metabolizes Trans-4-Hydroxy-L-Proline." *Science* 355 (6325). https://doi.org/10.1126/science.aai8386.

Li, Dinghua, Chi-Man Liu, Ruibang Luo, Kunihiko Sadakane, and Tak-Wah Lam. 2015. "MEGAHIT: An Ultra-Fast Single-Node Solution for Large and Complex Metagenomics Assembly via Succinct de Bruijn Graph." *Bioinformatics* 31 (10): 1674–6.

Luo, Ruibang, Binghang Liu, Yinlong Xie, Zhenyu Li, Weihua Huang, Jianying Yuan, Guangzhu He, et al. 2012. "SOAPdenovo2: An Empirically Improved Memory-Efficient Short-Read De Novo Assembler." *GigaScience* 1 (1): 18.

Mäkinen, Veli, Leena Salmela, and Johannes Ylinen. 2012. "Normalized N50 Assembly Metric Using Gap-Restricted Co-Linear Chaining." *BMC Bioinformatics* 13 (October): 255.

Mikheenko, Alla, Vladislav Saveliev, and Alexey Gurevich. 2016. "MetaQUAST: Evaluation of Metagenome Assemblies." *Bioinformatics* 32 (7): 1088–90.

Mineeva, Olga, Mateo Rojas-Carulla, Ruth E. Ley, Bernhard Schölkopf, and Nicholas D. Youngblut. 2020. "DeepMAsED: Evaluating the Quality of Metagenomic Assemblies." *Bioinformatics* 36 (10): 3011–17.

Mount, David W. 2007. "Using the Basic Local Alignment Search Tool (BLAST)." *CSH Protocols* 2007 (July): db.top17.

Nakayama, H., T. Kinouchi, K. Kataoka, S. Akimoto, Y. Matsuda, and Y. Ohnishi. 1997. "Intestinal Anaerobic Bacteria Hydrolyse Sorivudine, Producing the High Blood Concentration of 5-(E)-(2-Bromovinyl)uracil That Increases the Level and Toxicity of 5-Fluorouracil." *Pharmacogenetics* 7 (1): 35–43.

Namiki, Toshiaki, Tsuyoshi Hachiya, Hideaki Tanaka, and Yasubumi Sakakibara. 2012. "MetaVelvet: An Extension of Velvet Assembler to De Novo Metagenome Assembly from Short Sequence Reads." *Nucleic Acids Research* 40 (20): e155.

Noguchi, Hideki, Takeaki Taniguchi, and Takehiko Itoh. 2008. "MetaGeneAnnotator: Detecting Species-Specific Patterns of Ribosomal Binding Site for Precise Gene Prediction in Anonymous Prokaryotic and Phage Genomes." *DNA Research: An International Journal for Rapid Publication of Reports on Genes and Genomes* 15 (6): 387–96.

Nurk, Sergey, Dmitry Meleshko, Anton Korobeynikov, and Pavel A. Pevzner. 2017. "metaSPAdes: A New Versatile Metagenomic Assembler." *Genome Research* 27 (5): 824–34.

Peng, Yu, Henry C. M. Leung, S. M. Yiu, and Francis Y. L. Chin. 2012. "IDBA-UD: A De Novo Assembler for Single-Cell and Metagenomic Sequencing Data with Highly Uneven Depth." *Bioinformatics* 28 (11): 1420–8.

Rafii, F., J. B. Sutherland, E. B. Hansen Jr., and C. E. Cerniglia. 1997. "Reduction of Nitrazepam by Clostridium Leptum, a Nitroreductase-Producing Bacterium Isolated from the Human Intestinal Tract." *Clinical Infectious Diseases: An Official Publication of the Infectious Diseases Society of America* 25 Suppl 2 (September): S121–2.

Rho, Mina, Haixu Tang, and Yuzhen Ye. 2010. "FragGeneScan: Predicting Genes in Short and Error-Prone Reads." *Nucleic Acids Research* 38 (20): e191.

Rowland, Ian, Glenn Gibson, Almut Heinken, Karen Scott, Jonathan Swann, Ines Thiele, and Kieran Tuohy. 2018. "Gut Microbiota Functions: Metabolism of Nutrients and Other Food Components." *European Journal of Nutrition* 57 (1): 1–24.

Saitta, Kyle S., Carmen Zhang, Kang Kwang Lee, Kazunori Fujimoto, Matthew R. Redinbo, and Urs A. Boelsterli. 2014. "Bacterial β-Glucuronidase Inhibition Protects Mice against Enteropathy Induced by Indomethacin, Ketoprofen or Diclofenac: Mode of Action and Pharmacokinetics." *Xenobiotica; the Fate of Foreign Compounds in Biological Systems* 44 (1): 28–35.

Sayers, Ellie, Alex MacGregor, and Simon R. Carding. 2018. "Drug-Microbiota Interactions and Treatment Response: Relevance to Rheumatoid Arthritis." *AIMS Microbiology* 4 (4): 642–54.

Scott, Timothy A., Leonor M. Quintaneiro, Povilas Norvaisas, Prudence P. Lui, Matthew P. Wilson, Kit-Yi Leung, Lucia Herrera-Dominguez, et al. 2017. "Host-Microbe Co-Metabolism Dictates Cancer Drug Efficacy in *C. elegans*." *Cell* 169 (3): 442–56.e18.

Sharma, Ashok K., Shubham K. Jaiswal, Nikhil Chaudhary, and Vineet K. Sharma. 2017. "A Novel Approach for the Prediction of Species-Specific Biotransformation of Xenobiotic/Drug Molecules by the Human Gut Microbiota." *Scientific Reports* 7 (1): 9751.

Sousa, Tiago, Ronnie Paterson, Vanessa Moore, Anders Carlsson, Bertil Abrahamsson, and Abdul W. Basit. 2008. "The Gastrointestinal Microbiota as a Site for the Biotransformation of Drugs." *International Journal of Pharmaceutics* 363 (1–2): 1–25.

Spanogiannopoulos, Peter, Elizabeth N. Bess, Rachel N. Carmody, and Peter J. Turnbaugh. 2016. "The Microbial Pharmacists within Us: A Metagenomic View of Xenobiotic Metabolism." *Nature Reviews Microbiology* 14 (5): 273–87.

Spatz, M., D. W. Smith, E. G. McDaniel, and G. L. Laqueur. 1967. "Role of Intestinal Microorganisms in Determining Cycasin Toxicity." *Proceedings of the Society for Experimental Biology and Medicine. Society for Experimental Biology and Medicine* 124 (3): 691–7.

Ulmer, Jonathan E., Eric Morssing Vilén, Ramesh Babu Namburi, Alhosna Benjdia, Julie Beneteau, Annie Malleron, David Bonnaffé, et al. 2014. "Characterization of Glycosaminoglycan (GAG) Sulfatases from the Human Gut Symbiont Bacteroides Thetaiotaomicron Reveals the First GAG-Specific Bacterial Endosulfatase." *The Journal of Biological Chemistry* 289 (35): 24289–303.

Wallace, Bret D., Hongwei Wang, Kimberly T. Lane, John E. Scott, Jillian Orans, Ja Seol Koo, Madhukumar Venkatesh, et al. 2010. "Alleviating Cancer Drug Toxicity by Inhibiting a Bacterial Enzyme." *Science* 330 (6005): 831–5.

Wang, Zeneng, Elizabeth Klipfell, Brian J. Bennett, Robert Koeth, Bruce S. Levison, Brandon Dugar, Ariel E. Feldstein, et al. 2011. "Gut Flora Metabolism of Phosphatidylcholine Promotes Cardiovascular Disease." *Nature* 472 (7341): 57–63.

Wilson, Ian D., and Jeremy K. Nicholson. 2017. "Gut Microbiome Interactions with Drug Metabolism, Efficacy, and Toxicity." *Translational Research: The Journal of Laboratory and Clinical Medicine* 179 (January): 204–22.

Wishart, David S., Yannick D. Feunang, An C. Guo, Elvis J. Lo, Ana Marcu, Jason R. Grant, Tanvir Sajed, et al. 2018. "DrugBank 5.0: A Major Update to the DrugBank Database for 2018." *Nucleic Acids Research* 46 (D1): D1074–82.

Yu, Chaoheng, Bailing Zhou, Xuyang Xia, Shuang Chen, Yun Deng, Yantai Wang, Lei Wu, et al. 2019. "Prevotella Copri Is Associated with Carboplatin-Induced Gut Toxicity." *Cell Death & Disease* 10 (10): 714.

Zheng, Xiaojiao, Aihua Zhao, Guoxiang Xie, Yi Chi, Linjing Zhao, Houkai Li, Congrong Wang, et al. 2013. "Melamine-Induced Renal Toxicity Is Mediated by the Gut Microbiota." *Science Translational Medicine* 5 (172): 172ra22.

Zhu, Wenhan, Alexandre Lomsadze, and Mark Borodovsky. 2010. "Ab Initio Gene Identification in Metagenomic Sequences." *Nucleic Acids Research* 38 (12): e132.

Zimmermann, Michael, Maria Zimmermann-Kogadeeva, Rebekka Wegmann, and Andrew L. Goodman. 2019a. "Separating Host and Microbiome Contributions to Drug Pharmacokinetics and Toxicity." *Science* 363 (6427). https://doi.org/10.1126/science.aat9931.

Zimmermann, Michael, Maria Zimmermann-Kogadeeva, Rebekka Wegmann, and Andrew L. Goodman. 2019b. "Mapping Human Microbiome Drug Metabolism by Gut Bacteria and Their Genes." *Nature* 570 (7762): 462–7.

Section IV

Metagenomics of Various Ecotypes

6

Earthworm Gut Microbiome

The Uncharted Microbiome

Rashi Miglani*, Nagma Parveen, Satpal Singh Bisht*
Laboratory of Earthworm Biotechnology and Microbial Metagenomics
Department of Zoology, D.S.B Campus, Kumaun University, Nainital
Amrita Kumari Panda
Department of Biotechnology, Sant Gahira Guru University, Ambikapur, Chhattisgarh
Monu Bala
Department of Zoology, P.G. College, Syalde, Almora
Jyoti Upadhyay
School of Health Science, University of Petroleum and Studies, Dehradun
Ankit Kumar
Department of Pharmaceutical Sciences, Kumaun University Bhimtal Campus, Bhimtal
Surajit De Mandal
Laboratory of Bio-Pesticide Innovation and Application of Guangdong Province, College of Agriculture, South China Agricultural University, Guangzhou, People's Republic of China

CONTENTS

6.1 Introduction

Earthworms are nature's best recyclers, play a significant role in many ecological processes such as the formation of soil structure, and maintain biogeochemical cycling (Gong et al. 2018). As a keystone species, earthworm represents the largest animal biomass of terrestrial ecosystem, up to 90% of soil fauna (Edward 1983; Lavelle and Spain 2001). The earthworm's gut has massive space and abundant bacterial diversity, which contributes to microbial survival. The earthworm gut contains ~89.5 million liters of soil residue, and within 10 to 40 years, ~50% and ~90% of soil can move through the earthworm gut (Drake and Horn 2007; Pass et al. 2015; Thakuria et al. 2010).

Earthworms have the propensity to consume soil, and the related material—indeed, the composition of soil—is significantly influenced by these voracious feeders. It is essential to understand the physiological, morphological, and behavioral effects of earthworm on soil function (Arrow 1, Figure 6.1). Past studies suggest that soil microbial communities can mediate the impact of earthworms on soil functions (Arrow 2, Figure 6.1) (Medina-Sauza et al. 2019). Hence, different morpho groups, namely epigeic, endogeic, and anecic earthworms, which select soil microorganisms,

DOI: 10.1201/9781003042570-10

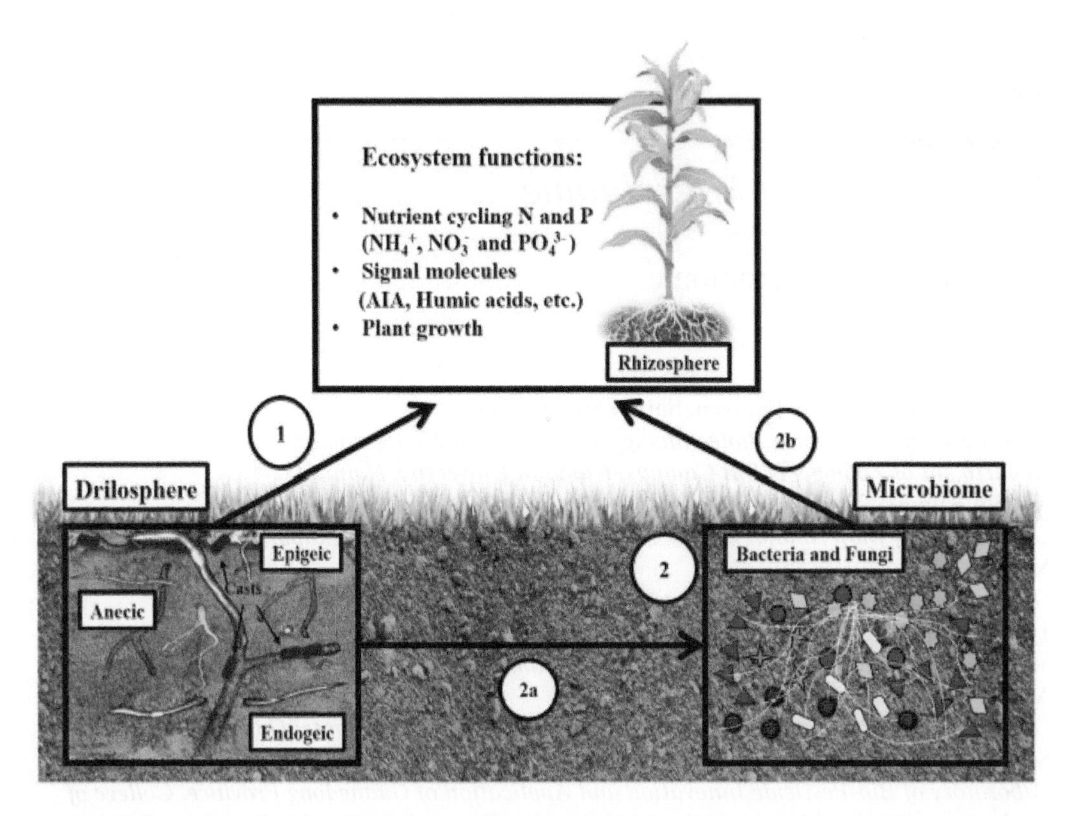

FIGURE 6.1 Microorganisms are indirectly arbitrated through earthworms' impact on plant growth and nitrogen cycling. The role of earthworms on nutrient cycling is either direct (1) or indirect (2). Earthworms control the functioning of the ecosystem by changing microbial groups (2a, 2b). The figure indicates both direct and indirect effects are significant in functioning of the ecosystem.

Source: Medina-Sauza et al. (2019)

mainly depend upon different kinds of species and show the effect on the soil microbial community (Byzov et al. 2015).

Earthworm behavior, physiology, and morphology are important to figure out their effect on soil functions (Arrow 1, Figure 6.1). The major purpose of this chapter is to highlight and examine the types of microbial populations that occur in the microbial biome of earthworm gut through metagenomics and to address the future prospective, including the potential impact of gut passage on soil microorganisms and other associated processes like nutrient cycling, change in soil microbiology, and geo-biology of terrestrial habitat.

6.2 Microbiome of Earthworm

The earthworm gut has a movable environment without oxygen to which the microorganisms of soils are subordinated (Drake and Horn 2007). Diversity in microorganisms of ingested soil and cast microorganisms are much influenced by the particular micro-environment of the earthworm gut (Drake et al. 2006). Earthworm nutrition plays a significant role in their effect on soil microbes. The biodiversity and complexity of microorganisms are also influenced by earthworm populations, which may be neutral, negative, or positive. It has been reported in recent research that there is a considerable increase in the population of specific bacterial groups in soils where earthworms are found (Medina-Sauza et al. 2019). Gong et al. (2018) found that in the long term, earthworms, by using their gut micro biota, modify microbes rather than eukaryotes in agriculture fields.

Regardless of the season, nearly 96% of the sequences were assigned only to 50 bacterial populations and three archaeal phyla, mainly constituting Proteobacteria, Acidobacteria, Actinobacteria, Nitrospirae, Bacteroidetes, and Firmicutes (Gong et al. 2018). Bacterial genera with the capacity to cope with high metal contamination were found in the research of Šrut et al. (2019), where they found *Dermacoccus, Rhizobium, Symbiobacterium, Rhodobacter*, and *Rathayibacter* to be the five most important bacteria of those that can survive easily in high metal contamination areas inside an earthworm gut. Some of the microbes isolated from various earthworm species are compiled in Table 6.1.

TABLE 6.1

Microbes (Bacteria) Found inside Earthworm Gut

Microbe Name	Family	Function	Reference
Dermacoccus	Dermacoccaceae	Heavy metal resistant (cadmium)	Šrut et al. 2019
Rathayibacter	Microbacteriaceae		
Sanguibacter	Sanguibacteraceae		
Chryseobacterium	Flavobacteriaceae		
Solibacillus	Planococcaceae		
Cohnella	Paenibacillaceae	Cellulolytic bacterial	
Streptomyces	Streptomycetaceae	Heavy metal resistant (cadmium); glucose isomerase activities	Thakuria et al. 2010; Šrut et al. 2019
Rhizobium	Rhizobiaceae	Bio-fertilizer	Hussain et al. 2016; Šrut et al. 2019
Symbiobacterium	Symbiobacteriaceae	NS	
Rhodobacter	Rhodobacteraceae	Perform anoxygenic photosynthesis, degrade insecticide (*Acephate*)	Drake and Horn 2007; Phugare et al. 2012; Šrut et al. 2019
Paenibacillus sp.	Paenibacillaceae	N_2O-producing	Ihssen et al. 2003; Šrut et al. 2019
Protochlamydia/Candidatus	Parachlamydiaceae	Symbionts of earthworms, heavy metal resistant (cadmium)	Nechitaylo et al. 2009; Šrut et al. 2019
Rummeliibacillus	Planococcaceae/ Caryophanaceae	NS	Šrut et al. 2019
Shewanella	Shewanellaceae	NS	
Pseudomonas sp.	Pseudomonadaceae	NO_2-producing	Drake and Horn 2007; Ihssen et al. 2003; Šrut et al. 2019
Roseococcus sp.	Acetobacteraceae	Endosulfan-degrading; chlorinated hydrocarbon-degrading	Šrut et al. 2019; Thakuria et al. 2010; Verma et al. 2006
Patulibacter	Patulibacteraceae	NS	Šrut et al. 2019
Yersinia	Enterobacteriaceae	NS	
Parvibaculum	Rhodobiaceae	NS	
Agrobacterium	Rhizobiaceae	Heavy metal resistant (cadmium)	
Ochrobactrum	Brucellaceae	NS	
Thermoactinomyces	Thermoactinomycetaceae	NS	
Exiguobacterium	Bacillaceae	Degrade insecticide (*Acephate*)	Phugare et al. 2012; Šrut et al. 2019
Uliginosibacterium	Rhodocyclaceae	NS	Šrut et al. 2019;
Paracoccus	Rhodobacteraceae	Denitrifying bacteria	Drake and Horn 2007; Karsten and Drake 1997; Šrut et al. 2019

(Continued)

TABLE 6.1 *(Continued)*

Microbe Name	Family	Function	Reference
Flavobacterium	Flavobacteriaceae	NO_2-producing, Hg-tolerant	Bratschen et al. 2020; Ihssen et al. 2003; Šrut et al. 2019
Phenylobacterium	Caulobacteraceae	NS	Šrut et al. 2019;
Achromobacter	Alcaligenaceae	Heavy metal resistant (cadmium), phosphate solubilizers	Brito-Vega and Espinosa-Victoria 2009
Leucobacter	Microbacteriaceae	Lignin-degrading activity	Bredon et al. 2018
Thermacetogenium	Thermoanaerobacteraceae	NS	Šrut et al. 2019
Bosea	Bradyrhizobiaceae	NS	
Nocardiodes	Nocardioidaceae	NS	
Bradyrhizobium	Bradyrhizobium	Improved distribution of nodules on soybean roots	Šrut et al. 2019; Pathma and Sakthive 2012
Dechloromonas	Rhodocyclaceae	NS	Drake and Horn 2007
Brucella	Brucellaceae	NS	
Ralstonia	Ralstoniaceae	NO_2-producing	Drake and Horn 2007; Ihssen et al. 2003
Photobacterium profundum	Vibrionaceae	NS	Drake and Horn 2007
Sinorhizobium	Rhizobiaceae	NO_2-producing	Ihssen et al. 2003; Drake and Horn 2007
Thiobacillus	Thiobacillaceae	NS	Drake and Horn 2007
Verrucomicrobia	Verrucomicrobiaceae	NS	Singh et al. 2015
Chloroflexi	Chloroflexaceae	NS	
Spirochaetes	Spirochaetaceae	NS	
Thermobifida	Nocardiopsaceae	NS	
Dehalobacter	Peptococcaceae	NS	
Prolixibacter	Prolixibacteraceae	NS	
Burkholderia sp.	Burkholderiaceae	Phosphate solubilizers	Brito-Vega and Espinosa-Victoria 2009; Hussain et al. 2016
Kluyvera ascorbata	Enterobacteriaceae	NS	
Bacillus sp.	Bacillaceae	Earthworm growth promoter	Hussain et al. 2016
Serratia sp.	Enterobacteriaceae	Phosphate solubilizers	Brito-Vega and Espinosa-Victoria 2009; Hussain et al. 2016
Bacillus cereus	Bacillaceae	Earthworm growth promoter	Hussain et al. 2016
Alphaproteobacteria	NA	Hg-tolerant	Bratschen et al. 2020
Planctomycetia	NA		
Actinobacteria	NA	Hg-tolerant, antifungal activity adjacent to *Colletotrichum coccodes, F. moliniforme, P. capsici, P. ultimum, R. solani*	Bratschen et al. 2020; Pathma and Sakthive 2012
Proteobacteria	NA		

NS= Not specified; NA= Not applicable

6.3 Gut Microbiome and Its Beneficial Effect in Earthworms

Soil is the growth promoter of diverse microbial flora. Bacteria present in the soil are the producers of several secondary metabolites, namely *Bacillus*, *Pseudomonas*, and *Streptomyces*, which act against various co-existing human pathogenic bacteria and phytopathogens like fungi (Pathma et al. 2011).

Traditionally, earthworms are the 'farmer's friend' and enhance microbial community and influence chemical and physical properties of soil. Through digesting and grinding of organic wastes with aerobic and anaerobic microbes, they induce the breakdown of large soil particles and leaf litter that affect the microbial population and turn organic wastes into vermicompost (Maboeta and Van Rensburg 2003). The activity of the earthworms enhances useful microflora and suppresses toxic pathogenic microbes. An important source of macro- and micro-nutrients and microbial enzymes are soil worm casts (Lavelle and Martin 1992).

The vermicomposting process is an effective nutrient recycling technique that involves earthworm harnessing, and they act as natural bioreactors for the decomposition of organic matter. Microbial activity–dependent vermicompost and richness in the availability of nutrients increase soil fertility, increase plant growth, and inhibit the pathogen population. Earthworms can transform organic waste into 'gold.' They are regarded as 'unheralded "soldiers" of mankind' by Charles Darwin and the 'intestine of earth' by Aristotle, as they are capable of digesting a variety of organic waste (Darwin and Seward 1903).

The area of influence of earthworms on microbial flora, fauna, and soil volume is called the 'drilosphere.' In soil volume, earthworm produces ground casts, middens, burrows, and diapause chambers (Brown et al. 2000). Earthworms participate in and influence soil functioning like carbon turnover, humus accumulation, and cellulose degradation. The intestines of earthworms contain various ranges of enzymes, microorganisms, and hormones that help in fast decomposition of semi-digested organic waste, transforming it into vermicompost in a shorter duration of time (approximately four to eight weeks) (Nagavellemma et al. 2004) in comparison to the traditional method of compositing, which involves only microbes and thereby increases the time duration (approximately 20 weeks) for compost production (Sánchez-Monedero et al. 2001). Through the earthworm gizzard, the organic waste travels and is ground into a fine powder; after this, the fermenting substances, enzymes, and microorganisms act and break it down it further into casts, which are further acted upon by microbes present in the earthworm gut, converting it into vermicompost (Dominguez and Edwards 2004).

The earthworm gut is a tube-like structure beginning from the mouth accompanied by the pharynx, which muscle tissues, esophagus, crop, gizzard, foregut, midgut, hindgut, accessory organs like digestive glands, and finally the anus. The gut of the earthworm comprises mucus made up of polysaccharides; proteins; mineral ions; organic matter; and microbial symbionts like bacteria, fungi, and protozoa. The high content of carbon, total nitrogen, organic carbon, and moisture content in the earthworm gut provides a suitable environment for the activation of microbes, which are dormant in nature. A wide range of enzymes like amylase, protease, cellulase, lipase, urease, and chitinase are found in the alimentary tract of earthworms. Enzymatic activity (cellulase and mannose) was found to be performed by gut microbes (Munnoli et al. 2010).

Earthworms break up the substrate and increase the surface area and active stage of vermicomposting through microbial degradation. The relation between the microbes and earthworm is reported to be complex, as certain categories of microbes are found in the diet of earthworms, shown by the presence of damaged microbes when they travel through the digestive system of the earthworm. A few categories of fungi like *Fusarium oxysporum*, microfungi, *Alternaria solani*, some protozoans, and yeast are digested by earthworms. *Serratia marcessens* and *Escherichia coli* were found to be fully eliminated, whereas *Bacillus cereus* var. *mycoides, Drawida calebi, Eisenia foetida*, and *Lumbricus terrestris* were reported to be decreased after passing through the gut of the earthworm (Edwards and Fletcher 1988).

Epigeic species of earthworm are phytophagous in nature. They feed, mineralize the surface litter, and contain an active gizzard that helps in fast conversion of organic waste into vermicompost. They are found to be effective bio-degraders and nutrient releasers. They include *Bimastus minusculus, Lumbricus rubellus, Eiseniella tetraedra, Eisenia foetida, Dendrodrilus rubidus*, and *Dendrodrilus octaedra.*

Endogeic species of earthworm form a wide range of horizontal burrows and are geophagous feeders on soil particulate and organic waste. They cause marked changes in the physical structure of soil and effectively utilize energy from deprived soils, hence increasing soil fertility. They include *Aporrectodea caliginosa, Aporrectodea trapezoids, Aporrectodea rosea, Allolobophora chlorotica, Millsonia anomala, Pontoscolex corethrurus, Octolasion cyaneum*, and *Octolasion lacteum.* They are further categorized as polyhumic endogeic, mesohumic endogeic, and oligohumic endogeic. Polyhumic endogeic worms feed on top soil; mesohumic worms feed on bulk soil, whereas oligohumic worms feed on poor and deep soil.

TABLE 6.2

Pollutant Transformation Ability of Earthworm Gut Bacteria

Pollutant Detoxified	Earthworm Species	Source	Bacteria	Reference
Heavy metals Cu^{2+} and Zn^{2+}	*Metaphire posthuma*	Intestine	*Bacillus licheniformia* strain KX 657843	Biswas et al. 2018a
		Gut	*Bacillus megaterium* strain MF589715, *B. licheniformia* strain MF589720, *Staphylococcus haemolyticus* strain MF589716	Biswas et al. 2018b
Pesticides 4,4-DDT helianthin B	*Eisenia fetida*	Gut	*Rhodococcus* sp., *Bacillus* sp.	Mudziwapasi et al. 2016
		Gut	*Pseudoxanthomonas*	Fu et al. 2019
Microplastics Low-density polyethylene	*Lumbricus terrestris*	Gut	*Microbacterium awajiense*, *Rhodococcus jostii*, *Mycobacterium vanbaalenii*, *Streptomyces fulvissimus* *Bacillus simplex*, *Bacillus* sp.	Lwanga et al. 2018
Antibiotics Ciprofloxacin	*Metaphire vulgaris*	Cast	*Flavobacterium*, *Turicibacter*	Pu et al. 2020

Another species of earthworm, aneceics, are phytogeophagous and feed on surface litter, forming permanent vertical burrows affecting air–water association. Anecic earthworms move from deeper layers to the surface, causing effective mixing of nutrients. They include *Lumbricus terrestris, Lumbricus polyphemus*, and *Aporrectodea longa* (Kooch and Jalilvand 2008). The earthworms act by loosening the soil and form a wide range of burrows, making soil porous. The pores in the soil improve water absorption and aeration and promote root penetration easily. They form soil aggregates, and their association with microbes in the burrows and casts plays a crucial role in the soil–air ecosystem. They speed up the soil recovery process and make it productive by retaining microbial flora that are beneficial for unproductive soil (Nakamura 1996).

The anoxic earthworm gut is a micro eco-zone where gut microbiota plays an essential role in maintaining the sustainability of the micro-ecozone and stability of soil ecosystem. The gut microbiome detoxifies various pollutants, such as heavy metal, organic pollutants, microplastics, and so on, through biodegradation, transformation, absorption, and bioaccumulation (Bhat et al. 2018). A few examples of the pollutant detoxification ability of earthworm gut bacteria are summarized in Table 6.2.

6.4 The Impact of Earthworm Gut Microbiome on Nutrient Cycling

Earthworms as decomposers are a keystone species in the soil food web. They release nutrients through digestion and excretion that directly concern soil fertility. Altering the microbial community structure affects nutrient cycling.

There are several recent reports on how the microbiome affects nutrient cycling.

- Numerous reports indicate carbon (C) and nitrogen (N) mineralization in the soil is predominantly performed by endogeic geophagous earthworms (Coq et al. 2007; Gopal et al. 2017; Lavelle et al. 1998), which affects the soil organic matter decomposition rate (Barois et al. 1987; Bernard et al. 2012). This priming effect also promotes the reutilizing of N, P, and other nutrients (Bertrand et al. 2015; Kuzyakov et al. 2000). According to one study, 70 microbial functions associated with plant microbe synthesis altered after introduction of the endogeic *Prontoscolex corethrurus* in the soil (Braga et al. 2016).
- In one study, the endogeic *Approtodea caliginosa* showed increases of two to three times in mineralized C in soil (Abail et al. 2017). Epigeic earthworms like *Eisenia fetida* and *Prontoscolex excavatus* enhance the decay rates of organic matter (Singh et al. 2015).

- Mineralization of soil organic matter (SOM) is directly associated with increased microbial activity of the intestinal microbiome (Barois and Lavelle 1986; Abail et al. 2017; Coq et al. 2007). Bernard et al. (2012) also reported increased growth in r strategic bacteria and their involvement in SOM mineralization.
- Proteobacteria growth is affected by nitrogen-rich intestinal mucus and fungal biomass decomposition in the intestinal passage of earthworms and produces labile C substrates. (Brown et al. 2000; Braga et al. 2016). SOM mineralization rates decrease in older casts compared to recent ones (Bertrand et al. 2015; Pulleman et al. 2005).
- The concentrations of mineral N increased five times in casts of *P. corethrurus* (Lavelle et al. 1992). Other reports indicated that *Eisenia fetida* also improved the mineralization of organic nitrogen in the rhizosphere of *Phormiumtenax* (Zhong et al. 2017). *P. corethrurus* (Wu et al. 2017) and *Lumbricus terrestris* (Athmann et al. 2017) both showed positive impact and increased nutrient mobilization.
- These outcomes hint at nutrient mineralization at both the drilosphere and rhizosphere region and can be considered microbial activity hotspots. *Pseudomonas* sp. is involved in the degradation of complex organic molecules and supports SOM decomposition (Bertrand et al. 2015; Fjøsne et al. 2018). Microbiota associated with earthworm release many enzymes that enhance the NO^3 and NH^{+4} in soil. For example, -N-acetylglucosaminidase enzyme increases in the presence of *Perionyx corethrurus* and is accessible via arbuscular mycorrhizal fungi (He et al. 2018).
- A further path to improved N mineralization through growth in NO_3-N leaching has been suggested to decrease microbial immobilization (Domínguez et al. 2004). Groffman et al. 2015 reported a decrease in N mineralization in the case of *Lumbricus rubellus*. However, it may be due to mineralized nitrogen that could be used more easily by plants, masking the available N concentrations (González and Zou 1999; Pashanasi et al. 1996; Wu et al. 2017).
- Soil N dynamics change according to earthworm species and different earthworm ecological groups (Aira et al. 2005; Clause et al. 2014; Decaëns et al. 1999; Groffman et al. 2015).
- Similar to nitrogen, phosphorus concentration is also high in earthworm casts (Jiménez et al. 2003; Kuczak et al. 2006; Ros et al. 2017; Vos et al. 2014) or in biopores fashioned by *L. terrestris* (Athmann et al. 2017) than in bulk soil. In *Lumbricus terrestris* casts, P concentration was around 30–1,000 times more than that found in bulk soil (Ros et al. 2017). *Prontoscolex corethruru*, an earthworm species (Chapuis-Lardy et al. 1998; Lopez-Hernandez et al. 1993; Patron et al. 1999), and *Eisenia fetida* (Cao et al. 2015) also reported enhancement in P.
- In soil or substrate, the P levels are initiated by epigeic earthworms, while the N level in soil is increased by endogeic worms rather than the other two groups of earthworms.
- Potassium (K) is another major nutrient used by plants (Amtmann et al. 2008; Chen et al. 2008; Sugumaran and Janarthanam 2007). Plants cannot make use of K in the form of silicates (Liu et al. 2006); therefore, earthworms help to free K from silicate minerals when they pass through their gut (Basker et al. 1994).
- The earthworm gut microbiome helps with weathering of minerals through decreasing pH and producing ion-complexing organic ligands (Sanz-Montero and Rodriguez-Aranda 2009), further confirmed by Teng et al. (2012) in *Metaphire tschiliensis tschiliensis* casts in clay soil.
- Enriched Ca content can contribute to an active calciferous gland in earthworms, which secretes calcium carbonate–rich mucus into the esophagus and leads to high concentration in casts (Drake and Horn 2007).
- The dominant bacterial phyla in the earthworm gut belong to Proteobacteria, Actinobacteria, Firmicutes, and Acidobacteria. However, more studies need to explore the link between gut microbiome and ecosystem functions. Several authors reported expression of bacterial genes involved in the N cycle (Hosseini Bai et al. 2015; Ribbons et al. 2018).
- The growth of bacterial denitrification genes (nirK and nosZ) in the rhizosphere is enhanced by the presence of *Prontoscolex corethrurus* (Braga et al. 2016; Chapuis-Lardy et al. 2010). Similar findings were noted by Nebert et al. (2011) for *Lumbricus rubellus*. de Menezes et al.

(2018) also reported an increase in the nosZ gene coding for nitrous oxide reductase (N_2O), suggesting the presence of highly denitrifying bacterial communities (Reverchon et al. 2015). Therefore, promoting the decomposition of SOM by earthworm promotes the soil conditions to support rich denitrified communities.

- For N_2O-producing bacteria, the earthworm gut is an ideal microenvironment that is probably derived from soil (Horn et al. 2003, 2006). According to Nebert et al. (2011), N_2O emissions in an abundance of the denitrification genes has no effect on *Approctodea caliginosa*. The *Maoridrilus crossalpinus* anecic species has been found to limit N_2O emissions associated with rhizobic bacteria thanks to aerobic environments that are unfavorable to denitrification due to burrowing habits (Kim et al. 2017).
- The incorporation of *Prontoscolex corethrurus* resulted in a larger amount of microbial functional genes linked to carbohydrate and lipid metabolism, biosynthesis processes, translation, oxidation, and cell proliferation (Braga et al. 2016). How earthworms alter microbial functional genes related to the P cycle is still to be investigated.

6.5 Genomics Approaches to Studying Gut Microbiology of Earthworm

The earthworm gut microbiota has a multifaceted interdependence with the host. Several conventional studies demonstrated that the earthworm gut has an equal microbial load parallel to soil. The earthworm is a predominant bio-indicator of the soil ecosystem and acts as an eco-toxicological assessment species for determination of the effects of pesticides on soil flora. The growth, reproduction, and mortality rate and cellular oxidative stress reactions of earthworms are early indicators of antagonistic effects of chemicals and pesticides. Advanced sequencing technologies and molecular approaches generate detailed insight into both culturable and uncultivatable microbial communities in different ecosystems including earthworm gut (Panda et al. 2017, 2015; De Mandal et al. 2015a, 2015b, 2016; Hong et al. 2011; Prochazkova et al. 2013).

A few functional metagenomic studies revealed biotechnologically important hydrolytic enzymes such as carboxyl/feruloyl esterase, glycosyl hydrolases, and platelet-activating factor acetyl-hydrolase, from earthworm egested matter (Nechitaylo et al. 2015). The earthworm gut is a transitory habitat for soil microbial flora, and it hosts a diverse range of soil prokaryotes, invertebrates, and vertebrates with varied phylogenies and roles. The change in microbial composition of the earthworm in response to arsenic gradient has newly been reported by 16S ribosomal RNA pyrosequencing (Pass et al. 2015). The surrounding environment affects the gut microbiome due to changed gut micro-habitat (anoxia, neutral pH, and increased carbon substrates). The draft genome sequences of three earthworm gut bacteria became available: 1) the whole genome of *Verminephrobacter eiseniae* msu (4.4 Mb) reported by Arumugaperumal et al. (2020) identified many symbiotic genes and pathways; 2) the 2.64-Mb genome of *Chryseomicrobium excrementi* sp. nov. isolated from *Eisenia fetida* cast identified novel genes involved in aromatic compound degradation (Saha et al. 2018); 3) similarly, the *Pradoshia eiseniae* gen. nov., sp. nov. genome (3.8 Mb) from gut of *Eisenia fetida* was studied by Saha et al. (2019).

The use of high-throughput sequencing technology for molecular screening of the gut microbial community will help in understanding the microbial functions associated with earthworm species. Singh et al. (2015) explored the taxonomic and functional diversity of the gut microbiome of *Eisenia foetida* and *Perionyx excavates* by employing a 16SrRNA amplicon sequencing approach. They observed a higher abundance of *Actinobacteria* and *Firmicutes* in the gut of *P. excavatus*. The 16S rDNA clonal library annotation revealed around 12–24% abundance of xylan degraders in both earthworm metagenomes.

Ammonia oxidizers (24.1%), chitin degraders (16.7%), and nitrogen fixers (7.4%) were comparatively higher in the *E. foetida* library, while sulfate oxidizers and reducers (12.1–29.6%) were abundant in the *P. excavates* amplicon library. In the *E. foetida* library, 3.7% of the clones were reported to be lignin degraders, whereas 1.7% of the clones were represented by cellulose degraders. The gut microbiomes comprise 17–22% dehalogenators and aromatic hydrocarbon degraders (1.7–5.6%), elucidating their bioremediation capabilities (Singh et al. 2015). Singh et al. (2016) reported the culturable gut microbiome of two epigeic earthworm, *Eisenia foetida* and *Perionyx excavates*, and revealed the significance of selection of earthworm substrate combinations for effective vermicomposting. Previous studies reported that the gut microbiota of *E. eugeniae* exploits fermented tannery waste mixtures and transforms it into nutrient-rich compost through enzyme activity (Ravindran et al. 2015).

The earthworm life forms, feeding mode, and genotype regulate gut microbiome diversity among different earthworm species (Sapkota et al. 2020). Sapkota et al. (2020) studied the gut microbiome of five earthworm species, *Aporrectodea longa, Aporrectodea caliginosa, Aporrectodea tuberculata, Allolobophora chlorotica,* and *Lumbricus herculeus*, by amplicon sequencing technique and concluded that gut microbial diversity is earthworm species specific. *Aporrectodea* sp. feeds on a variety of materials from horizon A (the top soil layer), such as humus, minerals, and decayed plant materials, and showed higher gut microbial diversity in comparison to *Lumbricus herculeus*, which feeds only on a narrow range of substrates (Sapkota et al. 2020). Many studies reported that microbiomes associated with the earthworm gut are molded by the life forms, feeding substrate, soil mineral, and other chemical content (Ma et al. 2017; Liu et al. 2018; Pass et al. 2015).

6.6 Future Perspectives

The characterization of earthworm gut microbiota by employing various molecular tools, including advanced genomics approaches, explores the diversity, relative abundance, and richness of gut microorganisms and simultaneously increases our understanding of how earthworm changes the soil microbiome. Earthworm alters the soil microbiome by stimulating certain microbial genera during the passage of soil through the gut and suppresses pathogenic organisms, thus playing a vibrant role as microbial engineer. Recent advances in next-generation sequencing (NGS) technologies have charted the structural and functional annotation of the gut microbiome of earthworm and explain how earthworm promotes various soil functions. Further studies are needed to understand how earthworm activates microorganisms and their interactions with other eukaryotic taxa present inside the gut to shape soil biogeochemistry. More investigations are required employing advanced sequencing technologies and bioinformatics tools to decipher the whole-genome sequence information of earthworm gut microbiota that may identify novel microbial genes and metabolic pathways required for pollutant transformation and other ecosystem services.

REFERENCES

Abail, Z., L. Sampedro, and J.K. Whalen. 2017. Short-term carbon mineralization from endogeic earthworm casts as influenced by properties of the ingested soil material. *Appl. Soil. Ecol.* 116: 79–86. DOI: 10.1128/JCM.44.4.1359-1366.2006

Aira, M., F. Monroy, and J. Domínguez. 2005. Ageing effects on nitrogen dynamics and enzyme activities in casts of *Aporrectodea caliginosa* (Lumbricidae). *Pedobiologia.* 49: 467–473. DOI: 10.1016/j.pedobi.2005.07.003

Amtmann, A., S. Troufflard, and P. Armengaud. 2008. The effect of potassium nutrition on pest and disease resistance in plants. *Physiol. Plant.* 133 (4): 682–691. DOI: 10.1111/j.1399-3054.2008.01075.x

Arumugaperumal, A., S. Paul, S. Lathakumari, R. Balasubramani, and S. Sivasubramaniam. 2020. The draft genome of a new *Verminephrobacter eiseniae* strain: a nephridial symbiont of earthworms. *Ann. Microbio.* 70: 1–18. DOI: 10.1186/s13213-020-01549-w

Athmann, M., T. Kautz, C. Banfield, S. Bauke, D.T. Hoang, M. Lüsebrink, J. Pausch, W. Amelung, Y. Kuzyakov, and U. Köpke. 2017. Six months of *L. terrestris* activity in root-formed biopores increases nutrient availability, microbial biomass and enzyme activity. *Appl. Soil. Ecol.* 120: 135–142. DOI: 10.1016/j.apsoil.2017.08.015

Barois, I., and P. Lavelle. 1986. Changes in respiration rate and some physicochemical properties of a tropical soil during transit through *Pontoscolex corethrurus* (Glossoscolecidae, Oligochaeta). *Soil. Biol. Biochem.* 18: 539–541. DOI: 10.1016/0038-0717(86)90012-X

Barois, I., B. Verdier, P. Kaiser, A. Mariotti, P. Rangel, and P. Lavelle. 1987. Influence of the tropical earthworm *Pontoscolex corethrurus* (Glossoscolecidae) on the fixation and mineralization of nitrogen. In: Omodeo, P., and Bonvicini, A.M. (eds) *On Earthworms.* Mucchi, Bologna, pp. 151–158.

Basker, A., J.H. Kirkman, and A.N. Macgregor. 1994. Changes in potassium availability and other soil properties due to soil ingestion by earthworms. *Biold. Fertil. Soils.* 17: 154–158. DOI: 10.1007/BF00337748

Bernard, L., L. Chapuis-Lardy, T. Razafimbelo, M. Razafindrakoto, A.L. Pablo, E. Legname, E.J. Poulain, T. Brüls, M. O'Donohue, A. Brauman, J-L, Chotte, and E. Blanchart. 2012. Endogeic earthworms shape bacterial functional communities and affect organic matter mineralization in a tropical soil. *ISME Journal.* 6: 213–222. DOI: 10.1038/ismej.2011.87

Bertrand, M., S. Barot, M. Blouin, J. Whalen, T. de Oliveira, and J.R. Estrade. 2015. Earthworm services for cropping systems. A review. *Agron. Sustain. Dev.* 35: 553–567. DOI: 10.1007/s13593-014-0269-7

Bhat, S.A., S. Singh, J. Singh, S. Kumar, and A.P. Vig. 2018. Bioremediation and detoxification of industrial wastes by earthworms: vermicompost as powerful crop nutrient in sustainable agriculture. *Bioresour. Technol.* 252: 172–179. DOI: 10.1016/j.biortech.2018.01.003

Biswas, J.K., A. Banerje, M. Rai, R. Naidu, B. Biswas, M. Vithanage, M.C. Dash, S.K. Sarkar, and E. Meers. 2018b. Potential application of selected metal resistant phosphate solubilizing bacteria isolated from the gut of earthworm (*Metaphire posthuma*) in plant growth promotion. *Geoderma.* 330: 117–124. DOI: 10.1016/j.*geoderma*.2018.05.034

Biswas, J.K., A. Banerjee, M.K. Rai, J. Rinklebe, S.M. Shaheen, S.K. Sarkar, M.C. Dash, A. Kaviraj, U. Langer, H. Song, M. Vithanage, M. Mondal, and N.K. Niazi. 2018a. Exploring potential applications of a novel extracellular polymeric substance synthesizing bacterium (*Bacillus licheniformis*) isolated from gut contents of earthworm (*Metaphire posthuma*) in environmental remediation. *Biodegradation.* 29: 323–337. DOI: 10.1007/s10532-018-9835-z

Braga, L.P., C.A. Yoshiura, C.D. Borges, M.A. Horn, G.G. Brown, H.L. Drake, and S.M. Tsai. 2016. Disentangling the influence of earthworms in sugarcane rhizosphere. *Sci. Rep.* 6: 38923. DOI: 10.1038/srep38923

Bratschen, J., S, Gygax, A. Mestrot, and A. Frossard. 2020. Soil Hg contamination impact on earthworms gut microbiome. *Appl. Sci.* 10 (7): 2565. DOI: 10.3390/app10072565

Bredon, M., J. Dittmer, C. Noei, B. Moumen, and D. Bouchon. 2018. Lignocellulose degradation at the holobiont level: teamwork in a keystone soil invertebrate. *Microbiome.* 6: 162. DOI: 10.1186/s40168-018-0536-y

Brito-Vega, H., and D. Espinosa-Victoria. 2009. Bacterial diversity in the digestive tract of earthworms (Oligochaeta). *J. Biol. Sci.* 9 (3): 192–199. DOI: 10.3923/jbs.2009.192.199

Brown, G.G., I. Barois, and P. Lavelle. 2000. Regulation of soil organic matter dynamics and microbial activity in the drilosphere and the role of interactions with other edaphic functional domains. *Eur. J. Soil. Biol.* 36 (3–4): 177–198. DOI: 10.1016/S1164-5563(00)01062-1

Byzov, B.A., V.V. Tikhonov, T.Y. Nechitailo, V.V. Deminc, and D.G. Zvyagintsev. 2015. Taxonomic composition and physiological and biochemical properties of bacteria in the digestive tracts of earthworms. *Eurasian J. Soil Sci.* 48: 268–275. DOI: 10.1134/S1064229315030035

Cao, J., Y. Huang, and C. Wang. 2015. Rhizosphere interactions between earthworms (*Eisenia fetida*) and arbuscular mycorrhizal fungus (*Funneliformis mosseae*) promote utilization efficiency of phytate phosphorus in maize. *Appl. Soil Ecol.* 94: 30–39. DOI: 10.1016/j.apsoil.2015.05.001

Chapuis-Lardy, L., A. Brauman, L. Bernard, A.L. Pablo, J. Toucet, M.J. Mano, L. Weber, D. Brunet, T. Razafimbelo, J.L. Chotte, and E. Blanchart. 2010. Effect of the endogeic earthworm *Pontoscolex corethrurus* on the microbial structure and activity related to CO_2 and N_2O fluxes from a tropical soil (Madagascar). *Appl. Soil Ecol.* 45: 201–208. DOI: 10.1016/j.apsoil.2010.04.006

Chapuis-Lardy, L., M. Brossard, P. Lavelle, and E. Schouller. 1998. Phosphorus transformations in a Ferralsol through ingestion by *Pontoscolex corethrurus*, a geophagous earthworm. *Eur. J. Soil. Biol.* 34: 61–67. DOI: 10.1016/S1164-5563(99)90002-X

Chen, S., B. Lian, and C.Q. Liu. 2008. *Bacillus mucilaginosus* on weathering of phosphorite and primary analysis of bacterial proteins during weathering. *Chin. J. Geochem.* 27: 209–216. DOI: 10.1007/s11631-008-0209-9

Clause, J., S. Barot, B. Richard, T. Decaëns, and E. Forey. 2014. The interactions between soil type and earthworm species determine the properties of earthworm casts. *Appl. Soil. Ecol.* 83: 149–158. DOI: 10.1016/j.apsoil.2013.12.006

Coq, S., B.G Barthès, R. Oliver, B. Rabary, and E. Blanchart. 2007. Earthworm activity affects soil aggregation and organic matter dynamics according to the quality and localization of crop residues—an experimental study (Madagascar). *Soil Biol. Biochem.* 39: 2119–2128. DOI: 10.1016/j.soilbio.2007.03.019

Darwin, F., and A.C. Seward. 1903. More letters of Charles Darwin. In: John, M. (ed) *A Record of His Work in Series of Hitherto Unpublished Letters*, vol 2. Hazell Watson and Viney, Ld. London, pp. 508.

De Mandal, S., H.T. Lalremsanga, and N.S. Kumar. 2015b. Bacterial diversity of Murlen National Park located in Indo-Burman Biodiversity hotspot region: A metagenomic approach. *Genom. Data* 5: 25–26.

De Mandal, S., A.K. Panda, S.S. Bisht, and N.S. Kumar. 2015a. First report of bacterial community from a bat guano using Illumina next-generation sequencing. *Genom. Data* 4: 99–101.

De Mandal, S., A.K. Panda, S.S Bisht, & N.S. Kumar. 2016. MiSeq HV4 16S rRNA gene analysis of bacterial community composition among the cave sediments of Indo-Burma biodiversity hotspot. *Environ. Sci. Pollut. Res.* 23 (12): 12216–12226.

De Menezes, A.B., P.M. Miranda Tendai, M. Lynne, T. Peter, B. Geoff, F. Mark, W. Tim, R. Alan, and T. Peter. 2018. Earthworm-induced shifts in microbial diversity in soils with rare versus established invasive earthworm populations. *FEMS Microbiol. Ecol.* 94 (5): 51. DOI: 10.1093/femsec/fiy051

Decaëns, T., A.F. Rangel, N. Asakawa, and R.J. Thomas. 1999. Carbon and nitrogen dynamics in ageing earthworm casts in grasslands of the eastern plains of Colombia. *Biol. Fertil. Soils.* 30: 20–28. DOI: 10.1007/s003740050582

Domínguez, J., P.J. Bohlen, and R.W. Parmelee. 2004. Earthworms increase nitrogen leaching to greater soil depths in row crop agroecosystems. *Ecosystems.* 7 (6): 672–685. DOI: 10.1007/s10021-004-0150-7

Dominguez, J., and C.A. Edwards. 2004. Vermicomposting organic wastes: A review. In: Shakir Hanna, S.H., and MikhaTl, W.Z.A., (eds) *Soil Zoology for Sustainable Development in the 21st Century.* Geocities, Cairo, pp. 369–395.

Drake, H.L., and M.A. Horn. 2007. As the worm turns: the earthworm gut as a transient habitat for soil microbial biomes. *Annu. Rev. Microbiol.* 61: 169–189. DOI: 10.1146/annurev.micro.61.080706.093139

Drake, H.L., A. Schramm, and M.A. Horn. 2006. Earthworms gut microbial biomes: their importance to soil microorganisms, denitrification, and the terrestrial production of the greenhouse gas N_2O. *Soil. Biol.* 6: 65–87. DOI: 10.1007/3-540-28185-1_3

Edwards, C.A. 1983. Earthworm ecology in cultured soils. In: Satchell, J.E. (ed) *Earthworm Ecology.* Springer, Dordrecht, pp. 123–137.

Edwards, C.A., and K.E. Fletcher. 1988. Interaction between earthworms and microorganisms in organic matter breakdown. *Agric. Ecosyst. Environ.* 20 (1–3): 235–249. DOI: 10.1007/s11356-017-8438-2

Fjøsne, T., F.D. Myromslien, R.C. Wilson, and K. Rudi. 2018. Earthworms are associated with subpopulations of *Gammaproteobacteria* irrespective of the total soil microbiota composition and stability. *FEMS Microbiol. Lett.* 365 (9). DOI: 10.1093/femsle/fny07

Fu, L., B. Yanan, Y.Z. Lu, J. Ding, D. Zhou, and R. Zeng. 2019. Degradation of organic pollutants by anaerobic methane-oxidizing microorganisms using methyl orange as example. *J. Hazard. Mater.* 364: 264–271. DOI: 10.1016/j.jhazmat.2018.10.036

Gong, X., Y. Jiang, Y. Zheng, X. Chen, H. Li, F. Hu, M. Liu, and S. Scheu. 2018. Earthworms differentially modify the microbiome of arable soils varying in residue management. *Soil Biol. Biochem.* 121: 120–129. DOI: 10.1016/j.soilbio.2018.03.011

González, G., and X. Zou. 1999. Earthworm influence on N availability and the growth of *Cecropia schreberiana* in tropical pastures and forest soils. *Pedobiologia.* 43 (6): 824–829. www.urbanfischer.de/journals/pedo

Gopal, M., S.S. Bhute, A. Gupta, S.R. Prabhu, G.V. Thomas, W.B. Whitman, and K. Jangid. 2017. Changes in structure and function of bacterial communities during coconut leaf vermicomposting. *Antonie Leeuwenhoek.* 110 (10): 1339–1355. DOI: 10.1007/s10482-017-0894-7

Groffman, P.G., T.J. Fahey, M.C. Fisk, J.B. Yavitt, R.E. Sherman, P.J. Bohlen, and J.C. Maerz. 2015. Earthworms increase soil microbial biomass carrying capacity and nitrogen retention in northern hardwood forests. *Soil Biol. Biochem.* 87: 51–58. DOI: 10.1016/j.soilbio.2015.03.025

He, X., Y. Chen, S. Liu, A. Gunina, X. Wang, W. Chen, Y. Shao, L. Shi, Q. Yao, J. Li, X. Zou, J.P. Schimel, W. Zhang, and S. Fu. 2018. Cooperation of earthworm and arbuscular mycorrhizae enhanced plant N uptake by balancing absorption and supply of ammonia. *Soil Biol. Biochem.* 116: 351–359. DOI: 10.1016/j.soilbio.2017.10.038

Hong, S.W., J.S. Lee, and K. Chung. 2011. Effect of enzyme producing microorganisms on the biomass of epigeic earthworms (*Eisenia fetida*) in vermicompost. *Biores. Tech.* 102 (10): 6344–6347. DOI: 10.1016/j.biortech.2011.02.096

Horn, M.A., H.L. Drake, and A. Schramm. 2006. Nitrous oxide reductase genes (nosZ) of denitrifying microbial populations in soil and the earthworm gut are phylogenetically similar. *App. Environ. Micro.* 72: 1019–1026. DOI: 10.1128/AEM.72.2.1019-1026.2006

Horn, M.A., A. Schramm, and H.L. Drake. 2003. The earthworm gut: an ideal habitat for ingested N_2O-producing microorganisms. *App. Environ. Micro.* 69 (3): 1662–1669. DOI: 10.1128/AEM.69.3.1662–1669.2003

Hosseini, B.S., F. Reverchon, C.Y. Xu, Z. Xu, T.J. Blumfield, H. Zhao, L.V. Zwieten, and H.M. Wallace. 2015. Wood biochar increases nitrogen retention in field settings mainly through abiotic processes. *Soil Biol. Biochem.* 90: 232–240. DOI: 10.1016/j.soilbio.2015.08.007

Hussain, N., A. Singh, S. Saha, M.V.S. Kumar, P. Bhattacharyya, and S.S. Bhattacharyya. 2016. Excellent N-fixing and P-solubilizing traits in earthworm gut-isolated bacteria: a vermicompost based assessment with vegetable market waste and rice straw feed mixtures. *Biores. Tech.* 222: 165–174. DOI: 10.1016/j.biortech.2016.09.115.

Ihssen, J., M.A. Horn, C. Matthies, A. Gobner, A. Schramm, and H.L. Drake. 2003. N_2O-producing microorganisms in the gut of the earthworm *Aporrectodea caliginosa* are indicative of ingested soil bacteria. *Appl. Environ. Micro.* 69 (3): 1655–1661. DOI: 10.1128/AEM.69.3.1655-1661.2003

Jiménez, J.J., A. Cepeda, T. Decaëns, A. Oberson, and D.K. Friesen. 2003. Phosphorus fractions and dynamics in surface earthworm casts under native and improved grasslands in a Colombian *Savanna oxisol. Soil Biol. Biochem.* 35 (5): 715–727. DOI: 10.1016/S0038-0717(03)00090-7

Karsten, G.R., and H.L. Drake. 1997. Denitrifying bacteria in the earthworm gastrointestinal tract and in vivo emission of nitrous oxide (N_2O) by earthworms. *Ameri. Soc. Microbiol.* 63 (5): 1878–1882. http://aem.asm.org/content/aem/63/5/1878.full.pdf

Kim, Y.N., B. Robinson, K.A. Lee, S. Boyer, and N. Dickinson. 2017. Interactions between earthworm burrowing, growth of a leguminous shrub and nitrogen cycling in a former agricultural soil. *Appl. Soil Eco.* 110: 79–87. DOI: 10.1016/j.apsoil.2016.10.001

Kooch, Y., and H. Jalilvand. 2008. Earthworm as ecosystem engineers and the most important detritivors in forest soils. *Pak. J. Biol. Sci.* 11 (6): 819–825. https://pubmed.ncbi.nlm.nih.gov/18814642

Kuczak, N.K., E.C.M. Fernandes, J. Lehmann, M.A. Rondon, and F.J. Luizão. 2006. Inorganic and organic phosphorus pools in earthworm casts (Glossoscolecidae) and a Brazilian rainforest oxisol. *Soil Biol. Biochem.* 38 (3): 553–560. DOI: 10.1016/j.soilbio.2005.06.007

Kuzyakov, Y., J.K. Friedel, and K. Stahr. 2000. Review of mechanisms and quantification of priming effects. *Soil Biol. Biochem.* 32: 1485–1498. DOI: 10.1016/S0038–0717(00)00084-5

Lavelle, P., and A. Martin. 1992. Small-scale and large-scale effects of endogeic earthworms on soil organic matter dynamics in soils of the humid tropics. *Soil Biol. Biochem.* 24 (12): 1491–1498. DOI: 10.1016/0038-0717(92)90138-N

Lavelle, P., G. Melendez, B. Pashanasi, and R. Schaefer. 1992. Nitrogen mineralization and reorganization in casts of the geophagous tropical earthworm *Pontoscolex corethrurus* (Glossoscolecidae). *Biol. Fert. Soil.* 14 (1): 49–53. DOI: 10.1007/BF00336302

Lavelle, P., B. Pashanasi, F.C.G. Charpentier, J.P. Rossi, L. Derouard, J. André, J. Ponge, and N. Bernier. 1998. Large-scale effect of earthworms on soil organic matter and nutrient dynamics. In: Edwards, C.A. (ed) *Earthworm Ecology.* St. Lucie Press, Boca Raton, FL, pp. 103–122. https://hal.archives-ouvertes.fr/hal-00505398

Lavelle, P., and A.V. Spain. 2001. *Soil Ecology.* Kluwer Scientific Publications, Amsterdam. DOI: 10.1007/978-94-017-5279-4

Liu, D., B. Lian, C. Wu, and P. Guo. 2018. A comparative study of gut microbiota profiles of earthworms fed in three different substrates. *Symbiosis.* 74 (1): 21–29. DOI: 10.1007/s13199-017-0491-6

Liu, W., S. Xu, Q. Wu, Y. Yang, and P.C. Luo. 2006. Decomposition of silicate minerals by *Bacillus mucilaginosus* in liquid culture. *Environ. Geochem. Heal.* 28 (1): 133–140. DOI: 10.1007/s10653-005-9022-0

Lopez-Hernandez, D., P. Lavelle, J.C. Fardeau, and M. Nino. 1993. Phosphorus transformations in two P-sorption contrasting tropical soils during transit through *Pontoscolex corethrurus* (Glossoscolecidae: Oligochaeta). *Soil Biol. Biochem.* 25 (6): 789–792. DOI: 10.1007/0038-0717 (93)90124-T

Lwanga, E.H., B. Thapa, X. Yang, H. Gertsen, T. Salánki, V. Geissen, and P. Garbeva. 2018. Decay of low-density polyethylene by bacteria extracted from earthworm's guts: a potential for soil restoration. *Sci. Total Environ.* 624: 753–757. DOI: 10.1016/j.scitotenv.2017.12.144

Ma, L., Y. Xie, Z. Han, J.P. Giesy, and X. Zhang. 2017. Responses of earthworms and microbial communities in their guts to Triclosan. *Chemosphere.* 168: 1194–1202. DOI: 10.1016/J.chemosphere.2016.10.079.

Maboeta, M.S., and L. Van Rensburg. 2003. Vermicomposting of industrially produced wood chips and sewage sludge utilizing. *Eisenia foetida. Ecotox. Environ. Safe.* 56 (2): 265–270. DOI: 10.1016/S0147-6513(02)00101-X

Medina-Sauza, R., M. Alvarez-Jimenez, A. Delhal, F. Reverchon, M. Blouin, J.A. Guerrero-Analco, C.R. Cerdan, R. Guevera, L. Villain, and I. Barosi. 2019. Earthworm building up soil microbiota, a review. *Fronti. Environ. Sci.* 7 (81): 1–20. DOI: 10.3389/fenvs.2019.00081

Mudziwapasi, M., S.S. Mlambo, N.L. Chigu, P.K. Kuipa, and W.T. Sanyika. 2016. Isolation and molecular characterization of bacteria from the gut of *Eisenia fetida* for biodegradation of 4,4 DDT. *J. App. Biol. Biotech.* 4 (5): 41–47. DOI: 10.7324/JABB.2016.40507

Munnoli, P.M., J.A.T. Da Silva, and B. Saroj. 2010. Dynamics of the soil-earthworm-plant relationship: a review. *Dyn. Soil, Dyn. Pla.* 4 (1): 1–2.

Nagavallemma, K.P., S.P. Wani, L. Stephane, V.V. Padmaja, C. Vineela, M. Babu Rao, and K.L. Sahrawat. 2004. Vermicomposting: recycling wastes into valuable organic fertiliser. Global Theme on Agrecosystems Report no.8. Patancheru 502324. International Crops Research Institute for the Semi-Arid Tropics, Andhra Pradesh, p. 20. http://oar.icrisat.org/id/eprint/3677

Nakamura, Y. 1996. Interactions between earthworms and microorganisms in biological control of plant root pathogens. *Farm. Jap.* 30 (6): 37–43.

Nebert, L.D., J. Bloem, I.M. Lubbers, and J.W. van Groeningen. 2011. Association of earthworm denitrifier interactions with increased emission of nitrous oxide from soil mesocosms amended with crop residue. *Appl. Environ. Microbio.* 77: 4097–4104. DOI: 10.1128/AEM.00033-11

Nechitaylo, T.Y., M. Ferrer, and P.N. Golyshin. 2015. Terrestrial invertebrate animal metagenomics, lumbricidae: activity-based metagenomics for mining new hydrolytic enzymes in microbial communities in earthworm-egested matter. *Encyc. Metagen.: Environ. Metagen.* 622–631. DOI: 10.1007/978-1-4614-6418-1

Nechitaylo, T.Y., K.N. Timmis, and P.N. Golyshin. 2009. *Candidatus lumbricincola*, a novel lineage of uncultured Mollicutes from earthworms of family Lumbricidae. *Environ. Micro.* 11 (4): 1016–1026. DOI: 10.1111/j.1462-2920.2008.01837.x

Panda, A.K., S.S. Bisht, B.R. Kaushal, S. De Mandal, N.S. Kumar, and B.C. Basistha. 2017. Bacterial diversity analysis of Yumthang hot spring, North Sikkim, India by Illumina sequencing. *Big Data Anal.* 2 (1): 1–7.

Panda, A.K., S.S. Bisht, N.S., Kumar, and S. De Mandal. 2015. Investigations on microbial diversity of Jakrem hot spring, Meghalaya, India using cultivation-independent approach. *Genom. Data* 4: 156–157.

Pashanasi, B., P. Lavelle, J. Alegre, and F. Charpentier. 1996. Effect of the endogeic earthworm *Pontoscolex corethrurus* on soil chemical characteristics and plant growth in a low-input tropical agroecosystem. *Soil Biol. Biochem.* 28 (2): 801–810. DOI: 10.1016/0038-0717(96)00018-1

Pass, D.A., A.J. Morgan, D.S. Read, D. Field, A.J. Weightman, and P. Kille. 2015. The effect of anthropogenic arsenic contamination on the earthworm microbiome. *Environ. Micro.* 17 (6): 1884–1896. DOI: 10.1111/1462-2920.12712

Pathma, J., G.R. Rahul, K.R. Kamaraj, R.R. Subashri, and N. Sakthivel. 2011. Secondary metabolite production by bacterial antagonists. *J. Biol. Control.* 25: 165–181.

Pathma, J., and N. Sakthivel. 2012. Microbial diversity of vermicompost bacteria that exhibit useful agricultural traits and waste management potential. *SpringerPlus.* 1 (1): 1–19. DOI: 10.1186/2193-1801-1-26

Patron, J.C., P. Sanchez, G.C. Brown, M. Brossard, I. Barois, and C. Gutierrez. 1999. Phosphorus in soil and *Brachiaria decumbens* plants as affected by the geophagous earthworm *Pontoscolex corethrurus* and P fertilization. *Pedobiologia.* 43 (6): 547–556. http://horizon.documentation.ird.fr/exl-doc/pleins_textes_7/b_fdi_53-54/010021114.pdf

Phugare, S.P., Y.B. Gaikwad, and J.P. Jadhav. 2012. Biodegradation of acephate using a developed bacterial consortium and toxicological analysis using earthworms (*Lumbricus terrestris*) as a model animal. *Int. Biodet. Biodeg.* 69: 1–9. DOI: 10.1016/j.ibiod.2011.11.013

Prochazkova, P., V. Sustr, J. Dvorak, R. Roubalova, F. Skanta, V. Pizl, and M. Bilej. 2013. Correlation between the activity of digestive enzymes and nonself recognition in the gut of *Eisenia andrei* earthworms. *J. Invert. Path.* 114 (3): 217–221. DOI: 10.1016/j.jip.2013.08.003

Pu, Q., H.T. Wang, T. Pan, H. Li, and J.Q. Su. 2020. Enhanced removal of ciprofloxacin and reduction of antibiotic resistance genes by earthworm *Metaphire vulgaris* in soil. *Sci. Total Env.* 742: 140409. DOI: 10.1016/j.scitotenv.2020

Pulleman, M.M., J. Six, A. Uyl, J.C.Y. Marinissen, and A.G. Jongmans. 2005. Earthworms and management affect organic matter incorporation and microaggregate formation in agriculture soils. *Appl. Soil Eco.* 29: 1–15. DOI: 10.1016/j.apsoil.2004.10.003

Ravindran, B., S.M. Contreras-Ramos, and G. Sekaran. 2015. Changes in earthworm gut associated enzymes and microbial diversity on the treatment of fermented tannery waste using epigeic earthworm *Eudrilus eugeniae*. *Ecol. Eng.* 74: 394–401. DOI: 10.1016/j.ecoleng.2014.10.014.

Reverchon, F., S.H. Bai, X. Liu, and T.J. Blumfield. 2015. Tree plantation systems influence nitrogen retention and the abundance of nitrogen functional genes in the Solomon Islands. *Front. Microbio.* 6: 1439–1445. DOI: 10.3389/fmicb.2015

Ribbons, R.R., S. Kepfer-Rojas, C. Kosawang, O.K. Hansen, P. Ambus, M. McDonald, S.J. Grayston, C.E. Prescott, and L. Vesterdal. 2018. Context-dependent tree species effects on soil nitrogen transformations and related microbial functional genes. *Biogeochemistry.* 140 (2): 145–160. DOI: 10.1007/s10533-018-0480-8

Ros, M.B., T. Hiemstra, J.W. van Groenigen, A. Chareesri, and G.F. Koopmans. 2017. Exploring the pathways of earthworm-induced phosphorus availability. *Geoderma.* 303: 99–109. DOI: 10.1016/j.geoderma.2017.05.012

Saha, T., B. Chakraborty, S. Das, N. Thakur, and R. Chakraborty. 2018. *Chryseomicrobium excrementi* sp. nov., a Gram-stain-positive rod-shaped bacterium isolated from an earthworm (Eisenia fetida) cast. *Int. J. Sys. Evol. Microbio.* 68 (7): 2165–2171. DOI: 10.1099/ijsem.0.002791

Saha, T., V.K. Ranjan, S. Ganguli, S. Thakur, B. Chakraborty, P. Barman, and R. Chakraborty, et al. 2019. *Pradoshia eiseniae* gen. nov., sp. nov., a spore-forming member of the family *Bacillaceae* capable of assimilating 3-nitropropionic acid, isolated from the anterior gut of the earthworm *Eisenia fetida*. *Int. J. Sys. Evol. Microbio.* 69 (5): 1265–1273. DOI: 10.1099/ijsem.0.003304

Sánchez-Monedero, M.A., A. Roig, C. Paredes, and M.P. Bernal. 2001. Nitrogen transformation during organic waste composting by the Rutgers system and its effects on ph, EC and maturity of the composting mixtures. *Biores. Technol.* 780 (3): 301–308. DOI: 10.1016/50960-8524(01)00031-1

Sanz-Montero, M.E., and J.P. Rodriguez-Aranda. 2009. Silicate bioweathering and biomineralization in lacustrine microbialites: ancient analogues from the Miocene Duero Basin, Spain. *Geo. Mag.* 146 (4): 527–539. DOI: 10.1017/S0016756808005906

Sapkota, R., S. Santos, P. Farias, P.H. Krogh, and A. Winding. 2020. Insights into the earthworm gut multi-kingdom microbial communities. *Sci. Total Environ.* 138301. DOI: 10.1016/j.scitotenv.2020.138301

Singh, A., D.P. Singh, R. Tiwari, K. Kumar, R.V. Singh, S. Singh, R. Prasanna, A.K. Saxena, and L. Nath. 2015. Taxonomic and functional annotation of gut bacterial communities of *Eisenia fetida* and *Perionyx excavatus*. *Microbiol. Res.* 175: 1–31. DOI: 10.1016/j.micres.2015.03.003

Singh, A., R. Tiwari, A. Sharma, A. Adak, S. Singh, R. Prasanna, and R.V. Singh. 2016. Taxonomic and functional diversity of the culturable microbiomes of epigeic earthworms and their prospects in agriculture. *J. Basic Microbiol.* 56 (9): 1009–1020. DOI: 10.1002/jobm.201500779

Šrut, M., S. Menke, M. Höckner, and S. Sommer. 2019. Earthworms and cadmium—heavy metal resistant gut bacteria as indicators for heavy metal pollution in soils? *Ecotox. Environ. Safe.* 171: 843–853. DOI: 10.1016/j.ecoenv.2018.12.102

Sugumaran, P., and Janarthanam, B., 2007. Solubilization of potassium containing minerals by bacteria and their effect on plant growth. *World J. Agri. Sci.* 3 (3): 350–355. DOI: 10.1007/s11631-016-0106-6

Teng, S.K., N.A.A. Aziz, M. Mustafa, S.A. Aziz, and Y.W. Yan. 2012. Evaluation on physical, chemical and biological properties of casts of geophagous earthworm *Metaphire tschiliensis tschiliensis*. *Scient. Res. Ess.* 7 (10): 1169–1174. DOI: 10.5897/SRE11.2233

Thakuria, D., O. Schmidt, D. Finan, D. Egan, and F.M. Doohan. 2010. Gut wall bacteria of earthworms: a natural selection process. *ISME J.* 4 (3): 357–366. DOI: 10.1038/ismej.2009.124

Verma, K., N. Agarwal, M. Farooq, R.B. Misra, and R.K. Hans. 2006. Endosulfan degradation by a Rhodococcus strain isolated from earthworm gut. *Ecotox. Environ. Safe.* 64 (3): 377–381. DOI: 10.1016/j.ecoenv.2005.05.14

Vos, H.M.J., M.B.H. Ros, G.F. Koopmans, and J.W. Van Groenigen. 2014. Do earthworms affect phosphorus availability to grass? A pot experiment. *Soil Biol. Biochem.* 79: 34–42. DOI: 10.1016/j.soilbio.2014.08.018

Wu, J., W. Zhang, Y. Shao, and S. Fu. 2017. Plant-facilitated effects of exotic earthworm *Pontoscolex corethrurus* on the soil carbon and nitrogen dynamics and soil microbial community in a subtropical field ecosystem. *Eco. Evol.* 7 (21): 8709–8718. DOI: 10.1002/ece3.3399

Zhong, H., Y.N. Kim, C. Smith, B. Robinson, and N. Dickinson. 2017. Seabird guano and phosphorus fractionation in a rhizosphere with earthworms. *Appl. Soil Eco.* 120: 197–205. DOI: 10.1016/j.apsoil.2017.08.00.

7

Metagenomics of Pollen-Borne Microbes and Gut Microbiota of the Honey Bee

Sampat Ghosh
Agriculture Science and Technology Research Institute, Andong National University, Republic of Korea
Saeed Mohamadzade Namin
Agriculture Science and Technology Research Institute, Andong National University, Republic of Korea; Department of Plant Protection, Faculty of Agriculture, Varamin-Pisahva Branch, Islamic Azad University, Varamin, Iran
Chuleui Jung*
Agriculture Science and Technology Research Institute, Andong National University, Republic of Korea; Department of Plant Medicals, Andong National University, Republic of Korea

CONTENTS

7.1 Introduction

Honey bee, as a pollinator, plays a vital role in global agriculture, as it is considered a pollinating powerhouse, accounting for 35% of global food production (Klein et al., 2007). Over the last few decades, significant loss of honey bee colonies has been reported from several parts of the world (Potts et al., 2010; vanEngelsdorp and Meixner, 2010; Dahle, 2010; van der Zee et al., 2012; Smith et al., 2013). The loss of honey bees can negatively affect the nutritional profile of humans, as well as increasing the burden of non-communicable diseases due to inadequate nutrients (Smith et al., 2015; Ghosh and Jung, 2018). Besides factors like loss of habitat, ecological degradation, overuse of insecticides (e.g. neonicotinoids), and the load of pathogens (Brown and Paxton, 2009; Goulson et al., 2015), poor nutrition has proven a significant factor responsible for population decline (Oldroyd, 2007; Naug, 2009).

Nectar and pollen are two main nutritional sources of the honey bee. Nectar, a carbohydrate source, serves as a source for the production of honey. On the other hand, pollen serves as the primary source of protein, essential lipids, minerals, and vitamins (Roulston et al., 2000; Human and Nicolson, 2006; Szczęsna, 2006; Morgano et al., 2012; Ghosh and Jung 2017). Nutritional quality plays a role in the selection and preference of pollen sources (Donkersley et al., 2017; Ghosh et al., 2020). However, in most cases, honey bees and other bee species do not rely only on one source; rather, they prefer to collect pollen from multiple floral sources. Ghosh and Jung (2020) reported the changes in the nutritional

DOI: 10.1201/9781003042570-11

composition from bee pollen to pollen patty, which is often used as a nutritional source for bees, although the pollen patties used in the study were prepared for bumblebees.

Crop (foregut) plays an important role in the honey bee's social as well as nutritional interface, as the crop serves as storage and is used for the transportation of nectar from the floral source to the hive, shares liquid nutrition to the nestmates, and selectively passes pollen into the midgut (Blatt and Roces, 2002; Corby-Harris et al., 2014). Forager bees use their forelegs and tongue to gain access to nectar and pollen. They accumulate all the pollens on the head and body hairs and consolidate using the forelegs, at the same time mixing with liquid sugars from the crop, and this sticky mixture is packed into a pollen basket on the hind leg, known as the corbiculae (Corby-Harris et al., 2014). Honey bees do not consume the just-collected fresh pollen; generally they collect pollen and store it for a few hours to days (Anderson et al., 2014). During the processing and storing of bee bread, the chemical composition of pollen is generally changed, presumably because of mixing bee pollen with nectar and microbial activity.

The microbiome consists of various microbial symbionts, where they are engaged in mutualistic, commensalist, and/or parasitic interactions with the host honey bee (Engel et al., 2016; Dharampal et al., 2019). Symbionts and other indigenous non-pathogenic microbial communities of the honey bee gut are essential for nutritional status as well as immunocompetence against invading pathogens (Anderson et al., 2011; Crotti et al., 2013; Zheng et al., 2017).

7.2 Composition of Honey Bee Gut Microbiota

Metagenomic analysis of honey bees helps in understanding the microbial communities in the honey bee gut as well as the hive environment, including pollen-borne microbes. The honey bee alimentary canal hosts many different bacteria. Anderson et al. (2011) proposed two major microbial niches and one occasional microbial niche throughout the hive and alimentary canal of honey bee. The first co-evolved with liquid transfer (polyethism) and food storage, the second co-evolved in the enzymatically active and nutrient-rich midgut, and the third is in the rectum, which generally provisions the niche with unused nutrients during the winter.

Table 7.1 represents a list of bacteria found in the honey bee gut and hive environment. The spatial composition of bacterial communities of the honey bee gut is represented in Figure 7.1. The isolated gut bacteria mostly belonged to three phyla, Firmicutes, Proteobacteria, and Actinobacteria. Firmicutes were found predominating, followed by Proteobacteria and Actinobacteria, and class γ-proteobacteria was the most abundant among proteobacteria (see Figure 7.2). Figure 7.3 represents a comparative view of the gut bacterial community between healthy honey bees and honey bees under stress. In general, firmicutes are Gram positive, while others are mostly Gram negative and classified as facultative anaerobes, are tolerant of acidic environments, and ferment sugar to produce lactic or acetic acid. Corby-Harris et al. (2014) showed that despite a very different diet, the forager honey bee gut contains core microbiota similar to that found in the gut of younger honey bees.

Similar to bacteria, fungi also have a mutualistic symbiotic effect with honey bees. However, in comparison to bacteria, the fungal community has not been extensively explored (Romero et al., 2019). Cox-Foster et al. (2007) carried out a metagenomics survey in order to explore the microflora in hives suffering from colony collapse disorder (CCD) and identified several pathogenic fungi like *Pandora delphacis*, *Nosema ceranae*, *Nosema apis*, and so on. Table 7.2 represents a list of non-pathogenic fungi. The fungal community might have the potential to contribute to the metabolic system of the honey bee. Division Ascomycota was found predominating, followed by Basidiomycota and Zygomycota (see Table 7.2). Pollen also provisions honey bees with specialized non-pathogenic microbes that ferment and improve digestibility, shelf-life, and nutritional content for honey bees (Dharampal et al., 2019). Most of the studies explored the taxonomic profiling of the fungal community associated with the honey bee

TABLE 7.1

List of Bacteria in the Gut of Honey Bee and Hive Environment

Taxonomic Position			Habitat				
Phylum	Class	Bacteria	Gut	Hive Environment	Characterization		Reference
Firmicutes	Bacilli	*Lactobacillus* spp.	L, Aw	BB, N/H	G^+	Ae-tol an.	1, 2, 3, 4, 6, 8, 9, 10, 11, 12
		Lactobacillus kunkeei	Aw	—	G^+	Ae-tol an.	17
		Bacillus spp.	L, Aw	Bb, N/H	G^+	F-an.	2, 3, 5, 7, 8
		Leuconostoc spp.	L, Aw	N/H	G^+	An	8
		Staphylococcus spp.	Aw	—	G^+	F-an.	19
		Enterococcus spp.	Aw	Bb	G^+	F-an.	2, 5, 11
		Planococcus maritimus	A	—	G^+	Ae	6
Actinobacteria	Actinobacteria	*Bifidobacterium* spp.	L, Aw	Bb, N/H	G^+	An	2, 4, 5, 6, 8, 9, 10, 12
		Propionibacterium spp.	L	—	G^+	An	6
Bacteroidetes	Sphingobacteriia	*Pedobacter africanus*	A	—	G^-	Ae	6
Proteobacteria	α-Proteobacteria	*Gluconobacter* spp.	Aw	N/H	$G^-(v)$	-	1, 12
		Bartonella spp.	Aw	—	G^-	An	4, 6, 9, 8, 12
		Bartonella apis	Aw	—	G^-	An	18
		Simonsiella spp.	L, Aw	—	G^-	Ae	4, 6, 9, 8, 12
		Ochrobactrum spp.	Aw	—	G^-	Ae	19
		Sphingomonas spp.	Aw	—	G^-	Ae	19
		Parasaccharibacter apium	Aw	—	G^-		13, 16, 14, 17
	β-Proteobacteria	*Ralstonia* spp.	Aw	—	G^-	Ae	19
		Snodgrassella alvi	Aw	—	G^-	Ae	15
	γ-Proteobacteria	*Pseudomonas* spp.	Aw	Bb	G^-	Ae or an	2, 5, 8
		Serratia spp.	Aw	—	G^-	An	4, 6, 8, 10
		Gilliamella apicola	Aw	—	G^-	-	15
		Frischella perrara	Aw	—	G^-	An	18

G^+ = Gram positive, G–= Gram negative; $G^-(v)$ = Gram variable but most likely Gram negative; Ae = Aerobic, An = Anaerobic, F-an = Facultative anaerobic; Ae-to = Aerotolerant; L = Larvae, Aw = Adult worker; Bb = Bee bread, N = Nectar, H = Honey

1 = Ruiz-Argueso and Rodriguez-Navarro 1975; 2 = Gilliam, 1997; 3 = Rada et al., 1997; 4 = Jeyaprakash et al., 2003; 5 = Kačániová et al., 2004; 6 = Mohr and Tebbe, 2006; 7 = Evans and Armstrong, 2006; 8 = Babendreier et al., 2007; 9 = Cox-Foster et al., 2007; 10 = Olofsson and Vásquez, 2008; 11 = Carina et al., 2011; 12 = Martinson et al., 2011; 13 = Anderson et al., 2013; 14 = Vojvodic et al., 2013; 15 = Kwong and Moran 2013; 16 = Anderson et al., 2014; 17 = Corby-Harris et al., 2014; 18 = Kwong and Moran 2016; 19 = Anjum et al., 2018

gut or hive environment, including bee bread or nectar. Although a symbiotic relationship of fungi as a food resource of hymenopteran species, like cultivating fungus as a food resource by leafcutter ant *Acromyrmex* and *Atta* (Currie et al., 1999) and cultivation of fungus *Monascus* sp. by stingless bee *Scaptotrigona depilis* inside the brood cell for provisioning food for larvae (Menezes et al., 2015) has been demonstrated, little is known about the functional role in honey bee fitness. Yun et al. (2018) reported that the gut and ovaries of queen honey bees were dominated by *Zygosaccharomyces*; on the other hand, *Saccharomyces* dominated the gut of nurse bees. Acetic acid tolerance of *Zygosaccharomyces* is presumably associated with the metabolism and transport of acetic acid in the presence of glucose (Sousa et al., 1996; Yun et al., 2018).

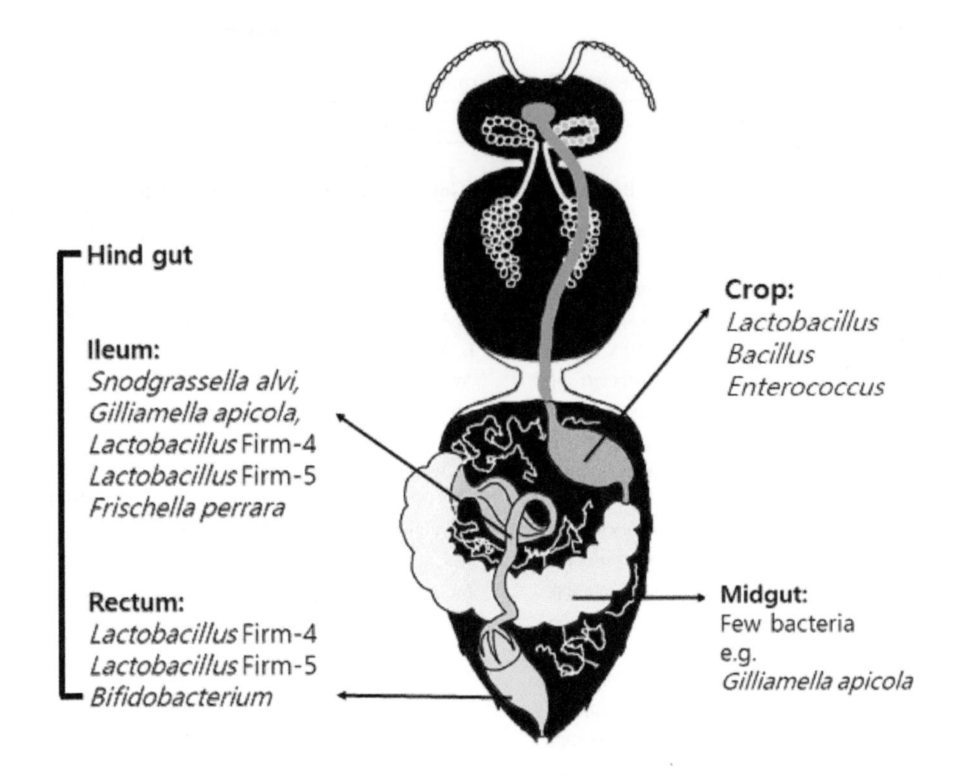

FIGURE 7.1 Spatial composition of bacterial communities in the honey bee gut.

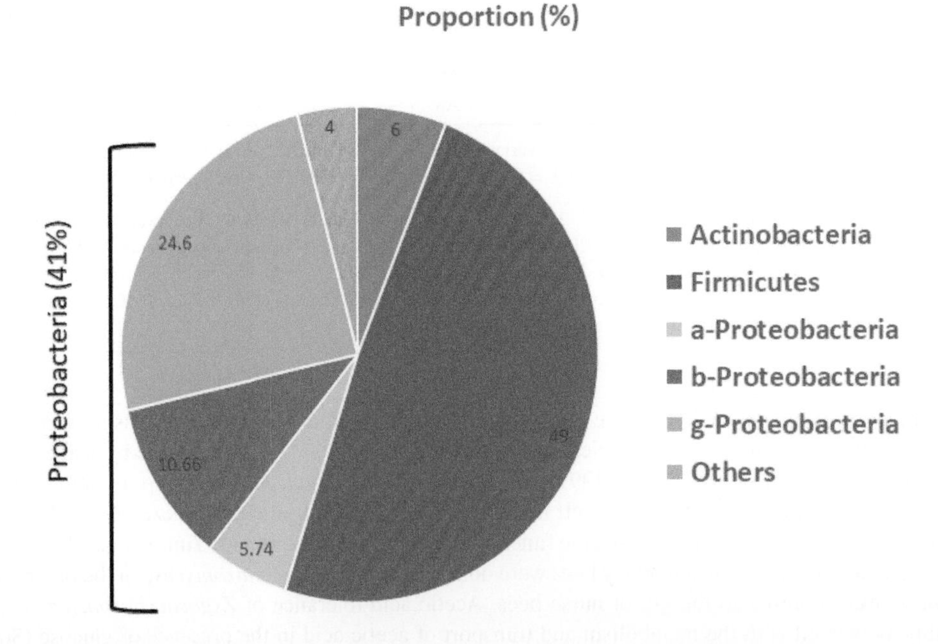

FIGURE 7.2 Phylum-wise distribution and relative abundance of gut microbial communities.

Source: Data obtained and calculated from Cox-Foster et al. (2007); Martinson et al. (2011); Martinson et al. (2012); Engel et al. (2012); Sabree et al. (2012); Moran et al. (2012); Corby-Harris et al. (2014); Kakumanu et al. (2016); Khan et al. (2017); Anjum et al. (2018).

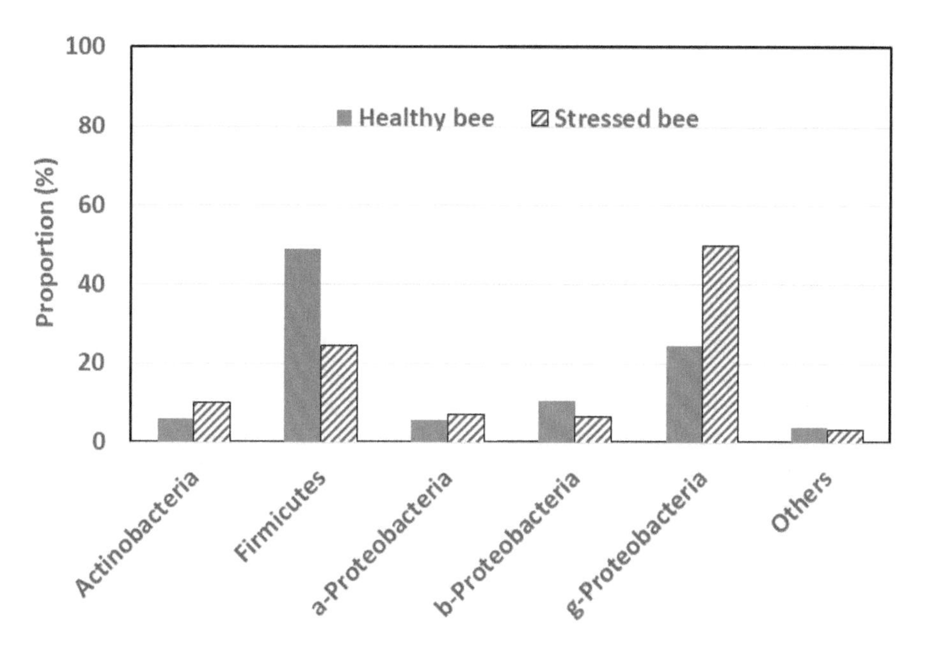

FIGURE 7.3 Comparison of gut bacterial community between healthy honey bees and honey bees under stress.

Source: Data obtained and calculated from Cox-Foster et al. (2007); Martinson et al. (2011); Martinson et al. (2012); Engel et al. (2012); Sabree et al. (2012); Moran et al. (2012); Corby-Harris et al. (2014); Kakumanu et al. (2016); Khan et al. (2017); Anjum et al. (2018).

TABLE 7.2

List of Fungi in the Gut of Honey Bee and Hive Environment

Taxonomic Position			Habitat		
Division	**Class**	**Fungi**	**Gut**	**Hive Environment**	**Reference**
Ascomycota	Eurotiomycetes	*Penicillium* spp.	L, Aw	Bb, N/H	2
		Aspergillus spp.	L, Aw	Bb, N/H	2
	Saccharomycetes	*Candida* spp.	Aw	Bb, N/H	1, 2
		Wickerhamomyces spp.	Aw	—	6
		Sachharomyces spp.	Aw	Bb, N/H	1, 2, 3, 4,
		Zygosaccharomyces spp.	Aw	—	5
	Dothideomycetes	*Cladosporium* spp.	-	pollen, Bb	7
Basidiomycota	Tremellomycetes	*Cryptococcus* spp.	Aw	Bb, N/H	1, 2
Zygomycota	Mucoromycetes	*Mucor* spp.	-	Bb	2, 4

L = Larvae, Aw = Adult worker; Bb = Bee bread, N = Nectar, H = Honey

1 = Sandhu and Waraich, 1985; 2 = Gilliam, 1997; 3 = Rada et al., 1997; 4 = Cox-Foster et al., 2007; 5 = Cornman et al., 2012; 6 = Yun et al., 2018; 7 = Disayathanoowat et al., 2020

7.3 Functional Role of Microbiota in Diet Improvement of the Honey Bee

Lactobacillus strains associated with honey bees mostly belong to two phylogenetically separate clades, Firm-4 and Firm-5, and a third clade, Firm-3, is occasionally found (Babendreier et al., 2007). Genome sequencing study revealed these two clades, especially Firm-5, have numerous phosphotransferase systems involving in the uptake of sugars (Ellegaard et al., 2015). The diet of honey bees is carbohydrate rich, as it consists of nectar, honey, and pollen. Therefore, the gut microbiota mostly include members like *Lactobacillus*, *Bacillus*, and *Bifidobacterium*, which have evolved to thrive on carbohydrate resources (Kwong and Moran, 2016). Genome-based investigation revealed *Gilliamella apicola* strains potentially digest the complex carbohydrates of pectin and xylan present in the pollen cell wall, which is otherwise indigestible by honey bees (Engel et al., 2012). Further, many sugars like mannose, arabinose, raffinose, galactose, and lactose are indigestible and potentially toxic to honey bees (Barker and Lehner, 1974). Some bacterial strains utilize those rarer sugars. A recent study demonstrates that gut symbiont *Gilliamella apicola* can metabolize mannose besides glucose and fructose, which improves dietary tolerances of honey bees (Zheng et al., 2016). In a recent investigation on digestion of pollen-derived polysaccharides of all core members of the bee gut microbes, it was found that *Bifidobacterium* and *Gilliamella* were primary degraders of hemicellulose and pectin (Zheng et al., 2019), which are the main components of pollen intine (Roulston and Cane 2000). Genome analysis indicated that *Bifidobacterium* species possessed many glycoside hydrolases (i.e. GH5, GH28, GH29, GH30, GH31, GH42, GH43, GH51, and GH78) in comparison to other genera like *Gilliamella* and *Lactobacillus*. On the other hand, *Gilliamella* possessed several polysaccharide lyases like PL1, PL9, PL22, and acetylase (CE12) (Zheng et al., 2019). In contrast, *Lactobacillus* played little or no role in polysaccharide digestion (Zheng et al., 2019). On the other hand, *Snodgrasella alvi* is a non-sugar fermenter (Kwong and Moran, 2013), providing an example of syntrophic interaction (Kwong and Moran, 2016). *S. alvi* relies on the aerobic oxidation of carboxylates such as citrate, malate, acetate, and lactic acid. The utilization of different resources enables a stable coexistence for sugar and non-sugar fermenters in the same gut environment. *Parasaccharibacter apium*, although a member of the Acetobacteraceae family, cannot produce acetic acid through the oxidation of sugars and alcohols. Interestingly, *P. apium* can thrive on royal jelly, unlike most other bacteria (Kwong and Moran, 2016). Kim et al. (2012) described the possibility of having pathogens of honey bees like *Serratia* from a predator like *Vespa*. However, the functions of bacteria belonging to genera *Acetobacter*, *Gluconobacter*, and *Bartonella* are not well known.

Saccharomyces is one of the most predominant fungi found in the gut and hive environment, including bee bread and nectar or honey. This fungus plays an important role in fermentation in the gut. Another fungus, *Metschnikowia*, is a relatively slow-growing fermenter and produces acid proteases (Kakumanu et al., 2016).

7.4 Role of Microbiota in the Physiology of the Honey Bee

Studies indicate that the presence of gut microbiota is necessary for normal body and gut-weight gain, especially during the time following the emergence from the pupal stage (Zheng et al., 2017). Non-sterilized pollen consumption increased the bodyweight of prepupae and took less developmental time in comparison to sterilized pollen consumption for the solitary bee *Osmia ribifloris* (Dharampal et al., 2019).

Zheng et al. (2017) demonstrated that ilp1 (insulin-like peptide 1) and Vg (vitellogenin) genes were expressed 5.8 and 4.9 times higher in normal honey bees than the honey bees without gut microbiota but both feeding on the same diet. Generally, ilp1 expresses the highest on a high-protein diet (Ihle et al., 2014). Therefore, the gut microbiota might have played a role in supplying protein or amino acids, resulting in higher expression of the ilp1 gene. Fatty acid represents a class of biomarkers that are commonly used to describe microbial trophic ecology. Zheng et al. (2017) demonstrated differences in the short-chain fatty acid (SCFA) content of the gut that suggested gut microbiota was the main producer of SCFAs. Acetate was found in very high concentrations in the ileum and rectum of honey bee adult workers with the normal gut microbiota compared to the germ-free honey bee. The abundance of free fatty acid (C12:0, C14:0; C14:1, C16:0, C16:1, and C18:1) was found to be about 15 times higher among larvae of *O. riniflorís* raised on natural pollen in comparison to larvae raised on sterile pollen (Dharampal et al., 2019). These fatty acids serve as a primary source of energy during non-feeding periods. Loper et al. (1980) found that corbicula pollen contained a higher amount of some fatty acids (lauric, palmitic, stearic, oleic, and arachidic acid) and cholesterol (unesterified and esterified) content than that found in hand-collected almond pollen.

7.5 Alteration of Gut Microbiota under Stress Conditions

Several scientific studies have demonstrated that the structure of gut microbiota is significantly affected or altered under stressed conditions like exposure to pesticides, the burden of pathogen load, and so on (Kakumanu et al., 2016; Tauber et al., 2019). Chlorothalonil caused significant changes to the bacterial communities of honey bee gut and thus functional potential like increasing oxidative phosphorylation and decreasing sugar metabolism (Kakumanu et al., 2016). A study by Li et al. (2017) demonstrated that honey bees whose microbiota had been eliminated by the application of antibiotics were found to be more susceptible to the *Nosema ceranae* infection. The study (Li et al., 2017) further demonstrated that the elimination of gut microbial community negatively impacted the functioning of the immune system, as the expression levels of genes encoding antimicrobial peptides like abaecin, defensin 1, and hymenoptaecin were low in the antibiotic treatment group. Another study by Forsgren et al. (2010) reported an antagonistic effect of lactic acid bacteria belonging to the genera *Lactobacillus* and *Bifidobacterium* originating from the honey stomach for *Paenibacillus larvae* (which causes American foulbrood disease for honey bee). Rouzé et al. (2019) showed that *Nosema ceranae* and chronic sub-lethal exposure to insecticides like coumaphos, fipronil, thiamethoxam, and imidacloprid decreased the abundance of *Bifidobacterium* and *Lactobacillus* regardless of the season. However, a relative abundance of *G. apicola* and *S. alvi* was found with the exposure of *N. ceranae* and a lethal dose of fipronil (Rouzé et al., 2019). Glyphosate, a widely used herbicide, caused a strong decrease in bacteria in *S. alvi*, a partial decrease in *G. apicola*, and an increase in *Lactobacillus* spp. (Blot et al., 2019). Another investigation revealed that yeast (*Wickerhamomyces anomalus*) augmentation for bees with developed microbiota appeared immunomodulatory and affected the microbial community (Tauber et al., 2019). Perturbation of the microbiota by pesticide, herbicide, and pathogen load could lead to a decline of the honey bee population. Wu et al. (2020) demonstrated that gut microbiota promoted the expression of the P450 enzyme in the midgut, which could help detoxification of pesticide exposures. The mortality rates of microbiota-deficient worker honey bees exposed to thiacloprid and tau-fluvalinate were significantly higher than those of worker bees with the conventional gut community. The study clearly suggests that gut symbionts also contribute to bee health through the modification of host xenobiotic detoxifying pathways.

7.6 Methodological Concern of Metagenomics Analysis of Gut and Pollen-Borne Microbiome

Methods involved in the microbiome analysis are represented schematically in Figure 7.4. Methods involving high-throughput techniques play an enormous role in the study of gut microbiome, such as taxonomic composition based on operational taxonomic units (OTUs), biodiversity, and so on. However, although marker gene studies are often limited to deducing the functional capabilities of microorganisms in terms of metabolism, 16s rRNA has extended its use to infer the functional contribution of particular community members (Langille et al., 2013), which is of particular interest to the microbiome study of honey bees.

7.7 Conclusion and Further Research Perspectives

Pollen-borne microbes and gut microbiota research on honey bees is an emerging field that could improve the understanding of the ecological and nutritional dynamics of the gut environment. For the last two decades, several valuable scientific investigations have contributed immense knowledge to the subject. High-throughput sequencing methods and associated bioinformatics have been employed for identification of the bacterial communities based on 16s rRNA and fungal communities based on ITS2. However, metabolic functions like diet improvement, physiological role, detoxifying activity, and immunomodulatory capacity of very few microbes have been explored so far. Immense scope remains for further investigation of in-detail functionality of microbiota in honey bee physiology and metabolism.

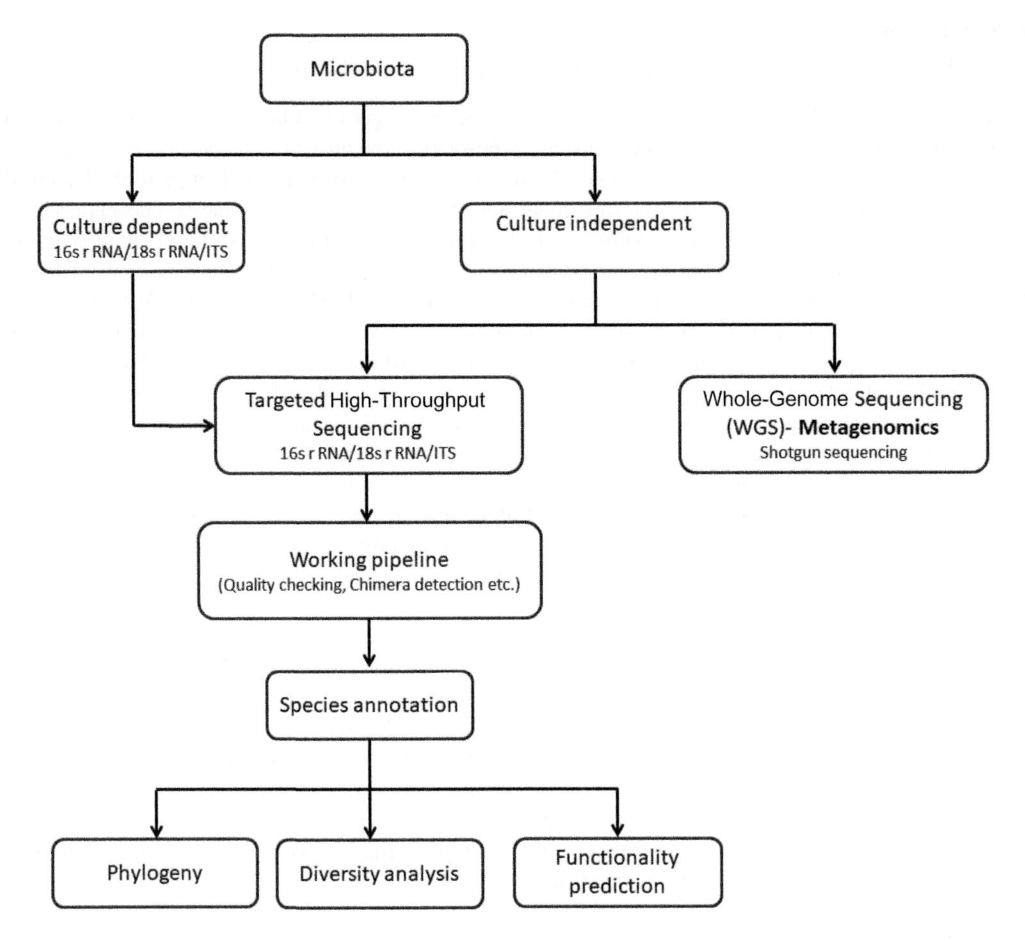

FIGURE 7.4 Schematic diagram representing the methods involved in microbiota analysis.

7.8 Acknowledgments

The authors are supported by the BSRP through the National Research Foundation of Korea (NRF), Ministry of Education (NRF—2018R1A6A1A03024862).

REFERENCES

Anderson, K. E., Mark J. Carroll, Tim Sheehan, Brendon M. Mott, Patrick Maes, and Vanessa Corby-Harris. "Hive-stored pollen of honey bees: many lines of evidence are consistent with pollen preservation, not nutrient conversion." *Molecular Ecology* 23, no. 23 (2014), 5904–5917. doi:10.1111/mec.12966.

Anderson, K. E., T. H. Sheehan, B. J. Eckholm, B. M. Mott, and G. DeGrandi-Hoffman. "An emerging paradigm of colony health: microbial balance of the honey bee and hive (*Apis mellifera*)." *Insectes Sociaux* 58, no. 4 (2011), 431–444. doi:10.1007/s00040-011-0194-6.

Anderson, Kirk E., Timothy H. Sheehan, Brendon M. Mott, Patrick Maes, Lucy Snyder, Melissa R. Schwan, Alexander Walton, Beryl M. Jones, and Vanessa Corby-Harris. "Microbial ecology of the hive and pollination landscape: bacterial associates from floral nectar, the alimentary tract and stored food of honey bees (*Apis mellifera*)." *PLoS One* 8, no. 12 (2013), e83125. doi:10.1371/journal.pone.0083125.

Anjum, Syed I., Abdul H. Shah, Muhammad Aurongzeb, Junaid Kori, M. K. Azim, Mohammad J. Ansari, and Li Bin. "Characterization of gut bacterial flora of *Apis mellifera* from north-west Pakistan." *Saudi Journal of Biological Sciences* 25, no. 2 (2018), 388–392. doi:10.1016/j.sjbs.2017.05.008.

Babendreier, Dirk, David Joller, Jörg Romeis, Franz Bigler, and Franco Widmer. "Bacterial community structures in honeybee intestines and their response to two insecticidal proteins." *FEMS Microbiology Ecology* 59, no. 3 (2007), 600–610. doi:10.1111/j.1574-6941.2006.00249.x.

Barker, Roy J., and Yolanda Lehner. "Acceptance and sustenance value of naturally occurring sugars fed to newly emerged adult workers of honey bees (*Apis mellifera* L.)." *Journal of Experimental Zoology* 187, no. 2 (1974), 277–285. doi:10.1002/jez.1401870211.

Blatt, Jasmina, and Flavio Roces. "The control of the proventriculus in the honeybee (*Apis mellifera carnica* L.) I. A dynamic process influenced by food quality and quantity?" *Journal of Insect Physiology* 48, no. 6 (2002), 643–654. doi:10.1016/s0022-1910(02)00090-2.

Blot, Nicolas, Loïs Veillat, Régis Rouzé, and Hélène Delatte. "Glyphosate, but not its metabolite AMPA, alters the honeybee gut microbiota." *PLoS One* 14, no. 4 (2019), e0215466. doi:10.1371/journal.pone.0215466.

Brown, Mark J., and Robert J. Paxton. "The conservation of bees: a global perspective." *Apidologie* 40, no. 3 (2009), 410–416. doi:10.1051/apido/2009019.

Carina Audisio, M., María J. Torres, Daniela C. Sabaté, Carolina Ibarguren, and María C. Apella. "Properties of different lactic acid bacteria isolated from *Apis mellifera* L. bee-gut." *Microbiological Research* 166, no. 1 (2011), 1–13. doi:10.1016/j.micres.2010.01.003.

Corby-Harris, Vanessa, Patrick Maes, and Kirk E. Anderson. "The bacterial communities associated with honey bee (*Apis mellifera*) foragers." *PLoS One* 9, no. 4 (2014), e95056. doi:10.1371/journal.pone.0095056.

Cornman, R. S., David R. Tarpy, Yanping Chen, Lacey Jeffreys, Dawn Lopez, Jeffery S. Pettis, Dennis VanEngelsdorp, and Jay D. Evans. "Pathogen webs in collapsing honey bee colonies." *PLoS One* 7, no. 8 (2012), e43562. doi:10.1371/journal.pone.0043562.

Cox-Foster, D. L., S. Conlan, E. C. Holmes, G. Palacios, J. D. Evans, N. A. Moran, P.-L. Quan, et al. "A metagenomic survey of microbes in honey bee colony collapse disorder." *Science* 318, no. 5848 (2007), 283–287. doi:10.1126/science.1146498.

Crotti, Elena, Luigi Sansonno, Erica M. Prosdocimi, Violetta Vacchini, Chadlia Hamdi, Ameur Cherif, Elena Gonella, Massimo Marzorati, and Annalisa Balloi. "Microbial symbionts of honeybees: a promising tool to improve honeybee health." *New Biotechnology* 30, no. 6 (2013), 716–722. doi:10.1016/j.nbt.2013.05.004.

Currie, Cameron R., James A. Scott, Richard C. Summerbell, and David Malloch. "Fungus-growing ants use antibiotic-producing bacteria to control garden parasites." *Nature* 398, no. 6729 (1999), 701–704. doi:10.1038/19519.

Dahle, Bjørn. "The role of *Varroa destructor* for honey bee colony losses in Norway." *Journal of Apicultural Research* 49, no. 1 (2010), 124–125. doi:10.3896/ibra.1.49.1.26.

Dharampal, Prarthana S., Caitlin Carlson, Cameron R. Currie, and Shawn A. Steffan. "Pollen-borne microbes shape bee fitness." *Proceedings of the Royal Society B: Biological Sciences* 286, no. 1904 (2019), 20182894. doi:10.1098/rspb.2018.2894.

Disayathanoowat, Terd, HuanYuan Li, Natapon Supapimon, Nakarin Suwannarach, Saisamorn Lumyong, Panuwan Chantawannakul, and Jun Guo. "Different dynamics of bacterial and fungal communities in hive-stored bee bread and their possible roles: a case study from two commercial honey bees in China." *Microorganisms* 8, no. 2 (2020), 264. doi:10.3390/microorganisms8020264.

Donkersley, Philip, Glenn Rhodes, Roger W. Pickup, Kevin C. Jones, Eileen F. Power, Geraldine A. Wright, and Kenneth Wilson. "Nutritional composition of honey bee food stores vary with floral composition." *Oecologia* 185, no. 4 (2017), 749–761. doi:10.1007/s00442-017-3968-3.

Ellegaard, Kirsten M., Daniel Tamarit, Emelie Javelind, Tobias C. Olofsson, Siv G. Andersson, and Alejandra Vásquez. "Extensive intra-phylotype diversity in lactobacilli and bifidobacteria from the honeybee gut." *BMC Genomics* 16, no. 1 (2015). doi:10.1186/s12864-015-1476-6.

Engel, P, Waldan K. Kwong, Quinn McFrederick, Kirk E. Anderson, Seth M. Barribeau, James A. Chandler, R. S. Cornman, et al. "The bee microbiome: impact on bee health and model for evolution and ecology of host-microbe interactions." *mBio* 7, no. 2 (2016). doi:10.1128/mbio.02164-15.

Engel, P., V. G. Martinson, and N. A. Moran. "Functional diversity within the simple gut microbiota of the honey bee." *Proceedings of the National Academy of Sciences* 109, no. 27 (2012), 11002–11007. doi:10.1073/pnas.1202970109.

Evans, Jay D., and Tamieka-Nicole Armstrong. "Antagonistic interactions between honey bee bacterial symbionts and implications for disease." *BMC Ecology* 6 (2006), 4–12. doi:10.11186/1472-6785/6/4.

Forsgren, Eva, Tobias C. Olofsson, Alejandra Vásquez, and Ingemar Fries. "Novel lactic acid bacteria inhibiting Paenibacillus larvae in honey bee larvae." *Apidologie* 41, no. 1 (2010), 99–108. doi:10.1051/apido/2009065.

Ghosh, Sampat, Hyejin Jeon, and Chuleui Jung. "Foraging behaviour and preference of pollen sources by honey bee (*Apis mellifera*) relative to protein contents." *Journal of Ecology and Environment* 44, no. 1 (2020). doi:10.1186/s41610-020-0149-9.

Ghosh, Sampat, and Chuleui Jung. "Nutritional value of bee-collected pollens of hardy kiwi, *Actinidia arguta* (Actinidiaceae) and oak, *Quercus* sp. (Fagaceae)." *Journal of Asia-Pacific Entomology* 20, no. 1 (2017), 245–251. doi:10.1016/j.aspen.2017.01.009.

Ghosh, Sampat, and Chuleui Jung. "Contribution of insect pollination to nutritional security of minerals and vitamins in Korea." *Journal of Asia-Pacific Entomology* 21, no. 2 (2018), 598–602. doi:10.1016/j.aspen.2018.03.014.

Ghosh, Sampat, and Chuleui Jung. "Changes in nutritional composition from bee pollen to pollen patty used in bumblebee rearing." *Journal of Asia-Pacific Entomology* 23, no. 3 (2020), 701–708. doi:10.1016/j.aspen.2020.04.008.

Gilliam, M. "Identification and roles of non-pathogenic microflora associated with honey bees." *FEMS Microbiology Letters* 155, no. 1 (1997), 1–10. doi:10.1016/s0378-1097(97)00337-6.

Goulson, D., E. Nicholls, C. Botias, and E.L. Rotheray. "Bee declines driven by combined stress from parasites, pesticides, and lack of flowers." *Science* 347, no. 6229 (2015), 1255957–1255957. doi:10.1126/science.1255957.

Human, Hannelie, and Sue W. Nicolson. "Nutritional content of fresh, bee-collected and stored pollen of *Aloe greatheadii* var. *davyana* (Asphodelaceae)." *Phytochemistry* 67, no. 14 (2006), 1486–1492. doi:10.1016/j.phytochem.2006.05.023.

Ihle, Kate E., Nicholas A. Baker, and Gro V. Amdam. "Insulin-like peptide response to nutritional input in honey bee workers." *Journal of Insect Physiology* 69 (2014), 49–55. doi:10.1016/j.jinsphys.2014.05.026.

Jeyaprakash, Ayyamperumal, Marjorie A. Hoy, and Michael H. Allsopp. "Bacterial diversity in worker adults of *Apis mellifera capensis* and *Apis mellifera scutellata* (Insecta: Hymenoptera) assessed using 16S rRNA sequences." *Journal of Invertebrate Pathology* 84, no. 2 (2003), 96–103. doi:10.1016/j.jip.2003.08.007.

Kačániová, M., R. Chlebo, M. Kopernický, and A. Trakovická. "Microflora of the honeybee gastrointestinal tract." *Folia Microbiologica* 49, no. 2 (2004), 169–171. doi:10.1007/bf02931394.

Kakumanu, Madhavi L., Alison M. Reeves, Troy D. Anderson, Richard R. Rodrigues, and Mark A. Williams. "Honey bee gut microbiome is altered by in-hive pesticide exposures." *Frontiers in Microbiology* 7 (2016). doi:10.3389/fmicb.2016.01255.

Khan, Khalid A., Mohammad J. Ansari, Ahmad Al-Ghamdi, Adgaba Nuru, Steve Harakeh, and Javaid Iqbal. "Investigation of gut microbial communities associated with indigenous honey bee (*Apis mellifera jemenitica*) from two different eco-regions of Saudi Arabia." *Saudi Journal of Biological Sciences* 24, no. 5 (2017), 1061–1068. doi:10.1016/j.sjbs.2017.01.055.

Kim, J., C. Jung, and Y. H. Jeon. "Possibility of honeybee infection of bacteria isolated from *Vespa mandarinia*." *Korean Journal of Apiculture* 27 (2012), 259–266.

Klein, Alexandra-Maria, Bernard E. Vaissière, James H. Cane, Ingolf Steffan-Dewenter, Saul A. Cunningham, Claire Kremen, and Teja Tscharntke. "Importance of pollinators in changing landscapes for world crops." *Proceedings of the Royal Society B: Biological Sciences* 274, no. 1608 (2007), 303–313. doi:10.1098/rspb.2006.3721.

Kwong, Waldan K., and Nancy A. Moran. "Cultivation and characterization of the gut symbionts of honey bees and bumble bees: description of *Snodgrassella alvi* gen. nov., sp. nov., a member of the family Neisseriaceae of the Betaproteobacteria, and *Gilliamella apicola* gen. nov., sp. nov., a member of Orbaceae fam. nov., Orbales ord. nov., a sister taxon to the order 'Enterobacteriales' of the Gammaproteobacteria." *International Journal of Systematic and Evolutionary Microbiology* 63, no. Pt_6 (2013), 2008–2018. doi:10.1099/ijs.0.044875-0.

Kwong, Waldan K., and Nancy A. Moran. "Gut microbial communities of social bees." *Nature Reviews Microbiology* 14, no. 6 (2016), 374–384. doi:10.1038/nrmicro.2016.43.

Langille, Morgan G., Jesse Zaneveld, J.G. Caporaso, Daniel McDonald, Dan Knights, Joshua A. Reyes, Jose C. Clemente, et al. "Predictive functional profiling of microbial communities using 16S rRNA marker gene sequences." *Nature Biotechnology* 31, no. 9 (2013), 814–821. doi:10.1038/nbt.2676.

Li, Jiang H., Jay D. Evans, Wen F. Li, Ya Z. Zhao, Gloria DeGrandi-Hoffman, Shao K. Huang, Zhi G. Li, Michele Hamilton, and Yan P. Chen. "New evidence showing that the destruction of gut bacteria by antibiotic treatment could increase the honey bee's vulnerability to Nosema infection." *PLoS One* 12, no. 11 (2017), e0187505. doi:10.1371/journal.pone.0187505.

Loper, G. M., L. N. Standifer, M. J. Thompson, and Martha Gilliam. "Biochemistry and microbiology of bee-collected almond (*Prunus dulcis*) pollen and bee bread. I-Fatty Acids, Sterols, Vitamins and Minerals." *Apidologie* 11, no. 1 (1980), 63–73. doi:10.1051/apido:19800108.

Martinson, Vincent G., Bryan N. Danforth, Robert L. Minckley, Olav Rueppell, Salim Tingek, and Nancy A. Moran. "A simple and distinctive microbiota associated with honey bees and bumble bees." *Molecular Ecology* 20, no. 3 (2011), 619–628. doi:10.1111/j.1365-294x.2010.04959.x.

Martinson, Vincent G., Jamie Moy, and Nancy A. Moran. "Establishment of characteristic gut bacteria during development of the honeybee worker." *Applied and Environmental Microbiology* 78, no. 8 (2012), 2830–2840. doi:10.1128/aem.07810-11.

Menezes, Cristiano, Ayrton Vollet-Neto, Anita J. Marsaioli, Davila Zampieri, Isabela C. Fontoura, Augusto D. Luchessi, and Vera L. Imperatriz-Fonseca. "A Brazilian social bee must cultivate fungus to survive." *Current Biology* 25, no. 21 (2015), 2851–2855. doi:10.1016/j.cub.2015.09.028.

Mohr, Kathrin I., and Christoph C. Tebbe. "Diversity and phylotype consistency of bacteria in the guts of three bee species (Apoidea) at an oilseed rape field." *Environmental Microbiology* 8, no. 2 (2006), 258–272. doi:10.1111/j.1462-2920.2005.00893.x.

Moran, Nancy A., Allison K. Hansen, J. E. Powell, and Zakee L. Sabree. "Distinctive gut microbiota of honey bees assessed using deep sampling from individual worker bees." *PLoS One* 7, no. 4 (2012), e36393. doi:10.1371/journal.pone.0036393.

Morgano, Marcelo A., Márcia C. Martins, Luana C. Rabonato, Raquel F. Milani, Katumi Yotsuyanagi, and Delia B. Rodriguez-Amaya. "A comprehensive investigation of the mineral composition of Brazilian bee pollen: geographic and seasonal variations and contribution to human diet." *Journal of the Brazilian Chemical Society*, 2012. doi:10.1590/s0103-50532012000400019.

Naug, Dhruba. "Nutritional stress due to habitat loss may explain recent honeybee colony collapses." *Biological Conservation* 142, no. 10 (2009), 2369–2372. doi:10.1016/j.biocon.2009.04.007.

Oldroyd, Benjamin P. "What's killing American honey bees?" *PLoS Biology* 5, no. 6 (2007), e168. doi:10.1371/journal.pbio.0050168.

Olofsson, Tobias C., and Alejandra Vásquez. "Detection and identification of a novel lactic acid bacterial flora within the honey stomach of the honeybee *Apis mellifera*." *Current Microbiology* 57, no. 4 (2008), 356–363. doi:10.1007/s00284-008-9202-0.

Potts, Simon G., Stuart P. Roberts, Robin Dean, Gay Marris, Mike A. Brown, Richard Jones, Peter Neumann, and Josef Settele. "Declines of managed honey bees and beekeepers in Europe." *Journal of Apicultural Research* 49, no. 1 (2010), 15–22. doi:10.3896/ibra.1.49.1.02.

Rada, V., M. Máchová, J. Huk, M. Marounek, and D. Dušková. "Microflora in the honeybee digestive tract: counts, characteristics and sensitivity to veterinary drugs." *Apidologie* 28, no. 6 (1997), 357–365. doi:10.1051/apido:19970603.

Romero, S., A. Nastasa, A. Chapman, W. K. Kwong, and L. J. Foster. "The honey bee gut microbiota: strategies for study and characterization." *Insect Molecular Biology* 28, no. 4 (2019), 455–472. doi:10.1111/imb.12567.

Roulston, T'ai H., and James H. Cane. "Pollen nutritional content and digestibility for animals." *Plant Systematics and Evolution* 222, (2000), 187–209. doi:10.1007/BF00984102.

Roulston, T'ai H., James H. Cane, and Stephen L. Buchmann. "What governs protein content of pollen: pollinator preferences, pollen-pistil interactions, or phylogeny?" *Ecological Monographs* 70, no. 4 (2000), 617. doi:10.2307/2657188.

Rouzé, Régis, Anne Moné, Frédéric Delbac, Luc Belzunces, and Nicolas Blot. "The honeybee gut microbiota is altered after chronic exposure to different families of insecticides and infection by *Nosema ceranae*." *Microbes and Environments* 34, no. 3 (2019), 226–233. doi:10.1264/jsme2.me18169.

Ruiz-Argueso, T., and A. Rodriguez-Navarro. "Microbiology of ripening honey." *Applied Microbiology* 30, no. 6 (1975), 893–896. doi:10.1128/aem.30.6.893-896.1975.

Sabree, Zakee L., Allison K. Hansen, and Nancy A. Moran. "Independent studies using deep sequencing resolve the same set of core bacterial species dominating gut communities of honey bees." *PLoS One* 7, no. 7 (2012), e41250. doi:10.1371/journal.pone.0041250.

Sandhu, Dhanwant K., and Manjit K. Waraich. "Yeasts associated with pollinating bees and flower nectar." *Microbial Ecology* 11, no. 1 (1985), 51–58. doi:10.1007/bf02015108.

Smith, Kristine M., Elizabeth H. Loh, Melinda K. Rostal, Carlos M. Zambrana-Torrelio, Luciana Mendiola, and Peter Daszak. "Pathogens, pests, and economics: drivers of honey bee colony declines and losses." *EcoHealth* 10, no. 4 (2013), 434–445. doi:10.1007/s10393-013-0870-2.

Smith, Matthew R., Gitanjali M. Singh, Dariush Mozaffarian, and Samuel S. Myers. "Effects of decreases of animal pollinators on human nutrition and global health: a modelling analysis." *The Lancet* 386, no. 10007 (2015), 1964–1972. doi:10.1016/s0140-6736(15)61085-6.

Sousa, M. J., L. Miranda, M. Côrte-Real, and C. Leão. "Transport of acetic acid in *Zygosaccharomyces bailii*: effects of ethanol and their implications on the resistance of the yeast to acidic environments." *Applied and Environmental Microbiology* 62, no. 9 (1996), 3152–3157. doi:10.1128/aem.62.9.3152-3157.1996.

Szczęsna, T. "Protein content and amino acid composition of bee collected pollen from selected botanical origins." *Journal of Apicultural Science* 50 (2006), 81–90.

Tauber, James P., Vy Nguyen, Dawn Lopez, and Jay D. Evans. "Effects of a resident yeast from the honeybee gut on immunity, microbiota, and Nosema disease." *Insects* 10, no. 9 (2019), 296. doi:10.3390/insects10090296.

van der Zee, Romée, Lennard Pisa, Sreten Andonov, Robert Brodschneider, Jean-Daniel Charrière, Róbert Chlebo, Mary F. Coffey, et al. "Managed honey bee colony losses in Canada, China, Europe, Israel and Turkey, for the winters of 2008–9 and 2009–10." *Journal of Apicultural Research* 51, no. 1 (2012), 100–114. doi:10.3896/ibra.1.51.1.12.

vanEngelsdorp, Dennis, and Marina D. Meixner. "A historical review of managed honey bee populations in Europe and the United States and the factors that may affect them." *Journal of Invertebrate Pathology* 103 (2010), S80-S95. doi:10.1016/j.jip.2009.06.011.

Vojvodic, Svjetlana, Sandra M. Rehan, and Kirk E. Anderson. "Microbial gut diversity of Africanized and European honey bee larval instars." *PLoS One* 8, no. 8 (2013), e72106. doi:10.1371/journal.pone.0072106.

Wu, Yuqi, Yufei Zheng, Yanan Chen, Shuai Wang, Yanping Chen, Fuliang Hu, and Huoqing Zheng. "Honey bee (*Apis mellifera*) gut microbiota promotes host endogenous detoxification capability via regulation of P450 gene expression in the digestive tract." *Microbial Biotechnology* 13, no. 4 (2020), 1201–1212. doi:10.1111/1751-7915.13579.

Yun, Ji-Hyun, Mi-Ja Jung, Pil S. Kim, and Jin-Woo Bae. "Social status shapes the bacterial and fungal gut communities of the honey bee." *Scientific Reports* 8, no. 1 (2018). doi:10.1038/s41598-018-19860-7.

Zheng, Hao, Alex Nishida, Waldan K. Kwong, Hauke Koch, Philipp Engel, Margaret I. Steele, and Nancy A. Moran. "Metabolism of toxic sugars by strains of the bee gut symbiont *Gilliamella apicola*." *mBio* 7, no. 6 (2016). doi:10.1128/mbio.01326-16.

Zheng, Hao, Julie Perreau, J. E. Powell, Benfeng Han, Zijing Zhang, Waldan K. Kwong, Susannah G. Tringe, and Nancy A. Moran. "Division of labor in honey bee gut microbiota for plant polysaccharide digestion." *Proceedings of the National Academy of Sciences* 116, no. 51 (2019), 25909–25916. doi:10.1073/pnas.1916224116.

Zheng, Hao, J. E. Powell, Margaret I. Steele, Carsten Dietrich, and Nancy A. Moran. "Honeybee gut microbiota promotes host weight gain via bacterial metabolism and hormonal signaling." *Proceedings of the National Academy of Sciences* 114, no. 18 (2017), 4775–4780. doi:10.1073/pnas.1701819114.

8

Viral Metagenomics

Hugo G. Castelán-Sánchez, Sonia Dávila Ramos*
Centro de Investigación en Dinámica Celular, Instituto de Investigación en Ciencias Básicas y Aplicadas, Universidad Autónoma del Estado de Morelos
Xiaolong Liang
Department of Earth and Planetary Sciences, Washington University in St. Louis
María del Rayo Sánchez-Carbente
Centro de Investigación en Biotecnología, Universidad Autónoma del Estado de Morelos

CONTENTS

8.1 What Is Viral Metagenomics?

Viruses, the most abundant entities in the biosphere, are widely distributed in nature. For example, viruses are found in high concentrations as 10^7-10^8 particles per milliliter on the surface of seawater (Breitbart et al. 2008), 10^7 to 10^9 particles per gram on soil (Emerson et al. 2018), and ~10^5 particles m^{-3} in the air, and in the human body, the viral concentration varies according to the part of the body (10^8 per milliliter in oral, 10^5 in the blood, 10^6 per cm^2 in the skin) (Zárate et al. 2017). However, identifying viral diversity using classical techniques is a big challenge since many of these are uncultivated because the conditions and systems to allow their replication are unknown. New techniques such as next-generation sequencing (NGS) make it possible to recover partial or whole genomes present in a specific environment (Roux et al. 2019). The collection of all viruses present in the sample is defined as

DOI: 10.1201/9781003042570-12

the virome (Zárate et al. 2017). Viral metagenomics is essential for understanding the ecology and evolution of viruses, helping in the diagnosis of diseases, discovering biotechnologically useful enzymes, and discovering new viruses (Roux et al. 2019).

Within an ecology context, the viral community structure is important because it moves nutrients into the ecosystem (Dell'Anno, Corinaldesi, and Danovaro 2015). In this same context, it is possible to know the virus–host relationship, such as mutualism between plants and viruses (Roossinck 2015) and interactions among bacteria and bacteriophages (Maslov and Sneppen 2017), thus allowing us to know the importance of viruses in diverse ecosystems. It has allowed us to understand how genetic material is exchanged between viruses and their hosts through gene transduction mechanisms. The existence of viral auxiliary metabolic genes (AMGs) has been found, which could be responsible for increasing host metabolism; participating in metabolic reprogramming; being involved in biogeochemical cycles; and increasing the production of viral progeny, increasing their fitness (Dávila et al. 2019; Castelán-Sánchez et al. 2019), where the virus has taken genes of its host through the 'plunder and pillage' strategy (Warwick-Dugdale, Buchholz et al. 2019).

In health and disease surveillance, eukaryotic viruses mainly affect animals, plants, and arthropod species (Cadwell et al. 2015). Viral metagenomics can provide hints about viruses causing disease with the advantage of their genomic characterization. The same applies to animal and plant health; the main viruses that cause disease in plants and animals can be identified, where co-infections are known to exacerbate diseases (Bosch et al. 2013; Wilkinson et al., 2011).

Finally, viral metagenomics has made it possible to identify about 750,000 uncultivated virus genomes. Carrying out viral metagenomics is a big challenge, since no single molecular marker performs all virus identification. Furthermore, viruses have diverse genetic material between RNA viruses and DNA viruses. dsDNA viral metagenomes are perhaps the most studied, because the protocols for general metagenomics have been standardized for this type of genome. However, there is an evolution in the protocols to obtain genomic information from RNA viruses, since these viruses represent approximately 70% of all viruses on the planet (Woolhouse and Adair 2013).

8.2 Sample Preparation and Processing of the Virome

The study of genomes has its limitations in obtaining good quality and quantity of nucleic acids. It is even more complicated if those organisms cannot be cultivated, and viruses are not the exception. With the development of metagenomic strategies (Rosario and Breitbart 2011), the isolation and cultivation steps to study microorganisms can be omitted.

Undoubtedly, obtaining good quality and quantity of viral genetic material depends on a good experimental design, which has to consider several important criteria, like obtaining the sample, the required quantity, and the type of storage for its processing.

8.2.1 Treatment of Samples and Limitations

8.2.1.1 Viromes from Environmental Samples

8.2.1.1.1 Aquatic Environments

Most of the studies on viromes have been carried out in the aquatic environment (Moon et al. 2020), and direct calculations have been made on viral abundance utilizing microscopy with epifluorescence and flow cytometry, combined with other viability techniques; thus, the calculation of viral abundance varies from 10^6 to 10^8 particles ml-1 (Corinaldesi, Tangherlini, and Dell'Anno 2017). These fluctuations are due to the characteristics of the environment where the samples are taken. Observations have often been made where we find a higher viral abundance in lakes than in the ocean or on the coasts than in the open sea and on the surface compared to the sea or deep lakes (Gong et al. 2018). However, these abundances are subjected to seasonality and the physicochemical characteristics of the sampling site, the abundance of the hosts, and the replicative cycle in which the viruses are found (lytic, lysogenic, or temperate and chronic). In this kind of environment, it is necessary to consider the stratification factor mainly caused by light and O_2 as the last electron acceptor, which is the main component that differentiates populations in aquatic ecosystems (Wommack and Colwell 2000). Taking into account both the abundance and diversity of this type of environment, we can consider some hypotheses and design methodologies that allow us to know the microbiome and virome, reducing as far as possible the methodological biases in taking samples to obtain consistent results. Several diversity studies have been reported in aquatic environments, mainly in marine oceans; however, only some of them focused on viromes. It is not surprising, since viruses, as we well know, are very diverse from their genomic composition (ssDNA, dsDNA, ssRNA, dsRNA) to the number of genes that compose and encode by them. There are no molecules such as the 16S rRNA gene for prokaryotes or 18S rRNA for eukaryotes that allow us to explore the virome present in a particular environment (Nooij et al. 2018).

The development of metagenomics, particularly of the shotgun type, made it possible for us to know the viral populations of an environment without the need to have prior knowledge of the sequences that made them up. From the studies carried out, we can observe that a large proportion of viral families belong to bacteriophages (viruses that infect bacteria) and to a lesser extent to archaeal viruses (viruses that infect archaea) and, lastly, if it is the case, viruses that infect fungi, algae, and organisms are present in those ecosystems. In extreme aquatic environments such as those of thermophiles or halophiles, the ratio of bacteriophages to that of the archaea virus is usually reversed (Ramos-Barbero et al. 2019). This is relevant in the methodological design to favor the extraction processes of the genomes of interest. For the extraction of viromes from aquatic environments, the samples go through a pre-filtration or centrifugation that eliminates contaminants from the virome and cellular debris of the microorganisms that belong to that ecosystem. Most of it is necessary to concentrate the samples obtained, ranging from 10s to 100s of liters, which is usually a bottleneck due to the limitations of the methods used, which are mostly unreliable, expensive, and not very efficient (Figure 8.1a). The concentration methods for smaller volumes pressure filtered the samples (7 mm Hg) through a 0.22-mm pore size polyethersulfone membrane filter cartridge (Sterivex; Millipore, Billerica, MA) using a peristaltic pump. This favors the presence of viral sequences, eliminating the other abundant cells, like prokaryotic cells. The next step was filtration by a 0.02-mm aluminum oxide filter (Anotop; Whatman, Middlesex, United Kingdom). Performing filtration without prior filtration of 0.22 mm has been effective for profiling metagenomics where PCR is performed with degenerate probes for specific genes of the desired group (Steward and Culley 2010). Another alternative is ultracentrifugation; the protocol consists, for example, of volumes of 210 mL (six tubes of 12 mL filled with 11.7 mL of samples per run) in three runs of 3 h 30 min each at 120,000 Å~g (i.e., 28,000 rpm), 4°C with an SW 40 with Ti rotor (LE80 K, Beckman). Viral pellets were then resuspended and incubated under agitation for at least 10 h at 4°C in an SM buffer [0.1 M NaCl, 8 mM $MgSO_4$–$7H_2O$, 50 mM Tris–HCl, and 0.005% (wt/vol) glycerol) adjusted at neutral pH] before analysis (final volume equal to about 400 μL) (Colombet et al. 2007).

For samples that come from large volumes, greater than 10 L of water, the most accepted concentration strategies are tangential flow filtration (TFF); 50 L subsamples were separately concentrated using large-scale 100-kDa TFF (Amersham Biosciences, Westborough, MA, USA; Cat. # UFP-100-C-9A) to 0.65–1 l followed by small-scale 100-kDa TFF (Millipore, Billerica, MA, USA, Cat. #PXB100C50) to 12–14 ml; viruses were collected in the retentate after a final washing step (Wommack et al. 2010). The other strategy, and the most accepted recently, is chemical concentration, such as flocculation with $FeCl_3$. For the $FeCl_3$ method, triplicate 20 L subsamples were subjected to a chemistry-based concentration method, where $FeCl_3$ creates a flocculus of the virus with iron, which precipitates and can be collected on 1.0-μm polycarbonate filters (GE Water and Process Technologies, Trevose, PA, USA; Cat. #K10CP14220) and resuspended in a magnesium-EDTA-ascorbate buffer (0.1 M Mg2EDTA, 0.2 M ascorbic acid, pH 6.0) using 1 ml of buffer per 1 L of seawater. Resuspension was allowed to go overnight, rotating in the dark at 4°C, and the filters were transferred to fresh tubes and centrifuged for 5 min at low speed to collect the remaining fluid (John et al. 2011).

8.2.1.1.2 Terrestrial Environments

For terrestrial environment samples, we face other problems; one of the most important is the diversity of existing soils and the composition of their biomes that create their particular properties and the fact that independent islands or microhabitats can be formed in soils. We could consider that each sample could have unique characteristics even taken in the same area, such as the rhizosphere or the surface. In these situations, microscale sampling could be the most suitable for estimating the viral abundance, which has been calculated to vary in ranges from 10^7 to 10^{10} viruses per gram of dry soil (Graham et al. 2019). This makes us see that viral diversity in this type of environment is practically unknown. However, the difficulties of extracting genetic material in optimal conditions from soils have not been entirely successful. Some of the soils considered explicitly diverse are the tundra, the taiga, the temperate forest, the tropical rainforest, the grassland, and the desert. Each of these has limitations in obtaining viral genomes, for instance, the problems of adherence of viral particles to the soil in the processing of the samples (Pratama and Elsas 2018).

In general, it is not practical to work with soil samples larger than 50 gr (Figure 8.1a). Once the sample is taken, it is resuspended in a buffer to maintain the integrity of the sample. There are some variations regarding the resuspension buffer of 1% potassium citrate (KC), amended 1% potassium citrate (AKC), and amended 5 mM sodium pyrophosphate (PP) adjustments in the buffer, which depends on the different types of soil. Two strategies are used to separate the viral particles from the soil, mainly sonication, which has been carried out on ice and without it, and another variant is both sonication and rest time. The other strategy used is vortex agitation, which has been seen to reduce tail breaking by up to 20% compared to sonication. Finally, the sample's storage conditions can be at 4°C or −80°C (Trubl et al. 2016, 2018). As can be seen, the combination of methods could be better adapted to one sample than another, so there is no universal protocol for obtaining virome from different soils.

8.2.1.2 Viromes from Biological Systems

8.2.1.2.1 Plants

Studies carried out to find how viruses interact with plants have focused mainly on the isolation and cultivation of viruses that cause disease in plants with economic interest for humans, with viral infections in crop plants the most studied; however, little is known about the importance of other viral interactions in these and other plants in general. There are several limitations to the study of viruses in plants, and, as in any sample, obtaining good-quality genomic material is one of them. In this case, the quantity is another issue, since plants' genome and associated bacteria are orders of magnitude higher than viruses. Although we can find that ssDNA or dsDNA viruses in plants are not frequent, we know that, for instance, the main viruses that infect plants have genomes of ssRNA with positive chains, although there are others with less frequency, so the methodology to obtain the virome could be focused on these viruses and favor the extraction of RNA from the sample (Muthukumar et al. 2009). Different strategies have been followed to obtain plant viruses, the most common of which are as follows. 1) Virus-like particles (VLP) enrichment once the sample has been homogenized and clarified. The treatment consists of centrifugation steps and ultracentrifugation with 20% sucrose pads; the pellet is resuspended for the extraction procedures and subsequent amplification of viral sequences by PCR (Wang et al. 2002). 2) The use of siRNA through short sequence assemblers (Kreuze et al. 2009). This method consists of extracting total RNA from the samples by selecting RNA bands corresponding to the size range 20–30 nts, ligation of 3′ and 5′ adapters to the RNA, cDNA synthesis followed by acrylamide gel purification in all steps, and a final step of PCR amplification to generate a DNA colony template library for Illumina sequencing. 3) Enrichment of dsRNA by precipitation once the tissue has been treated with liquid N2 and homogenized; the molecules are amplified by reverse transcriptase reactions (Roossinck 2015). This last strategy has made it possible to increase the knowledge of viral sequences that have not yet been obtained or reported in the databases.

8.2.1.2.2 Animal, Insect, and Human Virome

The study of animal and insect-related viruses has aroused great interest due to the relationship that these have with humans and evidently due to the transmission of diseases of zoonotic origin, due to fauna–livestock–human interface areas, where contact and emerging infections caused by it are becoming more frequent (Kwok et al. 2020). For animal and human samples (Conceição-Neto et al. 2015), the sample's origin must be taken into account, that is, if it is a sample of any secretion tissue or organ. The procedure for processing samples of secretions, for example, nasal or fecal, uses swabs that are resuspended in an elution medium and vortexed (Figure 8.1a); after a centrifugation step, the cells are eliminated and treated with nucleases to remove genomic material that is not of interest for the subsequent treatment of obtaining the viral genome (Blomström et al. 2018). In the case of tissue or organ samples, these are homogenized and filtered. The next step is to deal with DNases and RNases to eliminate most of the genetic material passed through the filter to later extract the viral genetic material from both DNA and RNA (Figure 8.1b). In the case of mosquito viromes, it has been increasingly relevant. These insects transmit vectors of important human diseases, so it is increasingly necessary to know the viruses that could be circulating or those that could become potential emerging infections (Shi et al. 2019). It is very common to use around five mosquitoes to obtain a good amount of viral genome. For this type of sample, homogenization is the limiting step to avoid clogging of the filters and loss of viral particles; the use of resuspension beds for this purpose seems to damage the integrity of the VLPs. Posterior centrifugation to eliminate the mosquito cells does not have to be at low speed and for a short time, and filtration has been recommended up to 0.45-mm filters to be able to recover viruses such as mimiviruses that are larger than the average (Figure 8.1a); finally, the use of chloroform in low concentrations works for the elimination of bacteria, although it can affect the abundance of viruses involved.

8.2.2 Viral Genome Extraction

The first step is to resuspend samples; the next step is the removal of cellular genomic material, which, in measure of the amount that is strained to the next stages, could represent an essential part of the sequences obtained at the end of the process. A DNase I digestion and RNase A digestion are performed previously to the concentration and broken steps (Sathiamoorthy et al. 2018) (Figure 8.1b). Here concentration steps are recommended; the principal methods include CsCl, sucrose, or Cs_2SO_4 gradients by ultracentrifugation (Trubl et al. 2019). Additionally, once the viral particles have been concentrated, the capsids or viral envelopes have to be broken to release the viral genomes, generally with Proteinase K or formamide (Trubl et al. 2019). One of the most recent strategies is to perform the purification of DNA and RNA genomes in parallel (Sathiamoorthy et al. 2018). However, the capsid or envelope has to be taken into account to break the viral particles and their genome's adequate extraction (Figure 8.1b). In the RNA viral genomes, a retro-transcription and amplification with random PCR steps are necessary; the random priming single-primer amplification (SISPA) method is frequently used. Nevertheless, it has been observed that amplifications lead to a bias in the sequence results of the structure of the virome in the samples (Greninger 2018) (Figure 8.1c).

8.2.3 Virome Sequencing

Virome sequencing is through next-generation sequencing; since there is not a molecular marker, the genetic material from the virus is enriched through the single-primer amplification method (random PCR), linker amplification shotgun libraries (LASLs), displacement amplification (MDA), PCR-based sequence independent or without basis amplification and is dependent on the DNA concentration. In the case of RNA genomes, a reverse transcriptase reaction is generally performed, and cDNA is required to sequence (Figure 8.1c) (Parras-Moltó et al. 2018). SISPA uses pseudo-degenerate oligonucleotides with a length of 6–12 random nucleotides, which are ligations of blunt-ended DNA molecules. However, this technique also presents bias on the conserved parts of the primer, amplifying these regions, and low-abundance viruses are captured (Parras-Moltó et al. 2018). The LASL library protocol was developed to capture complete phage genomes that use randomly amplified templates for PCR (Parras-Moltó et al. 2018). MDA is used under isothermal conditions by DNA polymerase Φ29 and random primers, where the template can be amplified 10,000-fold. However, this method has a bias in the enrichment of viruses, mainly ssDNA viruses (Figure 8.1c) (Parras-Moltó et al. 2018). In the case of human

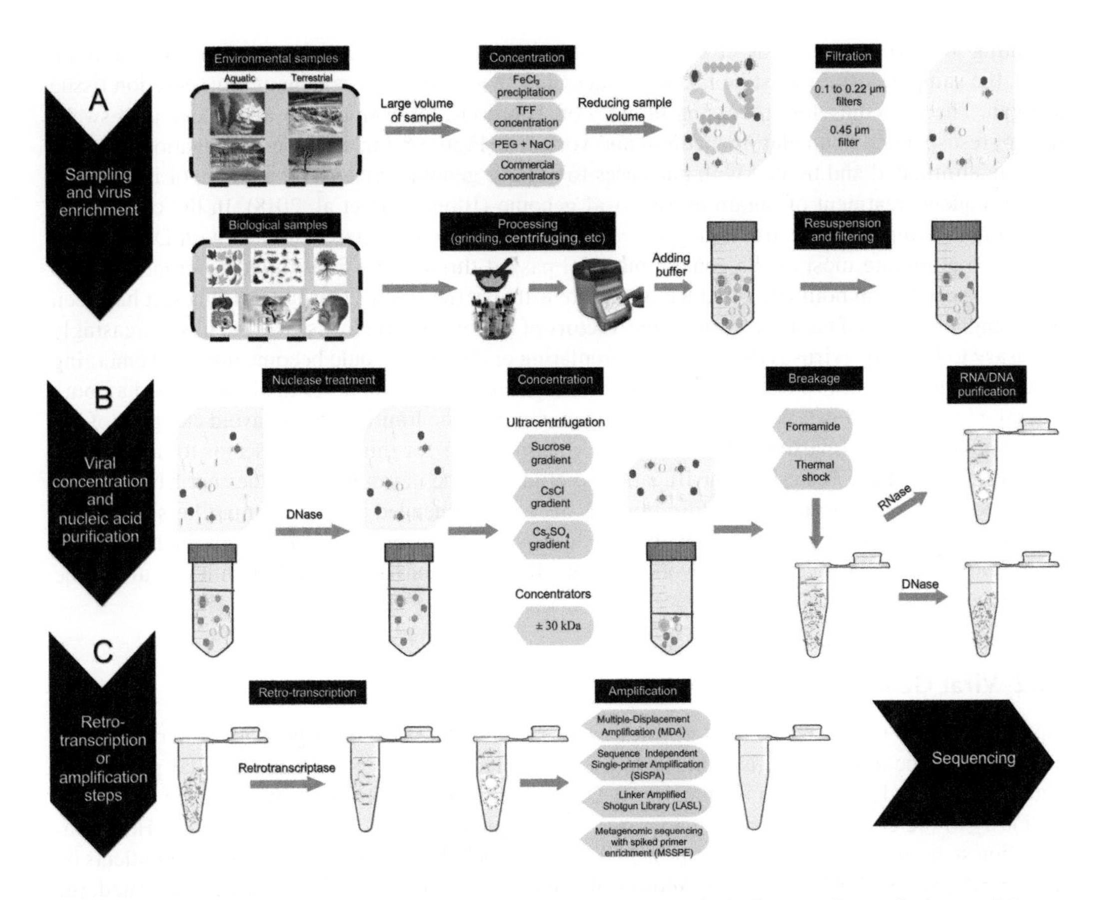

FIGURE 8.1 Wet lab pipeline for obtaining viral metagenomes. (a) Methods for sampling and virus enrichment. (b) Methods for viral concentration and nucleic acid purification. (c) Retro-transcription or amplification steps.

metagenomics, spiked primer enrichment (MSSPE) has been developed, where, through an analysis of the virus databases, *k*-mers were obtained, and primers were designed to be able to capture virus RNA (Deng et al. 2019). There are currently different platforms for sequencing, such as Illumina platforms (MiSeq, NextSeq, HiSeq), MinION, Pacbio, and so on. Platform selection depends on the performance and depth that one wants to recover for the viruses. However, as previously mentioned, the concentration of viruses in the environmental sample depends on many factors; based on the previous reports, a performance of 30 million reads on the Miseq Illumina platform is enough to recover complete viruses; lower than these yields, it is unlikely one would obtain complete genomes from metagenomes. New platforms such as MinION and Pacbio have been used in viral metagenomics; both platforms have the advantage that there is no pre-amplification of the genetic material prior to sequencing. The advantage of sequencing in MinION is that complete viral genomes can be obtained, and a wide variety of viruses can be captured; however, the disadvantage of these platforms is the high error rates (5–10%) (Warwick-Dugdale, Solonenko et al. 2019).

8.3 Bioinformatic Analysis to Viral Metagenomics

There are many bioinformatics workflows to find viruses, from single amplified genome (SAG) to metagenome-assembled genome (MAG). These programs can be divided into pipelines, programs that recover viruses from the assembly with different search approaches (alignment-based methods, gene-homology–based methods, free-alignment or deep learning architectures), and programs for taxonomic assignment (Ren et al. 2020). The workflows that need the assembled files or MAG as input to be able to identify viruses take specific characteristics of the viruses to extract the viral contigs from assemblies, and within this approach, there are many strategies such as those based on a comparison with reference sequences, hidden Markov models (HMMs), and using machine learning approaches. Finally, taxonomic assignment software has tools that taxonomically cluster viruses based on sequence similarity estimates. A good practice that is generally performed in metagenomic assemblies is MAG recovery; these recovered genomes are evaluated for quality and estimated completeness with different tools. In the case of the viral metagenome, viral genome assembly is challenging and ensures that these contigs are free from contamination by their host; thus, there is an effort to develop tools to assess the quality of recovered viral contigs (Nayfach et al. 2020). An important aspect that users of bioinformatics tools should know is that not all programs are used to find all types of viruses. Since viruses have genetic material diversity, there are specialized programs in DNA viruses and RNA viruses. Therefore, the appropriate tool must be selected to carry out searches and correct comparisons.

8.3.1 Pipelines to Recover Viruses from Metagenomes

The analysis pipelines to recover viral genomes are focused on viruses that infect humans and from environmental samples. These have the objective of carrying out analysis from the raw data without any pretreatment or aligning short reads to a reference database. Generally, the pipelines begin analysis from raw NGS data using quality control, then perform trim sequences, eliminate short reads, and remove adapter sequences. Some tools map or align the reads against genome reference hosts to eliminate host contamination. Some workflows remove duplicate sequences (deduplication) and unresolved nucleotides (Ns) and subsequently perform an assembly of the reads for a taxonomic classification according to a database such as NCBI, and the viruses are classified according to International Committee on Taxonomy of Viruses (ICTV) criteria. However, each pipeline uses particular strategies to recover virus genomes within a viral metagenome. The analysis of the bioinformatic sequences changes according to the type of sample that is being analyzed; in the case of working with a human virome or another host, the corresponding sequences must be eliminated, while in samples of environmental origin, it is not necessary to remove the host.

Many pipelines are designed to search for viruses that cause disease in humans, for example, VirusHunter (Zhao et al. 2013), Sequence-Based Ultrarapid Pathogen Identification (SURPI) (Naccache et al. 2014) Virus Identification Pipeline (VIP) (Y. Li et al. 2016), VirusSeeker (Zhao et al. 2017), ViromeScan (Rampelli et al. 2016), Viral Genome-Targeted Assembly Pipeline (VirusTAP) (Yamashita, Sekizuka, and Kuroda 2016), virMine (Garretto, Hatzopoulos, and Putonti 2019), DisCVR (Maabar et al. 2019), and VirAnnotOTU (Lefebvre et al. 2019). These programs usually map the reads against the human genome, and then the filtered reads are used in subsequent analyses. For environmental viruses, there are VirAmp (Wan et al. 2015), Holovir (Laffy et al. 2016), and FastViromeExplorer (Tithi et al. 2018), while other programs are focused on the recovery of RNA viruses, such as short RNA subtraction and assembly (SRSA), and others focus on RNA viruses in plants, such as VirFind (Ho and Tzanetakis 2014), Kodoja (Baizan-Edge et al. 2019) and VirusDetect (Zheng et al. 2017). However, these programs have different approaches; for example, VirFind uses BLAST, and Kodoja uses *k*-mer analysis strategy from data RNA-seq. VirusDetect is used to search for viruses by assembling total small RNAs (Zheng et al. 2017). Many of these programs identify viruses, but few perform a taxonomic classification; some are VirAnnotOTU, VIP, SRSA, HoloVir, and ViromeScan. The use of other tools is necessary for those that do not make this taxonomic assignment, and these will be discussed later.

8.3.2 Retrieval of Viruses from Contig Assembly

Assembly is a crucial step in characterizing the virome. For assembling viral sequences, assemblers are designed to generally assemble short reads specifically from metagenomes. One of the main assemblers is the de Bruijn algorithm, and the main disadvantage of de Bruijn assemblers is *k*-mer selection. *k*-mer size is essential to reconstruct genomes or metagenomes and not have suboptimal assemblies (Roux et al.

2017; Sutton et al. 2019). Common short-read assemblers of metagenomics are SPAdes (Bankevich et al. 2012), IDBA-UD (Peng et al. 2012), Megahit (D. Li et al. 2015), MetaVelvet (Namiki et al. 2012), and MetaVelvet-SL (Afiahayati, Sato, and Sakakibara 2015).

A benchmarking of assemblers for metagenomes has been developed for retrieval of viruses, and read preprocessing has been reported to have minimal impact on assembly production, except for poorly covered or represented genomes. In this comparison, Megahit was best at avoiding the construction of chimeric contigs; metaSPAdes was best at assembling low-coverage genomes within a single large contig and had fewer false positives when retrieving circular contigs or complete genomes. On the other hand, heterogeneity between the viral genomes is important; also, population shape and strain diversity can involve assembly failures. Strains closely related to each other are probably not assembled (Roux et al. 2017).

However, there are assemblers developed specifically to viral sequences. For example, Detect and Reconstruct Known Viral Genomes from Metagenomes (drVM), developed specifically to recover eukaryotic viruses, uses SPAdes for assembly in addition to annotating sequences using BLAST and mapping with the reference sequence (Lin and Liao 2017). VrAP (www.rna.uni-jena.de/research/software/vrap-viral-assembly-pipeline/) is a pipeline based on the SPAdes assembler, which also can classify viral contigs from host contigs depending on their density scores. Currently, MetaviralSPAdes has been developed, a tool based on the SPAdes assembler that allows the assembly of viral genomes and identifies viral contigs (Antipov et al. 2020).

Another method used for the recovered virus from raw reads is fragment recruitment (FR) (Cobián Güemes et al. 2016), which consists of mapping metagenomic reads against a database, which can be viral reference sequences or a set of genes of interest. Depending on the program, they use different algorithms, such as SMALT (Ponstingl and Ning 2010) and Bowtie2 (Langmead and Salzberg 2012). Currently, there are some tools to carry out FR from metagenomes such as FR-HIT (Niu et al. 2011), but FRAP and VR-MG-FRAP (Cobían Güemes 2019) specialize in viral sequences, and Tanoti (https://bioinformatics.cvr.ac.uk/software/tanoti/) is software to map reads through a guided reference using BLAST. Once the sequences have been assembled with any of the tools mentioned previously, the obtained contigs that correspond to the viral sequences can be classified or selected.

8.3.2.1 Gene Homology-Based Methods

These methods search for viruses from contigs by aligning the sequences with viral gene databases. This comparison generally uses BLAST search against a database. The comparison is made in some cases with an amino acid sequence, given that they are more evolutionarily conserved. Some of the most popular tools focus on searching for environmental viruses, such as VirSorter (Roux et al. 2015), which takes particular characteristics of viruses to be able to identify them. For example, VirSorter aligns against 'hallmark' viral genes and searches based on similarity in a previously built database using BLASTP and HMMs (Roux et al. 2015). With this program, new viruses can be identified. However, it is mainly focused on the search for bacteriophages. Within this category are the webservers and web tools. This kind of web tool makes work more friendly for people with little knowledge of bioinformatics. The pipelines mentioned previously are pipelines such as VirAmp (Wan et al. 2015), and web servers that receive contigs as input data are Metavir2 (Roux et al. 2014) and Viral Informatics Resource for Metagenome Exploration (VIROME) (Wommack et al. 2012).

These tools are designed to search for environmental viruses such as bacteriophages;, web-tools such as Phaster (Arndt et al. 2016) compare query sequences with a database of viral genes. Bioinformatics tools to make a taxonomic classification in metagenomes could work to identify viral sequences. These are not specialized tools for viruses, and based on our experience, the identification of new viruses is limited. Besides that, their viral database is little enriched in viral sequences or has biases for only viruses cultivable in the lab, and the latest viruses have no homologs in the databases. Some examples of these programs are Taxonomer (Flygare et al. 2016), CLARK (Ounit et al. 2015), Kaiju (Menzel, Ng, and Krogh 2016), Metaphlan2 (Truong et al. 2015), and Centrifuge (Kim et al. 2016). Thus, we do not recommend them when the coverage or number of viral sequences is low in the metagenome or for discovering new viruses.

There are a few tools to perform a functional annotation in viral sequences; they also accept contigs as input. For example, Viral Genome Organizer (VGO) (Upton et al. 2000), VMGAP (Viral MetaGenome Annotation Pipeline) (Lorenzi et al. 2011), Viral Genome Annotation (Vgas) (Zhang et al. 2019), PHANOTATE (McNair et al. 2019), and Viral Genome ORF Reader (VIGOR specifically for small viral genomes) use contigs to perform functional prediction (Wang, Sundaram, and Spiro 2010), and some tools that are not specialized in virus functional annotation but still can do it are Metagenomics-Rapid Annotation using Subsystem Technology (MG-RAST) (Keegan, Glass, and Meyer 2016) and GhostKoala (Kanehisa, Sato, and Morishima 2016).

8.3.2.2 Machine-Learning Architectures

Another strategy for classifying viruses is through machine-learning architectures; these approaches require an input sequence classified according to determined features. Machine-learning algorithms allow us to identify patterns or characteristics within sequences and then make predictions through *k*-mer–based approaches, random forest machine learning, and a trained convolutional neural network (Ponsero and Hurwitz 2019). One of the most popular programs that use this type of approach is Virfinder (Ren et al. 2017), which identifies viral signatures in contigs through machine learning, showing superior identification in identifying viral genomes compared to VirSorter. However, Virfinder is a classifier model that is trained with sequences from the RefSeq database with which it is possible to identify certain viral groups; however, if this algorithm is delivered with viruses representative of marine environments, it does not adequately identify viruses from other ecosystems. So, it is necessary to have a high set of viral viromes and genomes to have a good set of data to perform algorithm training and find more new viruses (Ponsero and Hurwitz 2019).

Recently, new tools have been developed with different classification algorithms based on machine learning; for example, the approach of machine learning using RSCU (Bzhalava et al. 2018). MARVEL is random forest machine learning that predicts virus sequences through genomic characteristics such as gene density and chain changes and then compares them with a database (Amgarten et al. 2018). DeepVirFinder uses convolutional neural networks for identification of viruses (Ren et al. 2020); ViraMiner is a random forest model to identify viruses in human samples, mapping viral genes to protein databases (Tampuu et al. 2019). PPR-Meta is a neural network architecture able to identify plasmid contigs and phages from the assembly (Fang et al. 2019). VIBRANT uses neural networks that are trained with a set of a protein family, which is capable of identifying dsDNA, ssDNA, and RNA viruses (Kieft, Zhou, and Anantharaman 2020).

Comprehensive tools such as What the Phage (WtP) have also been developed, which uses and integrates different identification tools in parallel to detect and annotate phages (Marquet et al. 2020). A total of nine different tools are used in parallel: VirSorter v1.0.6 83 (with and without virome mode) (Roux et al. 2015), Sourmash v2.0.1 (Pierce et al. 2019), Metaphinder with default database and own database (Jurtz et al. 2016), VirFinder v1.1 (Ren et al. 2017), MARVEL v0.2 (Amgarten et al. 2018), VirNet v0.1 (Abdelkareem et al. 2018), PPR-Meta v1.1 (Fang et al. 2019), Vibrant v1.2.1 (Kieft, Zhou, and Anantharaman 2020), and DeepVirFinder v1.0 (Ren et al. 2020).

8.3.3 Check of Quality of Virus Assembly

The minimum information standards on a non-cultured virus genome (MIUViG) have reported some considerations on the integrity of the virus genome; these are contigs with a length ≥10 kb, and the contig has a nucleotide identity ≥90% with ≥75% covered (Roux et al. 2019) and could be considered a complete virus. In the case of circular genomes, sequences must have the terminal repeats inverted, and for segmented viral genomes, a closely related reference genome and additional experiments in a wet lab are required (Roux et al. 2017). To assess the recovered viral genomes' completeness, CheckV has been recently developed (Nayfach et al. 2020). This program first removes host contamination by identifying non-viral genes using HMMs. The second step estimates the genome's completeness through average amino acid identity (ANI) compared with a database. It estimates the length of the new genome based on the hits in the database. In the last step, direct terminal repeats (DTRs), inverted terminal repeats (ITRs), and provirus integration sites are inferred in the sequences to assess the circular genome (Nayfach et al. 2020).

8.3.4 Taxonomic Tool Classification

The taxonomic classification of viruses is generally complicated, since there is no molecular marker to identify viruses. However, efforts have been made to carry out classification, for which various tools have been developed (Nooij et al. 2018). One way to classify viruses is through the reconstruction of phylogeny, and thus their possible origin can be inferred. Recently a new classification has been proposed where the virus can be classified from the genomic sequence into four mega-taxonomic realms: *Riboviria*, *Monodnaviria*, *Viridnaviria*, and *Duplodnaviria* (Koonin et al. 2020). Some tools for virus taxonomy are Phage Proteomic VICTOR (Meier-Kolthoff and Göker 2017), ViPTree (Nishimura et al. 2017), VicTree (Modha et al. 2018), and GRAViTy (Aiewsakun and Simmonds 2018). Another program for taxonomy classification is MEGAN (Huson et al. 2007), which uses the lowest common ancestor (LCA) assignment algorithm for classification of sequences. Another approach to taxonomic analysis are network-clustering approaches based on shared protein content between phages; one program that uses this approach is vConTACT (Bolduc et al. 2017).

8.3.5 Viral Host Prediction

There are different computational approaches to find potential virus hosts. According to MIUViG, the prediction of a viral host can be through sequence similarity with a reference virus genome (genetic homology); a comparison against a database with CRISPR-Cas spacers; through signatures such as G+C content, *k*-mer composition, and codon usage; and finally by comparing the abundance in host and virus sequences through spatial or temporal scales (Khot, Strous, and Hawley 2020). Other strategies have been developed based on machine-learning approaches, deep neural networks that do not depend on a comparison with reference genomes (Young, Rogers, and Robertson 2020). Another way to identify possible hosts is through an exact alignment match between the attP and attB sequence sites shared among lysogenic phages and bacteria (Khot, Strous, and Hawley 2020). Different tools have been developed, such as Hostfinder (Villarroel et al. 2016), WIsH (Galiez et al. 2017), Viral Host Predictor (Babayan, Orton, and Streicker 2018), VIDHOP (Mock et al. 2020), and VirHostMatcher (W. Wang et al. 2020).

8.3.6 Virus Databases

Within developed virus databases, these can be classified as general or specific databases. General databases contain information on viruses in general, while specific databases are characterized by having information on a particular virus. Many virus databases are focused on cultivating or isolating viruses in the laboratory. Nevertheless, in general, there are few virome databases. Some of the most consulted virus databases are GISAID database, Virus-Host DB (Mihara et al. 2016), viruSITE (Stano, Beke, and Klucar 2016), Virus Pathogen Resource (ViPR) (Blomström et al. 2018), virus capsid database (IRAM) (Almansour et al. 2019), VIROME (Wommack et al. 2012), Metavir2 (Roux et al. 2014, 2), VirusZone (Hulo et al. 2011) Virus RefSeq GenBank (Brister et al. 2015), and the ACLAME database (Leplae, Lima-Mendez, and Toussaint 2010). The most extensive database with the highest number of viruses is Integrated Microbial Genome/Virus (IMG/VR) (Paez-Espino et al. 2016). Other databases that contain viruses are the iMicrobe portal (http://imicrobe.us/, supported by iPlant) and Metagenomics-Rapid Annotation using Subsystem Technology (MG-RAST) (Keegan, Glass, and Meyer 2016).

8.4 Challenges and Applications in Human, Water, and Soil Viromes

Virology associated with human health has been extensively studied; however, environmental viruses have had relatively fewer research efforts than those in human virology. Despite the nano-scale size of viral particles, viruses may have critical roles in ecological and biogeochemical processes in all kinds of systems (Papaianni et al. 2020; Zimmerman et al. 2020). In the meantime, more and more researchers have recognized the importance of viruses, and more new remarkable capabilities have been credited to viruses. Thus, viral ecology research is entering the spotlight and experiencing incredible development. The basic information of viruses such as abundance, distribution, reproduction strategy, and community composition can be investigated via direct counting techniques and traditional molecular approaches (Liang, Zhuang et al. 2019; Liang, Zhang et al. 2020). While these traditional approaches provide important information and contribute to the research field, the resolution in examining specific scientific questions is limited and even incapable in these traditional methods. Therefore, more powerful techniques require a better understanding of viral ecology at higher altitudes. Continuing advances in molecular methods have provided promising approaches to understanding microbial community diversity and function in ecosystems. State-of-the-art sequencing technology generating large-scale datasets (e.g., genomics, proteomics, transcriptomics, metagenomics, metabolomics) has provided researchers opportunities to resolve scientific issues regarding the entire microbial community that could not be addressed before, for example, the correlation of microbial communities (specifically taxonomic composition, diversity, and metabolic functions) with environmental and biological factors (Coutinho et al. 2018). Specifically, viromics (viral metagenomics), focusing on addressing scientific questions associated with viral ecology, have demonstrated the effectiveness of techniques in characterizing viral communities in associated systems, such as human, water, and soil environments, and revealed the critical role of viruses in specific ecosystem functions (Emerson et al. 2018).

8.4.1 Human Virome

The human microbiome, the collection of body microbiota, and the overall genetic information have been shown to be an essential part of the well-being of human functionality and have extensive effects on people's health. Progress in human microbiome research has been greatly stimulated, especially with the launch of the Human Microbiome Project (Cho and Blaser 2012). As mentioned, viruses have critical roles in modulating the whole microbial community and may directly infect host cells, causing disease symptoms and, therefore, are correlated with human health and many processes in the body. The employment of NGS in the investigation of viromes of all types has led to tremendous discoveries and contributed to a better understanding of human viral ecology and the impact on human health (Ajami and Petrosino 2015; Santiago-Rodriguez and Hollister 2019).

The gut virome has been the leading topic in human virome research because of the utmost importance of gut virome to human health, and the human viromes that are characterized have mostly been derived from fecal samples (Deng et al. 2019; Ansari et al. 2020). DNA viruses were the most sought-after target in initial studies (Reyes et al. 2012). The human gut system is a more nutrient-enriched habitat for the microbiome than most natural systems like ocean and soil environments, and the virus–host interaction patterns showed notable differences compared with other environments (Silveira and Rohwer 2016; Shkoporov et al. 2019). Virome analysis has been a promising approach for studying viral dark matter, viral activities, and their function in sustaining human health or causing disease.

The human gut contains a diverse collection of viruses, with bacteriophages as the most represented viral members. Shkoporov et al. (2019) examined the temporal dynamics of fecal viral communities in healthy adults via longitudinal metagenomic analysis and showed that the viral community had high temporal stability and individual specificity, which is correlated with the co-occurring bacterial microbiome. Taxonomic annotation of viral genomes revealed that *Microviridae* and crass-like bacteriophages (associated with Bacteroidetes, predominant in the gut) were the most persistent colonizers in the gut. The correlation between stable viral communities and dominant bacterial taxonomic groups was also shown in CRISPR-based host prediction. Clear evidence has shown that the bacterial microbiome in the gut is closely related to human health, and the connection of bacteriophages with human health or diseases is less examined. Whole-virome analysis on inflammatory bowel disease (IBD) in conjunction with the internal dataset of ulcerative colitis was performed by Clooney et al. (2019), and the study showed that a stable group of virulent bacteriophages dominated in the healthy gut virome, while temperate bacteriophages replaced the virulent bacteriophages group in the gut with Crohn's disease. Equally important to the taxonomic annotation of human viromes, the functional annotation of human viral metagenomics allows the characterization of the roles of the virome in the host and ecosystem interactions. For example, Elbehery et al. (2018) built human virome protein cluster (HVPC) database viromes from human lung, gut (fecal), mouth, oropharynx, skin, and urine for a better functional annotation and diversity analysis of bronchoalveolar lavage viromes. The HVPC database also shed light on the potential impacts of phages on the horizontal gene transfer of bacterial virulence genes.

8.4.2 Water Virome

Viral ecology in aquatic environments has also been extensively studied and has provided important insight into the roles of viral communities in ecological and biogeochemical systems. The abundance of viruses, averaging 10 million per milliliter seawater on the ocean surface, is approximately tenfold higher than the co-occurring microorganisms of higher trophic levels (Breitbart et al. 2018). While microorganisms are directly engaged in Earth's biogeochemical cycles, viruses have important functions in modulating the biogeochemical processes via top-down control over microbial community and metabolic reprogramming (Hurwitz and U'Ren 2016; Howard-Varona et al. 2020). The advances and application of viromics have greatly boosted the field of environmental viruses, and great efforts have been made to elucidate viral community diversity, genetic composition, and associations with host populations. Continuous efforts have been centered on the distribution and variation of viral communities along spatial and temporal gradients for a better understanding of viral ecology in marine and other environments. To investigate the ecological drivers of viral community structure, Hurwitz and U'Ren (2016) introduced comparative metagenomics in the analysis of 32 viromes selected from the Pacific Ocean Virome dataset without the need for annotation and assembly, which combines shared *k*-mer analyses and regression modeling. The authors showed that geographic location, the distance from the shore, and depth were major predictors of community structure, and season, depth, and oxygen concentration were important factors shaping the viral communities at a specific ocean station. Similarly, a viromics study by Focardi et al. (2020) characterized bacteriophage community diversity in the dynamic east Australian current and showed distinct depth and regional patterns of viral diversity. In the meantime, diverse taxonomic groups of viruses have been reported, as well as large numbers of functional genes across different types of systems, all of which suggest the central roles of viral community in ecological and biogeochemical processes (Gregory et al. 2019; Ruiz-Perez et al. 2019).

8.4.3 Soil Virome

Research in soil viral ecology lags behind viral ecology research in other systems such as human and aquatic environments. However, soil virology has gained tremendous interest and has been a rapidly progressing scientific area. Only eight soil viromes and two Hypolite viromes were described in publications as of April 2017 (Williamson et al. 2017). However, more than 18 soil virome studies have been reported, since the number is still growing rapidly. The challenges in soil virome studies include extraction of viruses from soils and isolation of viral nucleotide acids (Trubl et al. 2019; Göller et al. 2020). Compared with the relatively uniform conditions in aquatic environments, soil systems are more complicated, considering the soils' heterogeneous structure (Ramirez et al. 2018; Roy et al. 2020; Liang, Regan et al., 2019). Previously limited soil virome studies have revealed diverse groups of ssDNA and dsDNA viruses were detected in soils and showed that the obtained viromes contained functional genes associated with host metabolism (e.g., membrane transport; iron acquisition and metabolism; cell signaling; and metabolism of nitrogen, sulfur, and aromatic compounds) (Han et al. 2017; Jin et al. 2019; Bi et al. 2020; Braga et al. 2020). In exploring the role of viruses in modulating soil bacterial community structure and function, Braga et al. (2020) showed that viruses were correlated with bacterial composition and diversity and had a positive relationship with the soil ammonium concentration of the potential roles of viruses in soil functions.

While NGS is a powerful technique in addressing fundamental questions, it is susceptible to constraints and biases (Lajoie and Kembel 2019). The technical limitations may be associated with the extraction/enrichment processes of viruses and nucleic acid, in which extractability and stability of nucleic acids differ. Moreover, introduced biases in the preparation of libraries and sequencing errors can also undermine the validity of the data. The comparability of gene abundance data among different experiments can be affected by high-level systematic variability in different experiments. The statistical challenges in comparative metagenomics include sequencing errors, read length-based biases, database biases, phylogenetic errors, average genome size, gene length, and PCR artifacts. Normalization in which these factors (systematic differences) are identified and removed is critical for data analysis. In the meantime, viromics only provides taxonomic information, and function prediction is limited depending on the content in the database. The insufficiency of viral sequence databases is also a big challenge for the application of viromics (Shkoporov and Hill 2019). The high concentration of viral DNA/RNA required for virome sequencing also poses challenges for low-biomass samples. Even specific functional genes are detected in viromes; they represent the potential for the specific functions that the viral community has, and further efforts are needed for ultimate verifications. Additionally, the coverage of discovered genes in metagenomic sequences is dependent on sequencing depth and community complexity, which may lead to an inability to completely sequence whole microorganisms in a designated sample. The high cost, potential contamination of DNA, and computationally intensive analysis can also affect the validity of this approach (Sutton, Clooney, and Hill 2020).

REFERENCES

Abdelkareem, Aly O., M. Khalil, M. Elaraby, H. Abbas, and A. Elbehery. 2018. "VirNet: Deep Attention Model for Viral Reads Identification." 2018 13th International Conference on Computer Engineering and Systems (ICCES). doi:10.1109/ICCES.2018.8639400.

Afiahayati, Kengo Sato, and Yasubumi Sakakibara. 2015. "MetaVelvet-SL: An Extension of the Velvet Assembler to a De Novo Metagenomic Assembler Utilizing Supervised Learning." *DNA Research: An International Journal for Rapid Publication of Reports on Genes and Genomes* 22 (1): 69–77. doi:10.1093/dnares/dsu041.

Aiewsakun, Pakorn, and Peter Simmonds. 2018. "The Genomic Underpinnings of Eukaryotic Virus Taxonomy: Creating a Sequence-Based Framework for Family-Level Virus Classification." *Microbiome* 6 (1): 38. doi:10.1186/s40168-018-0422-7.

Ajami, Nadim J., and Joseph F. Petrosino. 2015. "Chapter 9—Toward the Understanding of the Human Virome." In *Metagenomics for Microbiology*, edited by Jacques Izard and Maria C. Rivera, 135–43. Oxford: Academic Press. doi:10.1016/B978-0-12-410472-3.00009-9.

Almansour, Iman, Mazen Alhagri, Rahaf Alfares, Manal Alshehri, Razan Bakhashwain, and Ahmed Maarouf. 2019. "IRAM: Virus Capsid Database and Analysis Resource." *Database* 2019 (January). Oxford Academic. doi:10.1093/database/baz079.

Amgarten, Deyvid, Lucas P. P. Braga, Aline M. da Silva, and João C. Setubal. 2018. "MARVEL, a Tool for Prediction of Bacteriophage Sequences in Metagenomic Bins." *Frontiers in Genetics* 9. Frontiers. doi:10.3389/fgene.2018.00304.

Ansari, Mina Hojat, Mehregan Ebrahimi, Mohammad Reza Fattahi, Michael G. Gardner, Ali Reza Safarpour, Mohammad Ali Faghihi, and Kamran Bagheri Lankarani. 2020. "Viral Metagenomic Analysis of Fecal Samples Reveals an Enteric Virome Signature in Irritable Bowel Syndrome." *BMC Microbiology* 20 (1): 123. doi:10.1186/s12866-020-01817-4.

Antipov, Dmitry, Mikhail Raiko, Alla Lapidus, and Pavel A. Pevzner. 2020. "MetaviralSPAdes: Assembly of Viruses from Metagenomic Data." *Bioinformatics* 36 (14). Oxford Academic: 4126–29. doi:10.1093/bioinformatics/btaa490.

Arndt, David, Jason R. Grant, Ana Marcu, Tanvir Sajed, Allison Pon, Yongjie Liang, and David S. Wishart. 2016. "PHASTER: A Better, Faster Version of the PHAST Phage Search Tool." *Nucleic Acids Research* 44 (W1): W16–W21. doi:10.1093/nar/gkw387.

Babayan, Simon A., Richard J. Orton, and Daniel G. Streicker. 2018. "Predicting Reservoir Hosts and Arthropod Vectors from Evolutionary Signatures in RNA Virus Genomes." *Science* 362 (6414). American Association for the Advancement of Science: 577–80. doi:10.1126/science.aap9072.

Baizan-Edge, Amanda, Peter Cock, Stuart MacFarlane, Wendy McGavin, Lesley Torrance, and Susan Jones. 2019. "Kodoja: A Workflow for Virus Detection in Plants Using *k*-Mer Analysis of RNA-Sequencing Data." *Journal of General Virology* 100 (3): 533–42. doi:10.1099/jgv.0.001210.

Bankevich, Anton, Sergey Nurk, Dmitry Antipov, Alexey A. Gurevich, Mikhail Dvorkin, Alexander S. Kulikov, Valery M. Lesin, et al. 2012. "SPAdes: A New Genome Assembly Algorithm and Its Applications to Single-Cell Sequencing." *Journal of Computational Biology* 19 (5): 455–77. doi:10.1089/cmb.2012.0021.

Bi, Li, Dan-Ting Yu, Shuai Du, Li-Mei Zhang, Li-Yu Zhang, Chuan-Fa Wu, Chao Xiong, Li-Li Han, and Ji-Zheng He. 2020. "Diversity and Potential Biogeochemical Impacts of Viruses in Bulk and Rhizosphere Soils." *Environmental Microbiology* n/a (n/a). Accessed November 24. https://doi.org/10.1111/1462-2920.15010.

Blomström, Anne-Lie, Xingyu Ye, Caroline Fossum, Per Wallgren, and Mikael Berg. 2018. "Characterisation of the Virome of Tonsils from Conventional Pigs and from Specific Pathogen-Free Pigs." *Viruses* 10 (7). doi:10.3390/v10070382.

Bolduc, Benjamin, Ho Bin Jang, Guilhem Doulcier, Zhi-Qiang You, Simon Roux, and Matthew B. Sullivan. 2017. "VConTACT: An IVirus Tool to Classify Double-Stranded DNA Viruses That Infect Archaea and Bacteria." *PeerJ* 5 (May). doi:10.7717/peerj.3243.

Bosch, Astrid A. T. M., Giske Biesbroek, Krzysztof Trzcinski, Elisabeth A. M. Sanders, and Debby Bogaert. 2013. "Viral and Bacterial Interactions in the Upper Respiratory Tract." Edited by Tom C. Hobman. *PLoS Pathogens* 9 (1): e1003057. doi:10.1371/journal.ppat.1003057.

Braga, Lucas P. P., Aymé Spor, Witold Kot, Marie-Christine Breuil, Lars H. Hansen, João C. Setubal, and Laurent Philippot. 2020. "Impact of Phages on Soil Bacterial Communities and Nitrogen Availability under Different Assembly Scenarios." *Microbiome* 8 (1): 52. doi:10.1186/s40168-020-00822-z.

Breitbart, Mya, Chelsea Bonnain, Kema Malki, and Natalie A. Sawaya. 2018. "Phage Puppet Masters of the Marine Microbial Realm." *Nature Microbiology* 3 (7). Nature Publishing Group: 754–66. doi:10.1038/s41564-018-0166-y.

Breitbart, Mya, Matthew Haynes, Scott Kelley, Florent Angly, Robert A. Edwards, Ben Felts, Joseph M. Mahaffy, et al. 2008. "Viral Diversity and Dynamics in an Infant Gut." *Research in Microbiology* 159 (5): 367–73. doi:10.1016/j.resmic.2008.04.006.

Brister, J. Rodney, Danso Ako-adjei, Yiming Bao, and Olga Blinkova. 2015. "NCBI Viral Genomes Resource." *Nucleic Acids Research* 43 (D1): D571–77. doi:10.1093/nar/gku1207.

Bzhalava, Zurab, Ardi Tampuu, Piotr Bała, Raul Vicente, and Joakim Dillner. 2018. "Machine Learning for Detection of Viral Sequences in Human Metagenomic Datasets." *BMC Bioinformatics* 19 (1): 336. doi:10.1186/s12859-018-2340-x.

Cadwell, Ken. 2015. "The Virome in Host Health and Disease." *Immunity* 42 (5): 805–13. doi:10.1016/j.immuni.2015.05.003.

Castelán-Sánchez, Hugo G., Itzel Lopéz-Rosas, Wendy A. García-Suastegui, Raúl Peralta, Alan D. W. Dobson, Ramón Alberto Batista-García, and Sonia Dávila-Ramos. 2019. "Extremophile Deep-Sea Viral Communities from Hydrothermal Vents: Structural and Functional Analysis." *Marine Genomics* 46 (August): 16–28. doi:10.1016/j.margen.2019.03.001.

Cho, Ilseung, and Martin J. Blaser. 2012. "The Human Microbiome: At the Interface of Health and Disease." *Nature Reviews. Genetics* 13 (4): 260–70. doi:10.1038/nrg3182.

Clooney, Adam G., Thomas D. S. Sutton, Andrey N. Shkoporov, Ross K. Holohan, Karen M. Daly, Orla O'Regan, Feargal J. Ryan, et al. 2019. "Whole-Virome Analysis Sheds Light on Viral Dark Matter in Inflammatory Bowel Disease." *Cell Host & Microbe* 26 (6): 764–78.e5. doi:10.1016/j.chom.2019.10.009.

Cobían Güemes, Ana Georgina. 2019. "Exploring the Global Virome and Deciphering the Role of Phages in Cystic Fibrosis." UC San Diego. https://escholarship.org/uc/item/1103w639.

Cobián Güemes, Ana Georgina, Merry Youle, Vito Adrian Cantú, Ben Felts, James Nulton, and Forest Rohwer. 2016. "Viruses as Winners in the Game of Life." *Annual Review of Virology* 3 (1): 197–214. doi:10.1146/annurev-virology-100114-054952.

Colombet, J., A. Robin, L. Lavie, Y. Bettarel, H. M. Cauchie, and T. Sime-Ngando. 2007. "Virioplankton 'Pegylation': Use of PEG (Polyethylene Glycol) to Concentrate and Purify Viruses in Pelagic Ecosystems." *Journal of Microbiological Methods* 71 (3): 212–19. doi:10.1016/j.mimet.2007.08.012.

Conceição-Neto, Nádia, Mark Zeller, Hanne Lefrère, Pieter De Bruyn, Leen Beller, Ward Deboutte, Claude Kwe Yinda, et al. 2015. "Modular Approach to Customise Sample Preparation Procedures for Viral Metagenomics: A Reproducible Protocol for Virome Analysis." *Scientific Reports* 5 (1). Nature Publishing Group: 16532. doi:10.1038/srep16532.

Corinaldesi, Cinzia, Michael Tangherlini, and Antonio Dell'Anno. 2017. "From Virus Isolation to Metagenome Generation for Investigating Viral Diversity in Deep-Sea Sediments." *Scientific Reports* 7 (1). Nature Publishing Group: 8355. doi:10.1038/s41598-017-08783-4.

Coutinho, Felipe Hernandes, Gustavo Bueno Gregoracci, Juline Marta Walter, Cristiane Carneiro Thompson, and Fabiano L. Thompson. 2018. "Metagenomics Sheds Light on the Ecology of Marine Microbes and Their Viruses." *Trends in Microbiology* 26 (11). Elsevier: 955–65. doi:10.1016/j.tim.2018.05.015.

Dávila-Ramos, Sonia, Hugo G. Castelán-Sánchez, Liliana Martínez-Ávila, María del Rayo Sánchez-Carbente, Raúl Peralta, Armando Hernández-Mendoza, Alan D. W. Dobson, Ramón A. Gonzalez, Nina Pastor, and Ramón Alberto Batista-García. 2019. "A Review on Viral Metagenomics in Extreme Environments." *Frontiers in Microbiology* 10. Frontiers. doi:10.3389/fmicb.2019.02403.

Dell'Anno, Antonio, Cinzia Corinaldesi, and Roberto Danovaro. 2015. "Virus Decomposition Provides an Important Contribution to Benthic Deep-Sea Ecosystem Functioning." *Proceedings of the National Academy of Sciences* 112 (16). National Academy of Sciences: E2014–19. doi:10.1073/pnas.1422234112.

Deng, Ling, Ronalds Silins, Josué L. Castro-Mejía, Witold Kot, Leon Jessen, Jonathan Thorsen, Shiraz Shah, et al. 2019. "A Protocol for Extraction of Infective Viromes Suitable for Metagenomics Sequencing from Low Volume Fecal Samples." *Viruses* 11 (7). Multidisciplinary Digital Publishing Institute: 667. doi:10.3390/v11070667.

Elbehery, Ali H. A., Judith Feichtmayer, Dave Singh, Christian Griebler, and Li Deng. 2018. "The Human Virome Protein Cluster Database (HVPC): A Human Viral Metagenomic Database for Diversity and Function Annotation." *Frontiers in Microbiology* 9. Frontiers. doi:10.3389/fmicb.2018.01110.

Emerson, Joanne B., Simon Roux, Jennifer R. Brum, Benjamin Bolduc, Ben J. Woodcroft, Ho Bin Jang, Caitlin M. Singleton, et al. 2018. "Host-Linked Soil Viral Ecology along a Permafrost Thaw Gradient." *Nature Microbiology* 3 (8). Nature Publishing Group: 870–80. doi:10.1038/s41564-018-0190-y.

Fang, Zhencheng, Jie Tan, Shufang Wu, Mo Li, Congmin Xu, Zhongjie Xie, and Huaiqiu Zhu. 2019. "PPR-Meta: A Tool for Identifying Phages and Plasmids from Metagenomic Fragments Using Deep Learning." *GigaScience* 8 (6). Oxford Academic. doi:10.1093/gigascience/giz066.

Flygare, Steven, Keith Simmon, Chase Miller, Yi Qiao, Brett Kennedy, Tonya Di Sera, Erin H. Graf, et al. 2016. "Taxonomer: An Interactive Metagenomics Analysis Portal for Universal Pathogen Detection and Host MRNA Expression Profiling." *Genome Biology* 17 (1): 111. doi:10.1186/s13059-016-0969-1.

Focardi, Amaranta, Martin Ostrowski, Kirianne Goossen, Mark V. Brown, and Ian Paulsen. 2020. "Investigating the Diversity of Marine Bacteriophage in Contrasting Water Masses Associated with the East Australian Current (EAC) System." *Viruses* 12 (3). Multidisciplinary Digital Publishing Institute: 317. doi:10.3390/v12030317.

Galiez, Clovis, Matthias Siebert, François Enault, Jonathan Vincent, and Johannes Söding. 2017. "WIsH: Who Is the Host? Predicting Prokaryotic Hosts from Metagenomic Phage Contigs." *Bioinformatics* 33 (19). Oxford Academic: 3113–14. doi:10.1093/bioinformatics/btx383.

Garretto, Andrea, Thomas Hatzopoulos, and Catherine Putonti. 2019. "VirMine: Automated Detection of Viral Sequences from Complex Metagenomic Samples." *PeerJ* 7 (April). PeerJ Inc.: e6695. doi:10.7717/peerj.6695.

Göller, Pauline C., Jose M. Haro-Moreno, Francisco Rodriguez-Valera, Martin J. Loessner, and Elena Gómez-Sanz. 2020. "Uncovering a Hidden Diversity: Optimized Protocols for the Extraction of DsDNA Bacteriophages from Soil." *Microbiome* 8 (1): 17. doi:10.1186/s40168-020-0795-2.

Gong, Zheng, Yantao Liang, Min Wang, Yong Jiang, Qingwei Yang, Jun Xia, Xinhao Zhou, et al. 2018. "Viral Diversity and Its Relationship with Environmental Factors at the Surface and Deep Sea of Prydz Bay, Antarctica." *Frontiers in Microbiology* 9. Frontiers. doi:10.3389/fmicb.2018.02981.

Graham, Emily B., David Paez-Espino, Colin Brislawn, Russell Y. Neches, Kirsten S. Hofmockel, Ruonan Wu, Nikos C. Kyrpides, Janet K. Jansson, and Jason E. McDermott. 2019. "Untapped Viral Diversity in Global Soil Metagenomes." *BioRxiv*, March. Cold Spring Harbor Laboratory: 583997. doi:10.1101/583997.

Gregory, Ann C., Ahmed A. Zayed, Nádia Conceição-Neto, Ben Temperton, Ben Bolduc, Adriana Alberti, Mathieu Ardyna, et al. 2019. "Marine DNA Viral Macro- and Microdiversity from Pole to Pole." *Cell* 177 (5): 1109–123.e14. doi:10.1016/j.cell.2019.03.040.

Greninger, Alexander L. 2018. "A Decade of RNA Virus Metagenomics Is (Not) Enough." *Virus Research* 244 (January): 218–29. doi:10.1016/j.virusres.2017.10.014.

Han, Li-Li, Dan-Ting Yu, Li-Mei Zhang, Ju-Pei Shen, and Ji-Zheng He. 2017. "Genetic and Functional Diversity of Ubiquitous DNA Viruses in Selected Chinese Agricultural Soils." *Scientific Reports* 7 (1). Nature Publishing Group: 45142. doi:10.1038/srep45142.

Ho, Thien, and Ioannis E. Tzanetakis. 2014. "Development of a Virus Detection and Discovery Pipeline Using Next Generation Sequencing." *Virology* 471–473 (December): 54–60. doi:10.1016/j.virol.2014.09.019.

Howard-Varona, Cristina, Morgan M. Lindback, G. Eric Bastien, Natalie Solonenko, Ahmed A. Zayed, HoBin Jang, Bill Andreopoulos, et al. 2020. "Phage-Specific Metabolic Reprogramming of Virocells." *The ISME Journal* 14 (4). Nature Publishing Group: 881–95. doi:10.1038/s41396-019-0580-z.

Hulo, Chantal, Edouard de Castro, Patrick Masson, Lydie Bougueleret, Amos Bairoch, Ioannis Xenarios, and Philippe Le Mercier. 2011. "ViralZone: A Knowledge Resource to Understand Virus Diversity." *Nucleic Acids Research* 39 (Database issue): D576–82. doi:10.1093/nar/gkq901.

Hurwitz, Bonnie L., and Jana M U'Ren. 2016. "Viral Metabolic Reprogramming in Marine Ecosystems." *Current Opinion in Microbiology*, Environmental Microbiology * Special Section: Megaviromes 31 (June): 161–68. doi:10.1016/j.mib.2016.04.002.

Huson, D. H., A. F. Auch, J. Qi, and S. C. Schuster. 2007. "MEGAN Analysis of Metagenomic Data." *Genome Research* 17 (3): 377–86. doi:10.1101/gr.5969107.

Jin, Min, Xun Guo, Rui Zhang, Wu Qu, Boliang Gao, and Runying Zeng. 2019. "Diversities and Potential Biogeochemical Impacts of Mangrove Soil Viruses." *Microbiome* 7 (1): 58. doi:10.1186/s40168-019-0675-9.

John, Seth G., Carolina B. Mendez, Li Deng, Bonnie Poulos, Anne Kathryn M. Kauffman, Suzanne Kern, Jennifer Brum, Martin F. Polz, Edward A. Boyle, and Matthew B. Sullivan. 2011. "A Simple and Efficient Method for Concentration of Ocean Viruses by Chemical Flocculation." *Environmental Microbiology Reports* 3 (2): 195–202. doi:10.1111/j.1758-2229.2010.00208.x.

Jurtz, Vanessa Isabell, Julia Villarroel, Ole Lund, Mette Voldby Larsen, and Morten Nielsen. 2016. "MetaPhinder—Identifying Bacteriophage Sequences in Metagenomic Data Sets." *PLoS One* 11 (9). Public Library of Science: e0163111. doi:10.1371/journal.pone.0163111.

Kanehisa, Minoru, Yoko Sato, and Kanae Morishima. 2016. "BlastKOALA and GhostKOALA: KEGG Tools for Functional Characterization of Genome and Metagenome Sequences." *Journal of Molecular Biology* 428 (4). Computation Resources for Molecular Biology: 726–31. doi:10.1016/j.jmb.2015.11.006.

Keegan, Kevin P., Elizabeth M. Glass, and Folker Meyer. 2016. "MG-RAST, a Metagenomics Service for Analysis of Microbial Community Structure and Function." In *Microbial Environmental Genomics (MEG)*, edited by Francis Martin and Stephane Uroz, 207–33. Methods in Molecular Biology. New York, NY: Springer. doi:10.1007/978-1-4939-3369-3_13.

Khot, Varada, Marc Strous, and Alyse K. Hawley. 2020. "Computational Approaches in Viral Ecology." *Computational and Structural Biotechnology Journal* 18 (January): 1605–12. doi:10.1016/j.csbj.2020.06.019.

Kieft, Kristopher, Zhichao Zhou, and Karthik Anantharaman. 2020. "VIBRANT: Automated Recovery, Annotation and Curation of Microbial Viruses, and Evaluation of Viral Community Function from Genomic Sequences." *Microbiome* 8 (1): 90. doi:10.1186/s40168-020-00867-0.

Kim, Daehwan, Li Song, Florian P. Breitwieser, and Steven L. Salzberg. 2016. "Centrifuge: Rapid and Sensitive Classification of Metagenomic Sequences." *Genome Research* 26 (12): 1721–29. doi:10.1101/gr.210641.116.

Koonin, Eugene V., Valerian V. Dolja, Mart Krupovic, Arvind Varsani, Yuri I. Wolf, Natalya Yutin, F. Murilo Zerbini, and Jens H. Kuhn. 2020. "Global Organization and Proposed Megataxonomy of the Virus World." *Microbiology and Molecular Biology Reviews* 84 (2). American Society for Microbiology. doi:10.1128/MMBR.00061-19.

Kreuze, Jan F., Ana Perez, Milton Untiveros, Dora Quispe, Segundo Fuentes, Ian Barker, and Reinhard Simon. 2009. "Complete Viral Genome Sequence and Discovery of Novel Viruses by Deep Sequencing of Small RNAs: A Generic Method for Diagnosis, Discovery and Sequencing of Viruses." *Virology* 388 (1): 1–7. doi:10.1016/j.virol.2009.03.024.

Kwok, Kirsty T. T., David F. Nieuwenhuijse, My V. T. Phan, and Marion P. G. Koopmans. 2020. "Virus Metagenomics in Farm Animals: A Systematic Review." *Viruses* 12 (1). doi:10.3390/v12010107.

Laffy, Patrick W., Elisha M. Wood-Charlson, Dmitrij Turaev, Karen D. Weynberg, Emmanuelle S. Botté, Madeleine J. H. van Oppen, Nicole S. Webster, and Thomas Rattei. 2016. "HoloVir: A Workflow for Investigating the Diversity and Function of Viruses in Invertebrate Holobionts." *Frontiers in Microbiology* 7 (June). doi:10.3389/fmicb.2016.00822.

Lajoie, Geneviève, and Steven W. Kembel. 2019. "Making the Most of Trait-Based Approaches for Microbial Ecology." *Trends in Microbiology* 27 (10). Elsevier: 814–23. doi:10.1016/j.tim.2019.06.003.

Langmead, Ben, and Steven L. Salzberg. 2012. "Fast Gapped-Read Alignment with Bowtie 2." *Nature Methods* 9 (4): 357–59. doi:10.1038/nmeth.1923.

Lefebvre, Marie, Sébastien Theil, Yuxin Ma, and Thierry Candresse. 2019. "The VirAnnot Pipeline: A Resource for Automated Viral Diversity Estimation and Operational Taxonomy Units Assignation for Virome Sequencing Data." *Phytobiomes Journal* 3 (4). Scientific Societies: 256–59. doi:10.1094/PBIOMES-07-19-0037-A.

Leplae, Raphaël, Gipsi Lima-Mendez, and Ariane Toussaint. 2010. "ACLAME: A CLAssification of Mobile Genetic Elements, Update 2010." *Nucleic Acids Research* 38 (Database issue): D57–61. doi:10.1093/nar/gkp938.

Li, Dinghua, Chi-Man Liu, Ruibang Luo, Kunihiko Sadakane, and Tak-Wah Lam. 2015. "MEGAHIT: An Ultra-Fast Single-Node Solution for Large and Complex Metagenomics Assembly via Succinct de Bruijn Graph." *Bioinformatics* 31 (10): 1674–76. doi:10.1093/bioinformatics/btv033.

Li, Yang, Hao Wang, Kai Nie, Chen Zhang, Yi Zhang, Ji Wang, Peihua Niu, and Xuejun Ma. 2016. "VIP: An Integrated Pipeline for Metagenomics of Virus Identification and Discovery." *Scientific Reports* 6 (1). Nature Publishing Group: 23774. doi:10.1038/srep23774.

Liang, Xiaolong, Regan E. Wagner, Jie Zhuang, Jennifer M. DeBruyn, Steven W. Wilhelm, Fang Liu, Lu Yang, Margaret E. Staton, Andrew C. Sherfy, and Mark Radosevich. 2019. "Viral Abundance and Diversity Vary with Depth in a Southeastern United States Agricultural Ultisol." *Soil Biology and Biochemistry* 137 (October): 107546. doi:10.1016/j.soilbio.2019.107546.

Liang, Xiaolong, Yingyue Zhang, K. Eric Wommack, Steven W. Wilhelm, Jennifer M. DeBruyn, Andrew C. Sherfy, Jie Zhuang, and Mark Radosevich. 2020. "Lysogenic Reproductive Strategies of Viral Communities Vary with Soil Depth and Are Correlated with Bacterial Diversity." *Soil Biology and Biochemistry* 144 (May): 107767. doi:10.1016/j.soilbio.2020.107767.

Liang, Xiaolong, Jie Zhuang, Frank E. Löffler, Yingyue Zhang, Jennifer M. DeBruyn, Steven W. Wilhelm, Sean M. Schaeffer, and Mark Radosevich. 2019. "Viral and Bacterial Community Responses to Stimulated Fe(III)-Bioreduction during Simulated Subsurface Bioremediation." *Environmental Microbiology* 21 (6): 2043–55. https://doi.org/10.1111/1462-2920.14566.

Lin, Hsin-Hung, and Yu-Chieh Liao. 2017. "DrVM: A New Tool for Efficient Genome Assembly of Known Eukaryotic Viruses from Metagenomes." *GigaScience* 6 (2). doi:10.1093/gigascience/gix003.

Lorenzi, Hernan A., Jeff Hoover, Jason Inman, Todd Safford, Sean Murphy, Leonid Kagan, and Shannon J. Williamson. 2011. "TheViral MetaGenome Annotation Pipeline(VMGAP):An Automated Tool for the Functional Annotation of Viral Metagenomic Shotgun Sequencing Data." *Standards in Genomic Sciences* 4 (3): 418–29. doi:10.4056/sigs.1694706.

Maabar, Maha, Andrew J. Davison, Matej Vučak, Fiona Thorburn, Pablo R. Murcia, Rory Gunson, Massimo Palmarini, and Joseph Hughes. 2019. "DisCVR: Rapid Viral Diagnosis from High-Throughput Sequencing Data." *Virus Evolution* 5 (2). Oxford Academic. doi:10.1093/ve/vez033.

Marquet, Mike, Martin Hölzer, Mathias W. Pletz, Adrian Viehweger, Oliwia Makarewicz, Ralf Ehricht, and Christian Brandt. 2020. "What the Phage: A Scalable Workflow for the Identification and Analysis of Phage Sequences." Preprint. *Bioinformatics*. doi:10.1101/2020.07.24.219899.

Maslov, Sergei, and Kim Sneppen. 2017. "Population Cycles and Species Diversity in Dynamic Kill-the-Winner Model of Microbial Ecosystems." *Scientific Reports* 7 (1). Nature Publishing Group: 39642. doi:10.1038/srep39642.

McNair, Katelyn, Carol Zhou, Elizabeth A. Dinsdale, Brian Souza, and Robert A. Edwards. 2019. "PHANOTATE: A Novel Approach to Gene Identification in Phage Genomes." *Bioinformatics* 35 (22). Oxford Academic: 4537–42. doi:10.1093/bioinformatics/btz265.

Meier-Kolthoff, Jan P., and Markus Göker. 2017. "VICTOR: Genome-Based Phylogeny and Classification of Prokaryotic Viruses." Edited by Janet Kelso. *Bioinformatics* 33 (21): 3396–404. doi:10.1093/bioinformatics/btx440.

Menzel, Peter, Kim Lee Ng, and Anders Krogh. 2016. "Fast and Sensitive Taxonomic Classification for Metagenomics with Kaiju." *Nature Communications* 7 (1). Nature Publishing Group: 11257. doi:10.1038/ncomms11257.

Mihara, Tomoko, Yosuke Nishimura, Yugo Shimizu, Hiroki Nishiyama, Genki Yoshikawa, Hideya Uehara, Pascal Hingamp, Susumu Goto, and Hiroyuki Ogata. 2016. "Linking Virus Genomes with Host Taxonomy." *Viruses* 8 (3). Multidisciplinary Digital Publishing Institute: 66. doi:10.3390/v8030066.

Mock, Florian, Adrian Viehweger, Emanuel Barth, and Manja Marz. 2020. "VIDHOP, Viral Host Prediction with Deep Learning." *Bioinformatics*. Accessed November 24. doi:10.1093/bioinformatics/btaa705.

Modha, Sejal, Anil S. Thanki, Susan F. Cotmore, Andrew J. Davison, and Joseph Hughes. 2018. "ViCTree: An Automated Framework for Taxonomic Classification from Protein Sequences." *Bioinformatics (Oxford, England)* 34 (13): 2195–200. doi:10.1093/bioinformatics/bty099.

Moon, Kira, Jeong Ho Jeon, Ilnam Kang, Kwang Seung Park, Kihyun Lee, Chang-Jun Cha, Sang Hee Lee, and Jang-Cheon Cho. 2020. "Freshwater Viral Metagenome Reveals Novel and Functional Phage-Borne Antibiotic Resistance Genes." *Microbiome* 8 (1): 75. doi:10.1186/s40168-020-00863-4.

Muthukumar, Vijay, Ulrich Melcher, Marlee Pierce, Graham B. Wiley, Bruce A. Roe, Michael W. Palmer, Vaskar Thapa, Akhtar Ali, and Tao Ding. 2009. "Non-Cultivated Plants of the Tallgrass Prairie Preserve of Northeastern Oklahoma Frequently Contain Virus-Like Sequences in Particulate Fractions." *Virus Research* 141 (2). Plant Virus Epidemiology: Controlling Epidemics of Emerging and Established Plant Viruses—the Way Forward: 169–73. doi:10.1016/j.virusres.2008.06.016.

Naccache, Samia N., Scot Federman, Narayanan Veeraraghavan, Matei Zaharia, Deanna Lee, Erik Samayoa, Jerome Bouquet, et al. 2014. "A Cloud-Compatible Bioinformatics Pipeline for Ultrarapid Pathogen Identification from Next-Generation Sequencing of Clinical Samples." *Genome Research* 24 (7): 1180–92. doi:10.1101/gr.171934.113.

Namiki, Toshiaki, Tsuyoshi Hachiya, Hideaki Tanaka, and Yasubumi Sakakibara. 2012. "MetaVelvet: An Extension of Velvet Assembler to De Novo Metagenome Assembly from Short Sequence Reads." *Nucleic Acids Research* 40 (20): e155. doi:10.1093/nar/gks678.

Nayfach, Stephen, Antonio Pedro Camargo, Emiley Eloe-Fadrosh, Simon Roux, and Nikos Kyrpides. 2020. "CheckV: Assessing the Quality of Metagenome-Assembled Viral Genomes." Preprint. *Bioinformatics*. doi:10.1101/2020.05.06.081778.

Nishimura, Yosuke, Takashi Yoshida, Megumi Kuronishi, Hideya Uehara, Hiroyuki Ogata, and Susumu Goto. 2017. "ViPTree: The Viral Proteomic Tree Server." *Bioinformatics* 33 (15). Oxford Academic: 2379–80. doi:10.1093/bioinformatics/btx157.

Niu, Beifang, Zhengwei Zhu, Limin Fu, Sitao Wu, and Weizhong Li. 2011. "FR-HIT, a Very Fast Program to Recruit Metagenomic Reads to Homologous Reference Genomes." *Bioinformatics* 27 (12): 1704–705. doi:10.1093/bioinformatics/btr252.

Nooij, Sam, Dennis Schmitz, Harry Vennema, Annelies Kroneman, and Marion P. G. Koopmans. 2018. "Overview of Virus Metagenomic Classification Methods and Their Biological Applications." *Frontiers in Microbiology* 9. Frontiers. doi:10.3389/fmicb.2018.00749.

Ounit, Rachid, Steve Wanamaker, Timothy J. Close, and Stefano Lonardi. 2015. "CLARK: Fast and Accurate Classification of Metagenomic and Genomic Sequences Using Discriminative *k*-Mers." *BMC Genomics* 16 (1): 236. doi:10.1186/s12864-015-1419-2.

Paez-Espino, David, Emiley A. Eloe-Fadrosh, Georgios A. Pavlopoulos, Alex D. Thomas, Marcel Huntemann, Natalia Mikhailova, Edward Rubin, Natalia N. Ivanova, and Nikos C. Kyrpides. 2016. "Uncovering Earth's Virome." *Nature* 536 (7617). Nature Publishing Group: 425–30. doi:10.1038/nature19094.

Papaianni, Marina, Paola Cuomo, Andrea Fulgione, Donatella Albanese, Monica Gallo, Debora Paris, Andrea Motta, Domenico Iannelli, and Rosanna Capparelli. 2020. "Bacteriophages Promote Metabolic Changes in Bacteria Biofilm." *Microorganisms* 8 (4). doi:10.3390/microorganisms8040480.

Parras-Moltó, Marcos, Ana Rodríguez-Galet, Patricia Suárez-Rodríguez, and Alberto López-Bueno. 2018. "Evaluation of Bias Induced by Viral Enrichment and Random Amplification Protocols in Metagenomic Surveys of Saliva DNA Viruses." *Microbiome* 6 (1): 119. doi:10.1186/s40168-018-0507-3.

Peng, Y., H. C. M. Leung, S. M. Yiu, and F. Y. L. Chin. 2012. "IDBA-UD: A De Novo Assembler for Single-Cell and Metagenomic Sequencing Data with Highly Uneven Depth." *Bioinformatics* 28 (11): 1420–28. doi:10.1093/bioinformatics/bts174.

Pierce, N. Tessa, Luiz Irber, Taylor Reiter, Phillip Brooks, and C. Titus Brown. 2019. "Large-Scale Sequence Comparisons with Sourmash." *F1000Research* 8 (July). doi:10.12688/f1000research.19675.1.

Ponsero, Alise J., and Bonnie L. Hurwitz. 2019. "The Promises and Pitfalls of Machine Learning for Detecting Viruses in Aquatic Metagenomes." *Frontiers in Microbiology* 10. Frontiers. doi:10.3389/fmicb.2019.00806.

Ponstingl, H., and Zemin Ning. 2010. "SMALT—A New Mapper for DNA Sequencing Reads." doi:10.7490/F1000RESEARCH.327.1.

Pratama, Akbar Adjie, and Jan Dirk van Elsas. 2018. "The 'Neglected' Soil Virome—Potential Role and Impact." *Trends in Microbiology* 26 (8). Elsevier: 649–62. doi:10.1016/j.tim.2017.12.004.

Ramirez, Kelly S., Christopher G. Knight, Mattias de Hollander, Francis Q. Brearley, Bede Constantinides, Anne Cotton, Si Creer, et al. 2018. "Detecting Macroecological Patterns in Bacterial Communities across Independent Studies of Global Soils." *Nature Microbiology* 3 (2). Nature Publishing Group: 189–96. doi:10.1038/s41564-017-0062-x.

Ramos-Barbero, María Dolores, José M. Martínez, Cristina Almansa, Nuria Rodríguez, Judith Villamor, María Gomariz, Cristina Escudero, et al. 2019. "Prokaryotic and Viral Community Structure in the Singular Chaotropic Salt Lake Salar de Uyuni." *Environmental Microbiology* 21 (6): 2029–42. https://doi.org/10.1111/1462-2920.14549.

Rampelli, Simone, Matteo Soverini, Silvia Turroni, Sara Quercia, Elena Biagi, Patrizia Brigidi, and Marco Candela. 2016. "ViromeScan: A New Tool for Metagenomic Viral Community Profiling." *BMC Genomics* 17 (148). BioMed Central Ltd.

Ren, Jie, Nathan A. Ahlgren, Yang Young Lu, Jed A. Fuhrman, and Fengzhu Sun. 2017. "VirFinder: A Novel *k*-Mer Based Tool for Identifying Viral Sequences from Assembled Metagenomic Data." *Microbiome* 5 (1): 69. doi:10.1186/s40168-017-0283-5.

Ren, Jie, Kai Song, Chao Deng, Nathan A. Ahlgren, Jed A. Fuhrman, Yi Li, Xiaohui Xie, Ryan Poplin, and Fengzhu Sun. 2020. "Identifying Viruses from Metagenomic Data Using Deep Learning." *Quantitative Biology* 8 (1): 64–77. doi:10.1007/s40484-019-0187-4.

Reyes, Alejandro, Nicholas P. Semenkovich, Katrine Whiteson, Forest Rohwer, and Jeffrey I. Gordon. 2012. "Going Viral: Next-Generation Sequencing Applied to Phage Populations in the Human Gut." *Nature Reviews. Microbiology* 10 (9): 607–17. doi:10.1038/nrmicro2853.

Roossinck, Marilyn J. 2015. "Plants, Viruses and the Environment: Ecology and Mutualism." *Virology*, 60th Anniversary Issue 479–480 (May): 271–77. doi:10.1016/j.virol.2015.03.041.

Rosario, Karyna, and Mya Breitbart. 2011. "Exploring the Viral World through Metagenomics." *Current Opinion in Virology*, Vaccines/Viral genomics 1 (4): 289–97. doi:10.1016/j.coviro.2011.06.004.

Roux, Simon, Evelien M. Adriaenssens, Bas E. Dutilh, Eugene V. Koonin, Andrew M. Kropinski, Mart Krupovic, Jens H. Kuhn, et al. 2019. "Minimum Information about an Uncultivated Virus Genome (MIUViG)." *Nature Biotechnology* 37 (1). Nature Publishing Group: 29–37. doi:10.1038/nbt.4306.

Roux, Simon, Joanne B. Emerson, Emiley A. Eloe-Fadrosh, and Matthew B. Sullivan. 2017. "Benchmarking Viromics: An In Silico Evaluation of Metagenome-Enabled Estimates of Viral Community Composition and Diversity." *PeerJ* 5 (September). PeerJ Inc.: e3817. doi:10.7717/peerj.3817.

Roux, Simon, Francois Enault, Bonnie L. Hurwitz, and Matthew B. Sullivan. 2015. "VirSorter: Mining Viral Signal from Microbial Genomic Data." *PeerJ* 3 (May). PeerJ Inc.: e985. doi:10.7717/peerj.985.

Roux, Simon, Jeremy Tournayre, Antoine Mahul, Didier Debroas, and François Enault. 2014. "Metavir 2: New Tools for Viral Metagenome Comparison and Assembled Virome Analysis." *BMC Bioinformatics* 15 (1): 76. doi:10.1186/1471-2105-15-76.

Roy, Krishnakali, Dhritiman Ghosh, Jennifer M. DeBruyn, Tirthankar Dasgupta, K. Eric Wommack, Xiaolong Liang, Regan E. Wagner, and Mark Radosevich. 2020. "Temporal Dynamics of Soil Virus and Bacterial Populations in Agricultural and Early Plant Successional Soils." *Frontiers in Microbiology* 11. Frontiers. doi:10.3389/fmicb.2020.01494.

Ruiz-Perez, Carlos A., Despina Tsementzi, Janet K. Hatt, Matthew B. Sullivan, and Konstantinos T. Konstantinidis. 2019. "Prevalence of Viral Photosynthesis Genes along a Freshwater to Saltwater Transect in Southeast USA." *Environmental Microbiology Reports* 11 (5): 672–89. https://doi.org/10.1111/1758-2229.12780.

Santiago-Rodriguez, Tasha M., and Emily B. Hollister. 2019. "Human Virome and Disease: High-Throughput Sequencing for Virus Discovery, Identification of Phage-Bacteria Dysbiosis and Development of Therapeutic Approaches with Emphasis on the Human Gut." *Viruses* 11 (7). doi:10.3390/v11070656.

Sathiamoorthy, Sarmitha, Rebecca J. Malott, Lucy Gisonni-Lex, and Siemon H. S. Ng. 2018. "Selection and Evaluation of an Efficient Method for the Recovery of Viral Nucleic Acids from Complex Biologicals." *NPJ Vaccines* 3 (1). Nature Publishing Group: 1–6. doi:10.1038/s41541-018-0067-3.

Shi, Chenyan, Leen Beller, Ward Deboutte, Kwe Claude Yinda, Leen Delang, Anubis Vega-Rúa, Anna-Bella Failloux, and Jelle Matthijnssens. 2019. "Stable Distinct Core Eukaryotic Viromes in Different Mosquito Species from Guadeloupe, Using Single Mosquito Viral Metagenomics." *Microbiome* 7 (1): 121. doi:10.1186/s40168-019-0734-2.

Shkoporov, Andrey N., Adam G. Clooney, Thomas D. S. Sutton, Feargal J. Ryan, Karen M. Daly, James A. Nolan, Siobhan A. McDonnell, et al. 2019. "The Human Gut Virome Is Highly Diverse, Stable, and Individual Specific." *Cell Host & Microbe* 26 (4): 527–41.e5. doi:10.1016/j.chom.2019.09.009.

Shkoporov, Andrey N., and Colin Hill. 2019. "Bacteriophages of the Human Gut: The 'Known Unknown' of the Microbiome." *Cell Host & Microbe* 25 (2): 195–209. doi:10.1016/j.chom.2019.01.017.

Silveira, Cynthia B., and Forest L. Rohwer. 2016. "Piggyback-the-Winner in Host-Associated Microbial Communities." *NPJ Biofilms and Microbiomes* 2 (1). Nature Publishing Group: 1–5. doi:10.1038/npjbiofilms.2016.10.

Stano, Matej, G. Beke, and L. Klucar. 2016. "ViruSITE—Integrated Database for Viral Genomics." *Database J. Biol. Databases Curation.* doi:10.1093/database/baw162.

Steward, G., and Alexander I. Culley. 2010. "Extraction and Purification of Nucleic Acids from Viruses." doi:10.4319/mave.2010.978-0-9845591-0-7.154.

Sutton, Thomas D. S., Adam G. Clooney, and Colin Hill. 2020. "Giant Oversights in the Human Gut Virome." *Gut* 69 (7). BMJ Publishing Group: 1357–58. doi:10.1136/gutjnl-2019-319067.

Sutton, Thomas D. S., Adam G. Clooney, Feargal J. Ryan, R. Paul Ross, and Colin Hill. 2019. "Choice of Assembly Software Has a Critical Impact on Virome Characterisation." *Microbiome* 7 (1): 12. doi:10.1186/s40168-019-0626-5.

Tampuu, Ardi, Zurab Bzhalava, Joakim Dillner, and Raul Vicente. 2019. "ViraMiner: Deep Learning on Raw DNA Sequences for Identifying Viral Genomes in Human Samples." *PLoS One* 14 (9). doi:10.1371/journal.pone.0222271.

Tithi, Saima Sultana, Frank O. Aylward, Roderick V. Jensen, and Liqing Zhang. 2018. "FastViromeExplorer: A Pipeline for Virus and Phage Identification and Abundance Profiling in Metagenomics Data." *PeerJ* 6 (January): e4227. doi:10.7717/peerj.4227.

Trubl, Gareth, Ho Bin Jang, Simon Roux, Joanne B. Emerson, Natalie Solonenko, Dean R. Vik, Lindsey Solden, et al. 2018. "Soil Viruses Are Underexplored Players in Ecosystem Carbon Processing." *MSystems* 3 (5). American Society for Microbiology Journals. doi:10.1128/mSystems.00076-18.

Trubl, Gareth, Simon Roux, Natalie Solonenko, Yueh-Fen Li, Benjamin Bolduc, Josué Rodríguez-Ramos, Emiley A. Eloe-Fadrosh, Virginia I. Rich, and Matthew B. Sullivan. 2019. "Towards Optimized Viral Metagenomes for Double-Stranded and Single-Stranded DNA Viruses from Challenging Soils." *PeerJ* 7 (July). PeerJ Inc.: e7265. doi:10.7717/peerj.7265.

Trubl, Gareth, Natalie Solonenko, Lauren Chittick, Sergei A. Solonenko, Virginia I. Rich, and Matthew B. Sullivan. 2016. "Optimization of Viral Resuspension Methods for Carbon-Rich Soils along a Permafrost Thaw Gradient." *PeerJ* 4 (May). PeerJ Inc.: e1999. doi:10.7717/peerj.1999.

Truong, Duy Tin, Eric A. Franzosa, Timothy L. Tickle, Matthias Scholz, George Weingart, Edoardo Pasolli, Adrian Tett, Curtis Huttenhower, and Nicola Segata. 2015. "MetaPhlAn2 for Enhanced Metagenomic Taxonomic Profiling." *Nature Methods* 12 (10). Nature Publishing Group: 902–3. doi:10.1038/nmeth.3589.

Upton, Chris, Duncan Hogg, David Perrin, Matthew Boone, and Nomi L. Harris. 2000. "Viral Genome Organizer: A System for Analyzing Complete Viral Genomes." *Virus Research* 70 (1): 55–64. doi:10.1016/S0168-1702(00)00210-0.

Villarroel, Julia, Kortine Annina Kleinheinz, Vanessa Isabell Jurtz, Henrike Zschach, Ole Lund, Morten Nielsen, and Mette Voldby Larsen. 2016. "HostPhinder: A Phage Host Prediction Tool." *Viruses* 8 (5). doi:10.3390/v8050116.

Wan, Yinan, Daniel W. Renner, Istvan Albert, and Moriah L. Szpara. 2015. "VirAmp: A Galaxy-Based Viral Genome Assembly Pipeline." *GigaScience* 4: 19. doi:10.1186/s13742-015-0060-y.

Wang, David, Laurent Coscoy, Maxine Zylberberg, Pedro C. Avila, Homer A. Boushey, Don Ganem, and Joseph L. DeRisi. 2002. "Microarray-Based Detection and Genotyping of Viral Pathogens." *Proceedings of the National Academy of Sciences* 99 (24). National Academy of Sciences: 15687–92. doi:10.1073/pnas.242579699.

Wang, Shiliang, Jaideep P. Sundaram, and David Spiro. 2010. "VIGOR, an Annotation Program for Small Viral Genomes." *BMC Bioinformatics* 11 (1): 451. doi:10.1186/1471-2105-11-451.

Wang, Weili, Jie Ren, Kujin Tang, Emily Dart, Julio Cesar Ignacio-Espinoza, Jed A. Fuhrman, Jonathan Braun, Fengzhu Sun, and Nathan A. Ahlgren. 2020. "A Network-Based Integrated Framework for Predicting Virus–Prokaryote Interactions." *NAR Genomics and Bioinformatics* 2 (2). Oxford Academic. doi:10.1093/nargab/lqaa044.

Warwick-Dugdale, Joanna, Holger H. Buchholz, Michael J. Allen, and Ben Temperton. 2019. "Host-Hijacking and Planktonic Piracy: How Phages Command the Microbial High Seas." *Virology Journal* 16 (1): 15. doi:10.1186/s12985-019-1120-1.

Warwick-Dugdale, Joanna, Natalie Solonenko, Karen Moore, Lauren Chittick, Ann C. Gregory, Michael J. Allen, Matthew B. Sullivan, and Ben Temperton. 2019. "Long-Read Viral Metagenomics Captures Abundant and Microdiverse Viral Populations and Their Niche-Defining Genomic Islands." *PeerJ* 7 (April). doi:10.7717/peerj.6800.

Wilkinson, Katy, Wyn P. Grant, Laura E. Green, Stephen Hunter, Michael J. Jeger, Philip Lowe, Graham F. Medley, et al. 2011. "Infectious Diseases of Animals and Plants: An Interdisciplinary Approach." *Philosophical Transactions of the Royal Society B: Biological Sciences* 366 (1573): 1933–42. doi:10.1098/rstb.2010.0415.

Williamson, Kurt E., Jeffry J. Fuhrmann, K. Eric Wommack, and Mark Radosevich. 2017. "Viruses in Soil Ecosystems: An Unknown Quantity within an Unexplored Territory." *Annual Review of Virology* 4 (1): 201–19. doi:10.1146/annurev-virology-101416-041639.

Wommack, K. Eric, Jaysheel Bhavsar, Shawn W. Polson, Jing Chen, Michael Dumas, Sharath Srinivasiah, Megan Furman, Sanchita Jamindar, and Daniel J. Nasko. 2012. "VIROME: A Standard Operating Procedure for Analysis of Viral Metagenome Sequences." *Standards in Genomic Sciences* 6 (3). BioMed Central: 421–33. doi:10.4056/sigs.2945050.

Wommack, K. Eric, and Rita R. Colwell. 2000. "Virioplankton: Viruses in Aquatic Ecosystems." *Microbiology and Molecular Biology Reviews* 64 (1). American Society for Microbiology: 69–114. doi:10.1128/MMBR.64.1.69-114.2000.

Wommack, K. Eric, Télesphore Sime-Ngando, Danielle M. Winget, Sanchita Jamindar, and Rebekah R. Helton. 2010. "Filtration-Based Methods for the Collection of Viral Concentrates from Large Water Samples." In *Manual of Aquatic Viral Ecology (MAVE)*, 110–17. Advancing the Science for Limnology and Oceanography (ASLO). https://hal.archives-ouvertes.fr/hal-00528769.

Woolhouse, Mark E.J., and Kyle Adair. 2013. "The Diversity of Human RNA Viruses." *Future Virology* 8 (2): 159–71. doi:10.2217/fvl.12.129.

Yamashita, Akifumi, Tsuyoshi Sekizuka, and Makoto Kuroda. 2016. "VirusTAP: Viral Genome-Targeted Assembly Pipeline." *Frontiers in Microbiology* 7. Frontiers. doi:10.3389/fmicb.2016.00032.

Young, Francesca, Simon Rogers, and David L. Robertson. 2020. "Predicting Host Taxonomic Information from Viral Genomes: A Comparison of Feature Representations." *PLoS Computational Biology* 16 (5). Public Library of Science: e1007894. doi:10.1371/journal.pcbi.1007894.

Zárate, Selene, Blanca Taboada, Martha Yocupicio-Monroy, and Carlos F. Arias. 2017. "Human Virome." *Archives of Medical Research* 48 (8): 701–16. doi:10.1016/j.arcmed.2018.01.005.

Zhang, Kai-Yue, Yi-Zhou Gao, Meng-Ze Du, Shuo Liu, Chuan Dong, and Feng-Biao Guo. 2019. "Vgas: A Viral Genome Annotation System." *Frontiers in Microbiology* 10 (February). doi:10.3389/fmicb.2019.00184.

Zhao, Guoyan, Siddharth Krishnamurthy, Zhengqiu Cai, Vsevolod L. Popov, Hilda Guzman, Song Cao, Herbert W. Virgin, Robert B. Tesh, and David Wang. 2013. "Identification of Novel Viruses Using VirusHunter—An Automated Data Analysis Pipeline." *PLoS One* 8 (10): 11.

Zhao, Guoyan, Guang Wu, Efrem S. Lim, Lindsay Droit, Siddharth Krishnamurthy, Dan H. Barouch, Herbert W. Virgin, and David Wang. 2017. "VirusSeeker, a Computational Pipeline for Virus Discovery and Virome Composition Analysis." *Virology* 503 (March): 21–30. doi:10.1016/j.virol.2017.01.005.

Zheng, Yi, Shan Gao, Chellappan Padmanabhan, Rugang Li, Marco Galvez, Dina Gutierrez, Segundo Fuentes, Kai-Shu Ling, Jan Kreuze, and Zhangjun Fei. 2017. "VirusDetect: An Automated Pipeline for Efficient Virus Discovery Using Deep Sequencing of Small RNAs." *Virology* 500 (January): 130–38. doi:10.1016/j.virol.2016.10.017.

Zimmerman, Amy E., Cristina Howard-Varona, David M. Needham, Seth G. John, Alexandra Z. Worden, Matthew B. Sullivan, Jacob R. Waldbauer, and Maureen L. Coleman. 2020. "Metabolic and Biogeochemical Consequences of Viral Infection in Aquatic Ecosystems." *Nature Reviews Microbiology* 18 (1). Nature Publishing Group: 21–34. doi:10.1038/s41579-019-0270-x.

9

Freshwater Metagenomics

Overall Scenario

D. Patel, P. Chunarkar Patil*
Bharati Vidyapeeth (Deemed to be University), Pune, Maharashtra, India

CONTENTS

9.1 Introduction

Cognizance of the environment and its exploitable resources has been paramount in the advancement of humankind. Various methods have been used throughout history to understand these resources and their implications on human health. A significant contribution towards this was given by microbiology, enabling researchers to identify various etiological and pathogenic microorganisms inhabiting the environmental resources directly or indirectly involved in our daily activities. However, the culture-dependent nature of the techniques limited the scope of research (Handlesman 2004). This inspired the development of a culture-independent technique called metagenomics that 'reads' the entire genetic material derived from an environmental sample as a whole, followed by taxonomic classification and/or functional characterization. Metagenomics has since dominated the genomics industry to study the structure and role of microbial communities (microbiome) inhabiting environments like soil, aquatic systems, and even organs of the human body.

Over generations, scientists from various backgrounds have constantly made efforts to elucidate the mechanisms behind the role of water in maintaining diverse biotic and abiotic ecosystems and to find the factors responsible for a direct correlation between the quality of water and aspects of human life (healthcare, socio-economic challenges, cultural practices, and development). Waterborne diseases pose one of the biggest threats to humankind, with a global death toll of 1.8 million, out of which India alone

DOI: 10.1201/9781003042570-13

estimates 100,000 deaths (WHO 2004). Metagenomics made it feasible to uncoil the complexity of microbial biodiversity in water samples from both extreme (hot water springs, acid drains, polar ice caps, deep oceanic water) and moderate ecosystems (rivers, lakes, oceans, seas).

9.2 Freshwater Metagenomics Is the Desideratum

Despite their small size compared to marine ecosystems, freshwater systems bear the onus for most of the biogeochemical processes involved in mediating carbon flow. It has been observed that most nitrification and denitrification, methanogenesis and methanotrophy (through oxic surface waters), sulfate oxidation and reduction, carbon dioxide emission (estimated 100 million metric tons annually), and enzymatic reactions are the result of an array of metabolic processes carried out by freshwater microbial communities (Gutknecht et al. 2006). Identification and characterization of these microbial species, their metabolites, and the genes responsible for encoding them have critical applications like:

1. Discovering novel metabolites and characterizing them would extensively contribute to building the small molecule databases that are screened continuously and modified to find novel pharmacophores and bioactive agents (Minkiewicz et al. 2016).
2. Various bioinformatics portals maintain complex metabolic pathways of chemical reactions occurring within a cell. Discovery of unknown metabolites or the role of a known metabolite helps fill in missing data from the existing pathways (Cuadrat et al. 2018).
3. Various bacterial species thrive in freshwater conditions like very high or low oxygen or sulfur content, hot springs, high temperatures, high pressures, or extreme pH levels. Identification of genes and proteins contributing to their survival in such harsh conditions is an important research area (Marco 2010).
4. Microbial species in highly polluted environments have been known to develop mechanisms for the metabolism of hazardous pollutants. Deep functional screening of these mechanisms can reveal important agents that could be utilized to develop bioremediation processes (Devarapalli and Kumavat 2015).
5. Freshwater resources in the vicinity of industrial and urban areas are constantly exposed to sewage discharge, agricultural wastewater, and industrial contamination containing residual antibiotics. Microbial communities in such systems have developed antibiotic resistance. Identification of genes responsible for such resistance can expand our knowledge of factors responsible for antibiotic resistance (Nnadozie et al. 2019).
6. A large number of metabolites discovered are important biomarkers for physiological processes, pathogenic processes, gene mutation, polymorphisms, pharmacological responses, peptides, proteins, and other small molecules (Segata et al. 2011).

There are methodologies already developed under metagenomics that can deal with the kind of challenges posed while probing into the mentioned areas. Function-driven metagenomics has the potential to explore genes and their encoded proteins and metabolites and make important connections between them to deduce undergoing metabolic reactions. On the other hand, sequence-driven metagenomic studies can largely influence diversity studies.

Hence, it is clearly understood that activities of individual microbial species have micro-level influences on localized absorption of inorganic molecules and nutrients, while the community interactions among various species have a macro-level influence on the environment. Although environmental conditions have a major impact on the kind of microorganisms harbored in the respective community, the collective processes involved among the species of the community might exert their influence on the surrounding environment. Microbial communities are highly versatile in their activities, which range from seasonal changes and diurnal oscillations to long-term transformations suggesting evolution.

9.3 Technologies and Databases for Freshwater Metagenomics

Until the late 1970s, screening of microbial diversity was largely based on culture-dependent techniques. However, the popularity of such techniques declined because of some prominent disadvantages as it was realized that culture-dependent techniques were largely biased towards microbes that showed rapid growth, thereby ignoring the members of the community with low metabolic rates (Hugenholtz 2002). Realizing the limited approach of culture-dependent techniques, a search for new techniques based on amplification and sequencing of 16S rRNA sequences was needed for its universal presence over all microbial species. These techniques are called culture-independent techniques, and they encompass the state-of-the-art sequencing techniques like DGGE, T-RFLP, fluorescence in situ hybridization (FISH), and next-generation sequencing (NGS). In addition to sequencing technologies, metagenomics provides a framework of sequencing technologies, genome assembly, phylogenetic binning, read comparison, gene prediction, protein prediction pathway annotation, and data sharing platforms, as shown in Figure 9.1.

9.4 Freshwater Microbial Diversity

Freshwater bodies have an exclusive community that is consistent among similar ecosystems (like rivers, lakes, reservoirs, streams, etc.) but differs significantly from other biomes (like soil, sea, hot springs, etc.). Although diverse groups of microorganisms like protozoa, fungi, and archaea are present, bacterial species are the dominant inhabitants of freshwater bodies (~90%). A variety of bacterial species that exclusively inhabited freshwater systems were discovered as it was established that Gram-positive species of bacteria are also an active part of the respective microbiome (Schmeisser et al. 2003). Warnecke and his colleagues discovered that certain families of bacteria are universal to the freshwater systems, like Betaproteobacteria (Polynucleobacter, R-BT065, and GKS98), Bacteroides (SOL), and Actinobacteria (ACK M1) (2004). In many species, it has been noted that in different climatic zones, the cosmopolitan presence of certain families of bacteria shows adaptation to a different temperatures, but they are only

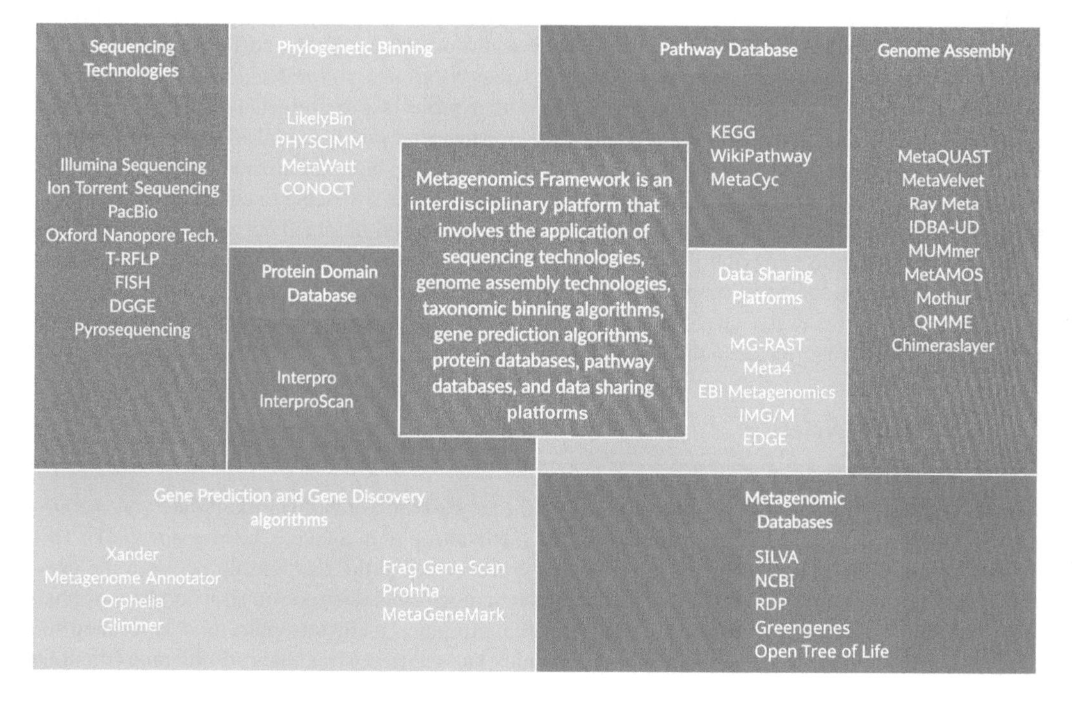

FIGURE 9.1 Metagenomics: framework of tools and technologies.

able to acclimatize to the particular temperature they are subjected to (Moon et al. 2004). To maintain the quality of water and pollutants, most of these community members are evidently involved directly. Analysis of such a diverse microbiome environment from a specific location can produce a great deal of information about the potential function, ecology, and diversity. The development of molecular techniques and analysis of metagenomic data from such freshwater sources can have considerable implications in industries involved in human health like food, agriculture, probiotics, and therapeutics.

Freshwater resources also include various pathogens from different taxa that are directly responsible for human waterborne diseases. Some microorganisms often isolated from water sources, including bacteria, protozoa, fungi, or viruses, are directly involved in pathogenic diseases like amoebiasis, cryptosporidiosis, SARS, polio, hepatitis, cholera, typhoid, leptospirosis, and others.

Moreover, most of the sequences obtained from freshwater sources are yet to be identified, since there are no matches to the existing reference databases. Classification of obtained sequences in respective phylogenetic 'bins' can provide inference on the unidentified species' taxonomy. However, it still has to be assumed that most of the species from freshwater bodies are yet to be identified.

9.5 Applications of Freshwater Metagenomics

Freshwater resources have been extensively sampled and sequenced over these years. Sequencing technologies have extensively contributed to the building of large and complex databases. Approximately 44,000 datasets have been analyzed and published in the EBI-EMBL database alone for freshwater microbiomes.

When metagenomics was first introduced around the 1990s, the disadvantages of shotgun sequencing limited its scope to a few studies like Schmeisser et al. (2003) surveying biofilms forming on rubber valves in drinking water networks, Warnecke et al. (2004) establishing the phylogenetic relationship of freshwater Actinobacterial colonies, and Moon et al. (2004) identifying the *bphC* gene that contributes to the degradation of aromatic pollutants in freshwater. However, the emergence of NGS and the global ocean sampling (GOS) expedition inspired a spurt of studies. Between 2005 and 2009, most of these projects were solely motivated by an objective of understanding the overall microbial diversity in urban freshwater bodies, like Cottrell et al. (2005) exploring the bacterial diversity in the Delaware River; Djikeng et al. (2009) investigating the RNA viruses of Lake Needwood in Maryland; and Debroas et al. (2009) identifying vital metabolic pathways of three major bacterial groups, *Actinobacteria, Proteobacteria*, and *Bactiroidetes*, in a mesotrophic lake in France. Other projects were targeted at micro-freshwater habitats, like Pope and Patel (2008) analyzing toxic freshwater cyanobacterial blooms and Breitbart et al. (2003) investigating freshwater microbialites to understand the biomineralization process in Cuatro Ciénegas, Mexico. Simultaneously, functional metagenomic studies were able to characterize 130 putative genes responsible for the maintenance of a toxic cyanobacterial bloom (Pope and Patel 2008) and identified a novel esterase enzyme (Wu and Sun 2009). Comparative metagenomics found its earliest applications in freshwater metagenomics as Sharma et al. (2009) established the metropolitan presence of Actinobacteria in freshwater lakes through consistency of ActR genes over various samples. Detailed applications of freshwater metagenomics are shown in Figure 9.2.

9.5.1 Ecological Cycles

9.5.1.1 Methane Cycle

An active role of freshwater Methylotenera species in one carbon metabolism like folate and methionine cycles was a pioneering work in understanding methylotrophy in a lake (Chistoserdov 2011). This was confirmed in the Lake Pavin study and further commented on the relationship between the presence of methyl-consuming (methylotrophic) Methylotenera and methane-consuming (methanotrophic) Methylobacter. A collective understanding of methanotrophic, methylotrophic, and methanogenic (Methanoregula) species and their metabolic dynamics has enormously enhanced our understanding of the methane cycle and the genes responsible for processes that involve nitrogen metabolism (nitrification, denitrification, nitrogen assimilation, fixation, and nitrate transport) (Biderre-Petit et al. 2018).

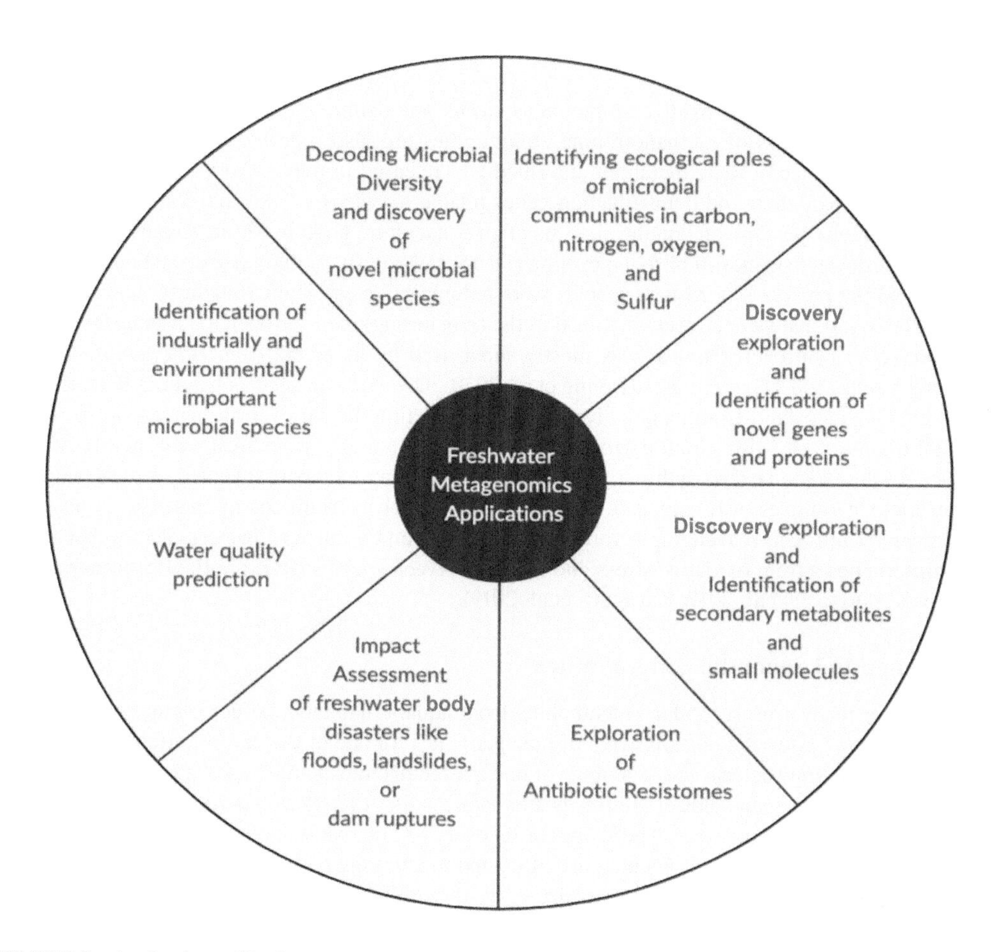

FIGURE 9.2 Applications of freshwater metagenomics.

Methane oxidation in Lake Erie freshwater wetlands was found to be majorly dominated by members of the methanotrophic bacterial taxa *Methylococcales*. Coupled with aerobic respiration of *Methylobacter*, *Methylococcales* can sustain energy production in low-oxygen oxic environments (Smith et al. 2018).

9.5.1.2 Sulfur Cycle

Similarly, dimethyl sulfate (DMS) is one of the most-emitted sulfur compounds in the atmosphere; hence, its role in the global sulfur cycle is inevitable. Although it was believed that marine gammaproteobacterial communities were involved in most global DMS degradation, the discovery of beta proteobacterial Methylophilaceae in lake sediments emphasized the contribution of freshwater bacterial communities in DMS degradation (Eyice et al. 2015). In a recent attempt, a curated database for microbial members responsible for the S cycle was created with a pool of 207 genes and 52 phyla. This database includes all the gene families responsible for S reduction, S oxidation, disproportionation, and transformation of sulfur compounds. Furthermore, the relative abundance of phyla involved in the sulfur cycle is also included (Yu et al. 2020). Simultaneously, a novel species from the chemoautotrophic bacterial order *Campylobacterales* was also observed to dominate sulfur oxidation in a metagenomic study performed on a geothermal spring in Greece (Meziti et al. 2020). Sulfur oxidation and sulfate reduction in Upper Mystic Lake coordinated by *Sulfuritalea hydrogenivorans* and *Desulfatirhabdium butyrativorans* strain HB1 provide further evidence for the presence of sulfur-cycling microbes in freshwater bodies (Williams et al. 2018).

9.5.1.3 Nitrogen Cycle

Due to their ability to nitrify (oxidize ammonia to nitrite) and denitrify (reduce nitrite to nitrous oxide or nitric oxide), members of methanotrophic communities are also reported to be critical factors in deciding the abundance of toxic metabolic intermediates released in the environment (Stein and Klotz 2011). Ammonia oxidation and denitrification genes related to *Nitrosomonas* and *Rhodocyclales* were recovered from functional metagenomic analysis of an aquaculture pond in China, suggesting an important role of freshwater species in nitrogen cycling (Deng et al. 2020). A study performed on mariculture water marked the presence of 25 new species from halophilic heterotrophic nitrification-aerobic denitrification (HNAD) that were actively involved in the transformation of nitrogen to gaseous form. It was also observed that nitrogen removal was mostly supervised by members from *Halomonas, Cobetia, Marinobacterium*, and *Marinomonas* (Huang et al. 2020). Denitrifying species from the Bacterioidetes group were also recovered from Upper Mystic Lake, suggesting their universal presence in freshwater bodies (Williams et al. 2018). Contrary to the direct participation of various freshwater microbes in the ecological cycles, some species indirectly influence other colony members involved in principal metabolic reactions. Examples of this include a diazotrophic cyanobacterium colony that fixes atmospheric nitrogen gas for nitrogen-scavenging aerobic proteobacteria and a range of bacteriophages that protect Actinobacteria hosts from oxidative stress and reactive oxygen species (ROS) in the freshwater aquatic environment (Driscoll et al. 2017; Kavagutti et al. 2019).

9.5.2 Genes and Secondary Metabolites

The process of recovering secondary metabolites from aquatic habitats and describing their potential function is another extensive research area of metagenomics. In one of the earliest attempts to identify novel chitin-degrading bacteria across a range of unexplored habitats, Cretoiu et al. (2012) reported high abundances of the *chiA* gene clusters in freshwater sponges like *Ephydatia fluviatilis* and its biochemical role in chitinoclastic processes. The scope for recovery of putative secondary metabolism genes has increased severalfold, with unprecedented use of taxonomic binning based on the probabilistic distance of genome abundance instead of a similarity detection approach that can only classify previously known sequences. Evidence for high productivity of such a system is observed in a recent study by Cuadrat et al. (2018) that successfully detected secondary metabolism genes like non-ribosomal peptide synthase (NRPS) and polyketide synthase (PKS), among many others. Functional screening for metabolites has further led to the recognition of various secondary metabolites that actively participate in microbial resistance against various commercially available drugs. A metagenomic study of fluoroquinolone-rich river sediment study in India revealed two novel secondary metabolism genes potentially involved in penicillin and carbapenem hydrolyzation (Marathe et al. 2018). Similarly, Lopez-Perez et al. (2013) characterized a putative open reading frame identical to the protein responsible for multidrug resistance in *Aeromonas salmonicida* from an industrial wetland in Mexico.

9.5.3 Antibiotic Resistomes

With the growing popularity of metagenomics in exploring antibiotic resistomes, researchers have invested even more interest in the urban rivers and lakes that are the most frequent receivers of pharmaceutical, agricultural, and domestic effluents. In one such antibiotic resistome analysis in a river catchment, Rowe et al. (2016) pinpointed up to 53 putative genes (like tetC, sul2, and tetW) contributing to microbial resistance towards seven classes of antimicrobials, including tetracycline and sulfonamides. These results were further confirmed by Chen et al. (2019) as they studied human-impacted river sediment habitat to discover around 20 antibiotic-resistant genes that contributed to the same degree of resistance against sulfonamide, tetracycline, and aminoglycosides. The major microbial species with potential antibiotic resistance genes for Bacitracin, beta-lactam, and multidrug were identified to be from the Polynucleobacter, Opitutus, and Flavobacterium taxonomic categories (Bai et al. 2019). However, despite our enhanced understanding of antibiotic resistomes by virtue of metagenomics, it has been challenging to discern the exact reason for their enrichment. Is human contamination an intervening component in the evolution of a resistome? Moreover, is it natural selection towards the resistance factor?

9.5.4 Impact Assessment

One of the most innovative applications of metagenomics has been observed in disaster management and impact assessment. A recent example of such an accomplishment of metagenomics can be observed in the case of the Fundao Dam rupture incident that possibly emanated >50 million m^3 of iron tailings into the Doce River of Brazil. In an attempt to analyze the impact of iron tailings on the peripheral coral reefs, Francini et al. (2019) expedited the process by metagenomic analysis of benthic microbial communities from reef systems spanning the path of the disaster. Wastewater, another important health-hazard agent, has been extensively studied using metagenomics to identify the inhabitant microbial communities and reveal interesting trends like prominent populations of certain organisms (Helicobacteraceae, Legionellaceae, Moraxellaceae, and Neisseriaceae) in highly contaminated samples (Garrido et al. 2017). Metagenomics has also been utilized in understanding the response of methanogenic species towards stressful situations like droughts and floods. The Zhalong wetland study reflects the various mechanisms and 'repair genes' used by methanogenic bacterial species to counter oxidative stress and maintain methane flux during dry conditions (Liu et al. 2021). A city-wide metagenomic project surveying six dam sites, two rivers, and four lakes was carried out in Pune to understand the impact of microbial communities on contamination, antibiotic resistomes, hydrological cycle, and seasonal change (Patil et al. 2020).

Detailed timeline of freshwater metagenomic research projects, key findings, and kind of sequencing technology used is summarized in Table 9.1.

TABLE 9.1

Timeline of Freshwater Metagenomic Research Projects, Key Findings, and Kind of Sequencing Technology Used

Biome Explored	Sequencing Technique	Key Findings	Year and Reference
Drinking-water biofilms	Snapshot genome sequencing	Microbial community closely related to Proteobacteria; 2,200 putative open reading frames; no pathogenic strains	Schmeisser et al. 2003
A cluster of seven freshwater systems	PCR amplification	Two exclusive monophyletic Actinobacterial clades	Warnecke et al. 2004
Several lakes in mid-Korea	T-RFLP metagenomics	*bphC* gene associated with degrading aromatic pollutants	Moon et al. 2004
Delaware River	PCR + FISH	Cytophaga-like members, Polynucleobacter and Betaproteobacteria colonies identified	Cottrell et al. 2005
Biofilms in drinking water	PCR amplification	Two novel esterase enzymes: EstCE1 and EstA3	Elend et al. 2006
Toxic cyanobacterial bloom	PCR cloning	Dominant members of a toxic freshwater bloom; 130 putative genes	Pope and Patel 2008
Mesotrophic lake, France	Gene amplification	Metabolic pathways of freshwater Actinobacteria, Proteobacteria, and Bacteroidetes	Debroas et al. 2009
Lake Nedwood, Maryland	Random priming mediated sequence Independent single-primer amplification (RP-SISPA) + MG-RAST + 454 pyrosequencing	Novel dsRNA viruses closely related to pathogenic Banna virus and Israeli acute paralysis virus	Djikeng et al. 2009

(Continued)

TABLE 9.1 *(Continued)*

Biome Explored	Sequencing Technique	Key Findings	Year and Reference
Living microbialites, Mexico	454 pyrosequencing	Redox-dependent communities; genes for the formation of exopolymer compounds and biofilms	Breitbart et al. 2003
Yangtze River, China	Subcloning	Novel esterase enzyme: EstY	Wu and Sun 2009
One pond and two lakes	Degenerate PCR primer amplification	Cosmopolitan freshwater Actinobacteria consumes rhodopsin	Sharma et al. 2009
Amazon River water column	454 pyrosequencing on GS-FLX titanium platform	Exclusively freshwater taxa like Polynucleobacter, sister SAR11 group, and Crenarchea; specialized role of mineralizing plant carbon	Ghai et al. 2011
Lake Washington	Whole-genome shotgun sequencing	Role of freshwater methylotenera species	Chistoserdov 2011
Sequences in OMeGA database	Available dataset	Microbes involved in the metabolism of toxic nitrogen species	Stein and Klotz. 2011
Lake Bourget and Lake Pavin	Pyrosequencing using 454 life sciences GS-FLX genome sequencer	Cluster and species richness higher in mesotrophic lake compared to oligotrophic lake	Roux et al. 2012
Flooded paddy field	PCR screening of existing libraries	Phylogenetic diversity of candidate division OP3	Glockner et al. 2010
Various lakes in China and North America	454 GS-FLX titanium sequencing + MG-RAST	Interplay between cyanobacterial and proteobacterial colonies in nitrogen assimilation across two continents	Steffen et al. 2012
Freshwater sponges	PCR-DGGE	Novel chitin-degrading gene *chiA*	Cretoiu et al. 2012
Industrial wetland in Mexico	PCR amplification	Putative ORF similar to the multidrug-resistant protein in *Aeromonas salmonicida*	Lopez-Perez et al. 2013
Lake Donghu, Wuhan, China (across three seasons: autumn, summer, and spring)	PCR amplification	Viral families are infecting bacterial and algal colonies	Ge et al. 2013
Lake Vattern, Lake Eholn, and Lake Erken	454 Pyrosequencing	Differences in marine and freshwater microbial pathways for carbon metabolism	Eiler et al. 2014
Tocil Lake	DNA stable isotope probing (SIP) metagenomics	Contribution of freshwater Methylophilaceae in DMS degradation	Eyice et al. 2015
Yellowstone Lake	Shotgun metagenomic sequencing	Three novel virophage genomes: YSLV5, YSLV6, and YSLV7	Zhou et al. 2015
Amazon basin: Lake Preta, Lake Poraque, Lake Anana, and Manacapura Great Lake	Illumina HiSeq 2500 sequencing platform	Quantitative composition of microbial phyla in the respective biomes	Toyama et al. 2016

Biome Explored	Sequencing Technique	Key Findings	Year and Reference
Tibetan mountain lake	Existing metagenomic sequences utilized	Eukaryotic and viral pathogens: coccolithophores and coccolithovirus	Oh et al. 2016
Agricultural and industrial wastewater	Illumina HiSeq 2500 sequencing	Fifty-three putative genes contributing to microbial resistome against seven classes of antimicrobial drugs	Rowe et al. 2016
Hot water spring, South Africa	Illumina MiSeq sequencing	dsDNA viral diversity	Zablocki et al. 2017
Upper Klamath Lake	PacBio shotgun sequencing + long-read metagenomics approach	Nitrogen-fixing diazotrophic cyanobacteria	Driscoll et al. 2017
Available metagenomic sequences	Dataset analysis	Contributing factors towards evolving resistome	Bengttson-Palme et al. 2017
Wastewater	Available sequences	Prominent populations of certain microorganisms in highly contaminated samples	Garrido et al. 2017
Lake Poraque, Amazon basin	Illumina HiSeq 1000 sequencing platform + MetaGeneMark + IDB-UD + MG-RAST + STAMP	Identified a freshwater beta-glucosidase	Toyama et al. 2018
Lake Stechlin, Germany	Illumina HiSeq 2500 sequencing + Anti-SMASH and NAPDOS workflow	Detected secondary metabolism genes like NRPS and PKS	Cuadrat et al. 2018
Indian river sediments	Sanger and PacBio RSII sequencing platform	Novel genes responsible for penicillin and carbapenem hydrolyzation	Marathe et al. 2018
Anoxic sulfidic spring	Illumina HiSeq 2500 sequencing platform	Four novel phylogenetically related genomes of Gram-negative flagellate-like organisms	Youssef et al. 2019
Pelagic freshwater habitat	Ultra-deep long-read nanopore sequencing	Virophage dynamics on the surface and deep waters; virophage pathogenic to prokaryotic organisms	Kavagutti et al. 2019
The existing collection of 14,000 metagenomes	Major capsid protein (MCP) prediction through hidden Markov model (HMM)	Increased the number of known virophage genomes by ten times	Paez-Espino et al. 2019
Lake Pavin	Illumina HiSeq 2000 sequencing	Metabolic potential of freshwater planktons in the methane cycle	Biderre-Petit et al. 2018
Human-impacted river sediments	Illumina HiSeq 2500 sequencing	Twenty antibiotic-resistant genes	Chen et al. 2019
Huaihe River basin, China	Illumina HiSeq 2500 sequencing	Microbial colonies with developed antibiotic resistance	Bai et al. 2019
Doce River, Brazil	PCR	Impact assessment of the Fundao Dam rupture disaster	Francini et al. 2019
Lake Redon, Pyrenees	Illumina MiSeq sequencing	Higher taxonomical diversity and functional potential at the deeper end of the lake	Llorens et al. 2020
Dams of Pune City, India	T-RFLP	Freshwater bacterial biodiversity	Patil et al. 2020

9.6 Conclusions

Freshwater metagenomics has been magnetic in terms of attracting funding and interest. Despite its evident success in the past decade, there are still certain lacunae in terms of technology and the scope of metagenomic research in freshwater environments. The former, that is, the technological aspects of metagenomics, have particularly faced a persistent challenge in the reproducibility of the results. Analytical techniques have often faced inconsistencies in generating a standard result for identical samples. Furthermore, the use of metagenomics has been limited to freshwater resources like lakes, rivers, and hot springs, while there is a parallel need to explore groundwater and rainwater microbiomes.

REFERENCES

Bai, Y., X. Ruan, X. Xie, and Z. Yan. 2019. "Antibiotic resistome profile based on metagenomics in raw surface drinking water source and the influence of environmental factor: A case study in Huaihe River Basin, China." *Environmental Pollution* 248: 438–447. doi: 10.1016/j.envpol.2019.02.057.

Bengttson-Palme, J., J. Larrson, and E. Kristiansson. 2017. "Using metagenomics to investigate human and environmental resistomes." *The Journal of Antimicrobial Chemotherapy* 72(10): 2690–2703. doi: 10.1093/jac/dkx199.

Biderre-Petit, C., N. Taib, H. Gardon, and C. Hochart. 2018. "New insights into the pelagic microorganisms involved in the methane cycle in the meromictic Lake Pavin through metagenomics." *FEMS Microbiology Ecology* 95(3): 183. doi: 10.1093/femsec/fiy183.

Breitbart, M., I. Hewson, B. Felts, J. Mahaffy, J. Nulton, P. Salamon, and F. Roher. 2003. "Metagenomic analyses of an uncultured viral community from human feces." *Journal of Bacteriology* 185(20): 6220–6223. doi: 10.1128/jb.185.20.6220-6223.2003.

Chen, H., X. Bai, L. Jing, R. Chen, and Y. Teng. 2019. "Characterization of antibiotic resistance genes in the sediments of an urban river revealed by comparative metagenomics analysis." *Science of the Total Environment* 653: 1513–1521. doi:10.1016/j.scitotenv.2018.11.052

Chistoserdov, L. 2011. "Methylotrophy in a lake: From metagenomics to single-organism physiology." *Applied and Environmental Microbiology* 77(14): 4705–4711. doi: 10.1128/AEM.00314-11.

Cottrell, M., L. Waidner, L. Yu, and D. Kirchman. 2005. "Bacterial diversity of metagenomic and PCR libraries from the Delaware River." *Environmental Microbiology* 7(12): 1883–1895. doi: 10.1111/j.1462-2920.2005.00762.x.

Cretoiu, M.S., A. Kielak, W.A. Al-Soud, S. Sorensen, and J.D. Van Elsas. 2012. "Mining of unexplored habitats for novel chitinases—chiA as a helper gene proxy in metagenomics." *Applied Microbiology and Biotechnology* 94: 1347–1358. doi: 10.1007/s00253-012-4057-5.

Cuadrat, R.C., D. Ionescu, A. Davila, and H.P.S. Grossart. 2018. "Recovering genomic clusters of secondary metabolites from lakes using genome-resolved metagenomics." *Frontiers in Microbiology* 9: 251. doi: 10.3389/fmicb.2018.00251

Debroas, D., J.F Humbert, F. Enault, G. Bronner, M. Faubladier, and E. Cornillot. 2009. "Metagenomic approach studying the taxonomic and functional diversity of the bacterial community in a mesotrophic lake (Lac du Bourget-France)." *Environmental Microbiology* 1 (9): 2412–2424. doi: 10.1111/j.1462-2920.2009.01969.x.

Deng, M., J. Hou, K. Song, J. Chen, J. Gou, D. Li, and X. He. 2020. "Community metagenomic assembly reveals microbes that contribute to the vertical stratification of nitrogen cycling in an aquaculture pond." *Aquaculture* 520: 734911. doi: 10.1016/j.aquaculture.2019.734911.

Devarapalli, P., and R. Kumavat. 2015. "Metagenomics—A technical drift in bioremediation." In *Advances in Bioremediation of Wastewater and Polluted Soil*, edited by Naofumi Shiomi, 73–91. London: Intechopen Limited.

Djikeng, A., R. Kuzmickaz, N. Anderson, and D. Spiro. 2009. "Metagenomic analysis of RNA viruses in a fresh water lake." *PLoS One* 4(9): e7264–e7278. doi: 10.1371/journal.pone.0007264.

Driscoll, C., T. Otten, N. Brown, and T. Dreher. 2017. "Towards long-read metagenomics: Complete assembly of three novel genomes from bacteria dependent on a diazotrophic cyanobacterium in a freshwater lake co-culture." *Standards in Genomic Sciences* 12(9). doi: 10.1186/s40793-017-0224-8.

Eiler, A., K.Z. Niedzwiedzka, M.M. Garcia, K. McMahon, R. Stepanauska, S. Anderrson, and S. Bertilsson. 2014. "Productivity and salinity structuring of the microplankton revealed by comparative freshwater metagenomics." *Environmental Microbiology* 16(9): 2682–2698. doi: 10.1111/1462-2920.12301.

Elend, C., C. Schmeisser, C. Leggewie, P. Babiak, J.D Carbarella, H.L. Steele et al. 2006. "Isolation and biochemical characterization of two novel metagenome-derived esterases." *Applied and Environmental Microbiology* 72(5): 3637–3645. doi: 10.1128/AEM.72.5.3637-3645.2006.

Eyice, O., M. Namura, Y. Chen, A. Mead, S. Samavedan, and H. Schafer. 2015. "SIP metagenomics identifies uncultivated Methylophilaceae as dimethyl sulphide degrading bacteria in soil and lake sediment." *The ISME Journal* 9: 2336–2348. doi: 10.1038/ismej.2015.37.

Francini F.R., M. Cordeiro, C. Omachi, A. Rocha, L. Bahiense, G. Garcia et al. 2019. "Remote sensing, isotopic composition and metagenomics analyses revealed Doce River ore plume reached the southern Abrolhos Bank Reefs." *Science of Total Environment* 697: 134038. doi: 10.1016/j.scitotenv.2019.134038.

Garrido C., J.A., M.P. Lopez, and I.O. Alberola. 2017. "Advanced microbial analysis for wastewater quality monitoring: Metagenomics trend." *Applied Microbiology and Biotechnology* 101(20): 7445–7458. doi: 10.1007/s00253-017-8490-3.

Ge, X., Y. Wu, M. Wang, J. Wang, L. Wu, X. Yang et al. 2013. "Viral metagenomics analysis of planktonic viruses in East Lake, Wuhan, China." *Viroligica Sinica* 28: 280–290. doi: 10.1007/s12250-013-3365-y.

Ghai, R., F.R. Valera, K. McMahon, D. Toyama, R. Rinke, T.C. Souza deOliviera, J. Garcia, F. Miranda, and I. Henrique-Silva. 2011. "Metagenomics of the water column in the pristine upper course of the Amazon River." *PLoS One* 6(8): e23785. doi: 10.1371/journal.pone.0023785.

Glockner, J., M. Kube, P. Shrestha, M. Weber, F. Glockner, R. Reinhardt, and W. Liesack. 2010. "Phylogenetic diversity and metagenomics of candidate division OP3." *Environmental Microbiology* 12(5): 1462–2912. doi: 10.1111/j.1462-2920.2010.02164.x.

Gutknecht, J.L.M., R. Goodman, and T. Balser. 2006. "Linking soil processes and microbial ecology in freshwater wetland ecosystems." *Plant and Soil* 289: 17–34. doi: 10.1007/s11104-006-9105-4

Handlesman, J. 2004. "Metagenomics: Application of genomics to uncultured microorganisms." *Microbiology and Molecular Biology Reviews* 68(4): 669–685. doi: 10.1128/MMBR.68.4.669-685.2004.

Huang, F., L. Pan, Z. He, M. Zhang, and M. Zhang. 2020. "Identification, interactions, nitrogen removal pathways and performances of culturable heterotrophic nitrification-aerobic denitrification bacteria from mariculture water by using cell culture and metagenomics" *Science of the Total Environment* 732: 123698. doi: 10.1016/j.scitotenv.2020.139268

Hugenholtz, P. 2002. "Exploring prokaryotic diversity in the genomic era." *Genome Biology* 3(2): reviews0003.1–reviews0003.8. doi: 10.1186/gb-2002-3-2-reviews0003.

Kavagutti, V.S., A.S. Andrei, M. Mehrshad, M. Salcher, and R. Ghai. 2019. "Phage-centric ecological interactions in aquatic ecosystems revealed through ultra-deep metagenomics." Microbiome 7(135). doi: 10.1186/s40168-019-0752-0.

Llorens M.T., J Catalan, and E. Casamayer. 2020. "Taxonomy and functional interactions in upper and bottom waters of an oligotrophic high-mountain deep lake (Redon, Pyrenees) unveiled by microbial metagenomics." *The Science of Total Environment* 707: 135929. doi: 10.1016/j.scitotenv.2019.135929.

Lopez-Perez, M., S. Mirete, E.J. Valadez, and J.E Gonzalez-Pastor. 2013. "Identification and modeling of a novel chloramphenicol resistance protein detected by functional metagenomics in a wetland of Lerma, Mexico." *International Microbiology: The Official Journal of the Spanish Society for Microbiology* 16(2): 103–111. doi: 10.2436/20.1501.01.185.

Marathe, N.P., A. Janzon, S. Kotsakis, C.F. Flach, M. Razavi, F. Berglund, E. Kristiansson, and J. Larrson. 2018. "Functional metagenomics reveals a novel carbapenem-hydrolyzing mobile beta-lactamase from Indian river sediments contaminated with antibiotic production waste." *Environment International* 112: 279–286. doi: 10.1016/j.envint.2017.12.036.

Marco, D. 2010. *Metagenomics: Theory, Methods and Applications*. Norfolk: Caister Academic Press. *International Microbiology* 13: 41–42.

Meziti A., E. Nikouli, K. Hatt, T. Konstantinidis, and K.A. Kormans. 2020. "Time series metagenomic sampling of the Thermopyles, Greece, geothermal springs reveals stable microbial communities dominated by novel sulfur-oxidizing chemoautotrophs." *Environmental Microbiology*. doi: 10.1111/1462-2920.15373.

Minkiewicz, P., M. Darewicz, A. Iwaniak, J. Bucholska, P. Starowicz, and E. Czyrko. 2016. "Internet databases of the properties, enzymatic reactions, and metabolism of small molecules—Search options and applications in food science." *International Journal of Molecular Science* 17(12): 2039–2063. doi: 10.3390/toxins6102912.

Moon, M.S., D.H. Lee, and C.K. Kim. 2004. "Identification of the bphC gene for meta-cleavage of aromatic pollutants from a metagenomic library derived from lake waters." *Biotechnology and Bioprocess Engineering* 9: 393–399. doi: 10.1007/BF02933064.

Nnadozie, C.F., and O. Odume. 2019. "Freshwater environments as reservoirs of antibiotic resistant bacteria and their role in the dissemination of antibiotic resistance genes." *Environmental Pollution* 254(Part B): 113067–113080. doi: 10.1016/j.envpol.2019.113067.

Oh, S., D. Yoo, and W.T. Liu. 2016. "Metagenomics reveals a novel virophage population in a Tibetan mountain lake." *Microbes and Environments* 31(2): 173–177. doi: 10.1264/jsme2.ME16003.

Paez-Espino, D., J. Zhou, S. Roux, S. Nayfach, G. Pavlopoulos, F. Schulz, K. McMahon, D. Walsch et al. 2019. "Diversity, evolution, and classification of virophages uncovered through global metagenomics." *Microbiome* 7(1): 157. doi: 10.1186/s40168-019-0768-5.

Patil, P.C., D.A. Patel, V. Tale. 2020. "Metagenomic analysis of dam reservoirs in Pune City for bacterial fingerprints through BLAST and Kaiju tool." *Biosci Biotech Res Asia* 17(4): 839–852.

Pope, P.B., and B.K.C. Patel. 2008. "Metagenomic analysis of a freshwater toxic cyanobacteria bloom." *FEMS Microbiology Ecology* 64(1): 9–27. doi: 10.1111/j.1574-6941.2008.00448.x.

Roux, S., F. Enault, A. Robin, V. Ravet, S. Personnic, S. Theil, J. Colombet, T.S. Ngando, and D. Debroas. 2012. "Assessing the diversity and specificity of two freshwater viral communities through metagenomics." *PLoS One* 7(3): e33641. doi: 10.1371/journal.pone.0033641.

Rowe, W., D.V. Jeffreys, C.B. Austin, J. Ryan, D. Maskell, and G. Pearce. 2016. "Comparative metagenomics reveals a diverse range of antimicrobial resistance genes in effluents entering a river catchment." *Water Science and Technology* 73(7): 1541–1549. doi: 10.2166/wst.2015.634.

Schmeisser, C., C. Stockigt, C. Raasch, J. Wingender, K.N. Timmis, D.F. Wenderoth, H.C. Flemming, H. Liesegang, R.A. Schmitz, K.E. Jaeger, and W.R. Streit. 2003. "Metagenome survey of biofilms in drinking-water networks." *Applied and Environmental Microbiology* 69(12): 7298–7309. doi: 10.1128/aem.69.12.7298-7309.2003.

Segata, N., J. Izard, L. Waldron, D. Gevers, L. Miropolsky, W. Garrett, and C. Huttenhower. 2011. "Metagenomic biomarker discovery and explanation." *Genome Biology* 12(6): R60–R78. doi: 10.1186/gb-2011-12-6-r60.

Sharma, A.K., K. Sommerfeld, G. Bullerjahn, A. Matteson, S. Wilhelm, J. Jazbera, U. Brandt, Ford Dolittle, and Martin Hahn. 2009. "Actinorhodopsin genes discovered in diverse freshwater habitats and among cultivated freshwater Actinobacteria." *ISME Journal* 3(6): 726–737. doi: 10.1038/ismej.2009.13.

Smith, G., J. Angle, L. Solden, M. Borton, R. Daly, M. Johnston, K. Stefanik, R. Wolfe, B. Gil, and K. Wrighton. 2018. "Members of the genus methylobacter are inferred to account for the majority of aerobic methane oxidation in oxic soils from a freshwater wetland." *mBio* 9(6): e00815–818. doi: 10.1128/mBio.00815-18.

Steffen Morgan, Zhou Li, Chad Effler, Loren Hauser, Gregory Boyer, and Steven Wilhelm. 2012. "Comparative metagenomics of toxic freshwater cyanobacteria bloom communities on two continents." *PLoS One* 7(8): e44002. doi: 10.1371/journal.pone.0044002.

Stein, Lisa Y., and Martin Klotz. 2011. "Nitrifying and denitrifying pathways of methanotrophic bacteria." *Biochemical Society Interactions* 39(6): 1826–2831. doi: 10.1042/BST20110712.

Toyama, D., M. Abrahão B. Morais, F.C. Ramos, L.M. Zanphorlin, C.C. Costa Tonoli, A.F. Balula et al. 2018. "A novel β-glucosidase isolated from the microbial metagenome of Lake Poraquê (Amazon, Brazil)." *Biochimica et biophysica Acta-Proteins and Proteomics* 1866(4): 569–579. doi: 10.1016/j.bbapap.2018.02.001.

Toyama, D., L.T. Kishi, C.D. Santos-Junior, A.S. Costa, T.C. Souza de Oliviera, F. Pellon de Miranda, and F.H. Silva. 2016. "Metagenomics analysis of microorganisms in freshwater lakes of the Amazon basin." *Genome Announcements* 4(6): e01440–16. doi: 10.1128/genomeA.01440-16.

Warnecke, F., R. Amann, and J. Pernthaler. 2004. "Actinobacterial 16S rRNA genes from freshwater habitats cluster in four distinct lineages." *Environmental Microbiology* 6(3): 242–253. doi: 10.1111/j.1462-2920.2004.00561.x.

Williams, K.A., S. Olesen, B. Scandella, K. Delwiche, S. Spencer, E. Myers, S. Abraham, A. Sooklal, and S. Preheim. 2018. "Dynamics of microbial populations mediating biogeochemical cycling in a freshwater lake." *Microbiome* 6(165). doi: 10.1186/s40168-018-0556-7

World Health Organisation. 2004. "The global burden of diseases." www.who.int/healthinfo/global_burden_disease/GBD_report_2004update_full.pdf?ua.

Wu, C., and B. Sun. 2009. "Identification of novel esterase from metagenomic library of Yangtze River." *Journal of Microbiology and Biotechnology* 19(2): 187–193. doi: 10.4014/jmb.0804.292.

Youssef, N.H., I. Farag, R. Hahn, H. Premathilake, E. Fry, M. Hart, K. Huffaker, E. Bird et al. 2019. "Candidatus *Krumholzibacterium zodletonense* gen. nov., sp nov, the first representative of the candidate phylum Krumholzibacteriota phyl. nov. recovered from an anoxic sulfidic spring using genome resolved metagenomics." *Systematic and Applied Microbiology* 42(1): 85–93. doi: 10.1016/j.syapm.2018.11.002.

Yu, X., J. Zhou, W. Song, M. Xu, Q. He, Y. Peng, and Z. He. 2020. "SCycDB: A curated functional gene database for metagenomic profiling of sulfur cycling pathways." *Molecular Ecological Resources*. doi: 10.1111/1755-0998.13306

Zablocki, O., L.J. van Zyl, B. Kirby, and M. Trindade. 2017. "Diversity of dsDNA viruses in a South African hot spring assessed by metagenomics and microscopy." *Viruses* 9(11): 348. doi: 10.3390/v9110348.

Zhou, J., D. Sun, A. Childers, and T. Mcdermott. 2015. "Three novel virophage genomes discovered from Yellowstone Lake metagenomes." Journal of Virology 89(2): 1278–1285. doi: 10.1128/JVI.03039-14.

Section V

Applications

10

Microbiomes in the Rice Ecosystem

Arabinda Mahanty*, Srikanta Lenka, Prabhukarthikeyan SR, Totan Adak, Raghu S., Prakash Chandra Rath
ICAR-National Rice Research Institute, Crop Protection Division, Cuttack, India
Koustav Kumar Panda
Department of Biochemistry, Biotechnology and Physiology, Paralakhemundi, India
Amrita Kumari Panda
Department of Biotechnology, Sarguja University, Ambikapur, India

CONTENTS

10.1 Introduction

Rice is one of the predominant staple foods for about 3.5 billion people across the world. In terms of volume of production, rice ranks third among agricultural commodities, after sugarcane and maize. It serves as a chief source of energy, providing 27% of the dietary energy supply in 33 developing countries. It also serves as a good source of dietary protein, providing 20% of the protein requirement, along with a fair amount of micronutrients, including zinc, riboflavin, niacin, and thiamine (FAO 2004).

Rice plants, like other crop plants, are exposed to a diverse range of biotic and abiotic factors which influence its productivity. Due to the ever-increasing population, there is a pressing need to increase food production. It has been projected that an increase of 25–70% above current production levels would be required to meet the crop demand in the year 2050 (Hunter et al. 2017). However, the resources would be limited, and achieving the target crop yield in an environmentally sustainable way is going to be one of the largest challenges of the 21st century. This would require all the available resources to be used to their fullest potential, which includes the use of rice varieties with higher yield potential and disease and pest resistance, improved agronomic practices, harnessing beneficial microbes, and control of microbial pathogens.

The rice plant is in constant interaction with microbes, and its physiology is highly influenced by this interaction. From the root in the rhizosphere to the leaf and seed, the rice plant is surrounded by a wide range of microbes. The collection of microbes surrounding a particular environment or niche is known as the microbiota. With the development of sequencing technologies, it is possible to sequence the genomes of these microbiota, and the sum total of the genomic elements in a particular environment is known as

DOI: 10.1201/9781003042570-15

the microbiome. Studying the microbiome not only provides an opportunity to identify and harness beneficial microbes but also provides a scope for eliminating harmful microbes, thereby increasing productivity.

The present chapter provides an overview of the microbiomes associated with different parts of the rice plant.

10.2 Strategies for Studying Microbiomes

Traditionally, microbial diversity has been studied using marker genes such as 16s rRNA sequencing for bacteria and 18s rNA or the internal transcribed spacer (ITS) region sequences of eukaryotes like fungi. Other than these, polymerase chain reaction-differential gel electrophoresis (PCR-DGGE) and terminal restriction fragment length polymorphism have been used for studying plant microbiomes (Kim and Lee 2020). However, with developments in high-throughput sequencing or next-generation sequencing (NGS) technology and other -omics technologies such as metagenomics, metatranscriptomics, and metabolomics, the study of the microbiome has been revolutionized.

The NGS platform uses three different approaches for microbiome exploration: 1) the amplicon sequencing approach, 2) the metagenome sequencing approach, and 3) the metatranscriptome sequencing approach.

The amplicon sequencing approach involves amplification of the targeted genomic region such as the 16S rDNA gene (for bacteria) and the 18S rRNA gene or the internal transcribed spacer region (for fungi). This region is amplified and sequenced to find the diversity of the microbial community. The metagenome sequencing approach involves sequencing of the complete DNA extracted from microbial communities present in a sample without any specific gene target. The sequences thus generated are compared with databases and genes are identified. The third and seldom-used approach for microbiome exploration is the metatranscriptomics approach, which involves the sequencing of the transcripts of genes of the microbial community (Sebastien et al. 2015).

The microbiomes in rice ecosystems mainly consist of bacteria, fungi, archaea, and viruses, which not only impact the health of the plant but also the health of the consumer and environment. The rhizospheric microbiome could play a role in improving the growth of the plant by increasing nutrient availability and production of plant growth hormones and substances that antagonize other harmful microbes. At the same time, these microbes could be important from a consumer health point of view, as these could influence the accumulation of toxic compounds in rice (University of Queensland 2020). Similarly, these could also influence the amount of methane produced during the rice cultivation process, thereby affecting environmental health (Liechty et al. 2020). Studying the microbiomes under different conditions provides an understanding of the influence of the microbes on the rice plant and the environment.

To study the microbiomes associated with different tissues of the rice plant in different conditions, all the mentioned techniques have been extensively used. For example, Moronta-Barrios et al. (2018) and Arjun and Harikrishnan (2011) used 16S rRNA sequencing–based microbiome analysis to study the rice root microbiota and rhizospheric microbiomes, respectively. Similarly, Imchen et al. (2019) used the 16S rRNA sequencing method to study the changes in the rhizospheric metagenome during various growth stages of the rice plant. The metagenome shotgun sequencing technique has been extensively used for exploring rice-associated microbiomes and is discussed in later sections.

Along with the 16S sequencing and metagenomics approaches, the metatranscriptomics approach has also been used for exploring rice-associated microbiomes. Masuda et al. (2018) used the metatranscriptomic approach to study the community structure of the microbes associated with methane metabolism in paddy soils. Similarly, Peng et al. (2018) used it to study the effect of differential temperature on structural and functional organization of food webs of rice field soil.

10.3 Microbiomes Surrounding Different Parts of the Rice Plant

The microbiomes surrounding a plant are broadly categorized into two categories; one which resides under the soil, which includes the rhizobiome (rhizospheric microbiome) and endospheric microbiome (Vukanti

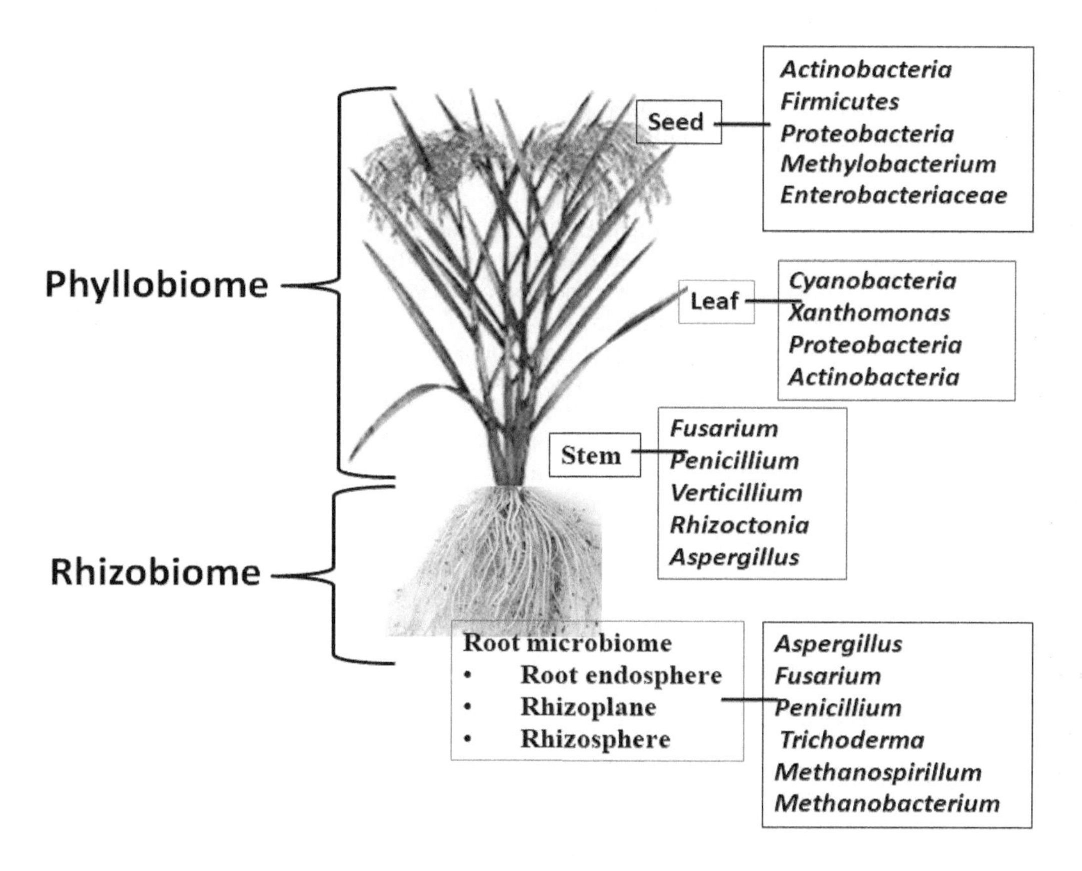

FIGURE 10. 1 Microbiomes associated with different parts of the rice plant, with important representatives.

2020), and the other one resides in the phyllospheric region (above the soil), sometimes called the phyllobiome (Sindelar and Bastola 2020) (Figure 10.1). The microbiomes associated with different parts of the rice plant and their significance are described in detail in later sections.

10.3.1 Root- and Rhizosphere-Associated Microbiomes

The interaction between the plant and its root associated microbiome plays an important role in the availability of nutrition and growth of the plant. In terms of space occupied, microbes could be categorized as endospheric (inside root), rhizoplane (root surface), and rhizospheric microbes (soil surrounding root surface) (Ding et al. 2019). Often, there exists a symbiotic relation between plants and microbes; microbes are dependent on plants for sugars, whereas plants are dependent on microbes for fixed nitrogen. Some of these microbes also play an important role in resisting pathogenic microbes and the development of induced systemic resistance. Moreover, these microbiomes also include methanogenic bacteria, which contributes significantly to global methane production. As the ultimate objective of agricultural scientists and agriculturists is to enhance rice production with minimal methane production, it is extremely important to understand the rice root and rhizospheric microbiome.

The rice plant secretes oxygen through its root aerenchyma, which leads to the formation of an oxygen-rich zone surrounded by anoxic bulk soil (Ding et al. 2019; Yuan et al. 2016; Zhao et al. 2018). This oxic–anoxic interface provides a niche for a diverse group of microbes which includes aerobic, anaerobic, or facultative anaerobic microbes (Ding et al. 2019; Li and Wang 2013).

Ding et al. (2019) extensively reviewed publications on root-associated microbiomes and compiled the different bacterial, fungal, and archaeal populations associated with the root. Numerous studies have uncovered a high diversity of rice endophytic microbiomes and taxonomic information on bacterial, archaeal, and fungal communities. The bacterial phyla commonly inhabiting the rice endosphere are Gammaproteobacteria, Firmicutes, Actinobacteria, and Bacteroidetes (Reinhold-Hurek and Hurek 2011; Sessitsch et al. 2012). The families Enterobacteriaceae and Pseudomonadaceae of the phylum Gammaproteobacteria phylum and the Bacillus genus of the Firmicutes phylum (Hallmann et al. 1997; Hardoim et al. 2008) are the predominant ones.

Among the fungi, *Aspergillus*, *Fusarium*, *Penicillium*, and *Trichoderma* have been found to be the predominant genera in the rice endosphere (Potshangbam et al. 2017; Santos-Medellin et al. 2017). Methanogenic archaea like *Methanospirillum* and *Methanobacterium* genera have also been frequently detected inside rice roots (Edwards et al. 2015).

However, the root and rhizosphere microbiome community structures are highly dynamic and are influenced by a number of biotic and abiotic factors which include the cultivar, soil condition, biogeography, and so on (Gaiero et al. 2013). Edwards et al. (2015) have shown that microbial communities vary by soil type, rice genotype, geographical effects, and fertilizer application. Ikeda et al. (2014) have shown that the root microbiome changes with the dosage of nitrogen fertilizer. They reported increase in abundances of the bacteria *Burkholderia*, *Bradyrhizobium*, and *Methylosinus* in soil where low-nitrogen fertilizers were added compared to soil in which a standard quantity (30 Kg h^{-1}) of nitrogen was added. Conversely, the abundance of methanogenic archaea was lower in a low-nitrogen field.

10.3.2 Phyllosphere Microbiomes

10.3.2.1 Stem-Associated Microbiomes

The above ground part, otherwise known as the phyllosphere, which includes the stem, provides harsher environmental conditions for microbial growth. It is exposed to ultraviolet radiation and rapid fluctuations in temperature on a daily basis. Moreover, it provides less access to nutrients (Kim and Lee 2020). Thus, the microbial density is comparatively less in the stem. Consequently, a limited number of studies have been carried out on the stem microbiome.

Wang et al. (2016) studied the rice sprout, stem, and root microbiomes, and they reported that the stem microbiome has a higher diversity of fungal operational taxonomic units (OTUs) and lower diversity of bacterial taxonomic units as compared to the root. They found 5,277 ITS-2 (ITS-2: internal transcribed spacer 2 regions of fungal rRNA genes) tags and 310 V3–V4 tags (V3–V4: hyper-variable regions of bacterial 16S rRNA genes) in stem. The predominant fungal genera in the stem included *Fusarium*, *Penicillium*, *Aspergillus*, *Trichosporon*, *Pestalotiopsis*, *Verticillium*, and *Cryptococcus*. The predominant bacterial genera included *Methylobacterium*, *Caulobacter*, *Sphingomonas*, *Brevundimonas*, *Devosia*, *Corynebacterium*, and *Pseudomonas*.

10.3.2.2 Leaf Microbiomes

The leaf is the primary site of photosynthesis in plants. Roman-Reyna et al. (2020) extensively studied the leaf microbiome of rice accessions growing in China and the Philippines.

The most abundant microbial phyla found in the leaves were Proteobacteria, Firmicutes, Actinobacteria, Cyanobacteria, Tenericutes, and Euryarchaeota. However, the microbiomes observed in Chinese and Philippine rice plant leaves starkly varied. In the case of Chinese plants, the abundance of the genera *Propionibacterium, Agrobacterium, Acidovorax*, and *Enterobacter* was much higher compared to the Philippine plants, whereas the reverse pattern was observed for *Xanthomonas* and *Serratia*. The authors attributed the genetic ability of the host plant for cellulose and salicylate accumulation as one of the major factors that affect microbiome composition.

Roman-Reyna et al. (2020) employed a unique approach to study the leaf microbiome of rice; they extracted non-plant sequence reads from the 3,000 rice genome project in which the total genome of the rice and associated resident microbial communities have been sequenced. They found that the genera *Pseudomonas, Xanthomonas, Mycoplasma*, and *Burkholderia* were the most abundant bacteria in the leaf microbiome.

10.3.2.3 Seed Microbiome

The seed microbiome plays a very important role in the life cycle of the rice plant. The genetic information in the seed not only comprises the parental genome but also includes the metagenome. The resident bacteria in the endophyte of the seed restrict the growth of other pathogens, which tend to attack seeds during the germination process, thereby promoting seedling development.

The seed microbiome of the rice plant has been studied both by culture-based and sequencing-based methods. Raj et al. (2019) explored the rice seed microbiome using both culture-based and metagenomics approaches. Their study revealed that among the bacterial species, Proteobacteria was the most dominant phylum, comprising five genera, eight species, (72%) sequences, followed by Actinobacteria (18% sequences) and Firmicutes (10% sequences). However, the microbiomes were found to vary with agroecosystem and cultivar. Similarly, Midha et al. (2016) reported the presence of the bacterial phyla Proteobacteria, Firmicutes, and Actinobacteria, which included 15 distinct genera and 29 species. Eyre et al. (2019) further studied microbiome distribution in different compartments of the seed, like the outer husk, husk, outer grain, and grain. It was observed that the microbiomes found in each compartment were fairly distinct from those in the other compartments. However, the genera *Shingomonas* and *Methylobacterium* and the family Enterobacteriaceae were the key bacteria found in all compartments of the seed. Similarly, the genera *Alternaria* and *Hannaella* and the phyla Pleosporales were important fungal species found in the seed.

Wang et al. (2020) characterized the microbiomes in seedlings directly originating from the seed and compared them with the microbiome of the seed in order to identify the core microbes associated with the seedling. The study revealed that *Rhizobium*, *Pantoea*, *Sphingomonas*, and *Paenibacillus* constituted the core bacterial genera associated with the seed.

10.4 Pathogenic Microbes of the Rice Plant

Diseases caused by pathogenic bacteria, fungi, and viruses are among the stumbling blocks for rice production. The important diseases of rice plant include blast, brown spot, sheath blight disease, false smut, bacterial leaf blight, and bacterial leaf streak (Table 10.1). Some diseases like blast may cause up to 80% yield loss in countries like India (http://agritech.tnau.ac.in/crop_protection/crop_prot_crop%20 diseases_cereals_paddy.html). Globally, it has been reported to cause 30% loss in rice production, which could be used for feeding 60 million people across the world (Nalley et al. 2016). Similarly, other diseases also cause significant loss in crop yield.

These disease-causing microbes are present in the rice ecosystem, are transmitted from affected areas to unaffected areas through various means, and infect plants when the environment is conducive to growth and/or the environmental conditions are stressful for the rice plant. Diseases are often controlled by the application of chemical pesticides. However, the proper identification of the disease and correct dosage of chemical agents are the two most important aspects of disease management. Wrong identification of disease could lead to inappropriate use of chemical pesticides, which may pose a number of environmental problems. Thus, it is necessary for farmers to correctly identify the disease and use appropriate chemical or biological agents to control them.

10.5 Microbes with Antagonistic Effects against Pathogens

The rice microbiome not only consists of pathogenic microbes, but it also includes bacteria and fungi that inhibit the growth of pathogens, and these microbes can induce systematic resistance to the rice plant and also promote growth. Some of these have been extensively used as biocontrol agents for controlling plant diseases. The most commonly used biocontrol agents include *Trichoderma* sp. and *Pseudomonas* sp. A list of fungi and bacteria species that have been found effective against rice disease pathogens is given in Table 10.2.

TABLE 10.1

Some Important Diseases of the Rice Plant and Their Causal Pathogens

Disease	Causal Pathogen	Symptoms	Image
Blast	*Magnaporthe grisea*	Small specks formed on leaves, which enlarge to spindle-shaped spots	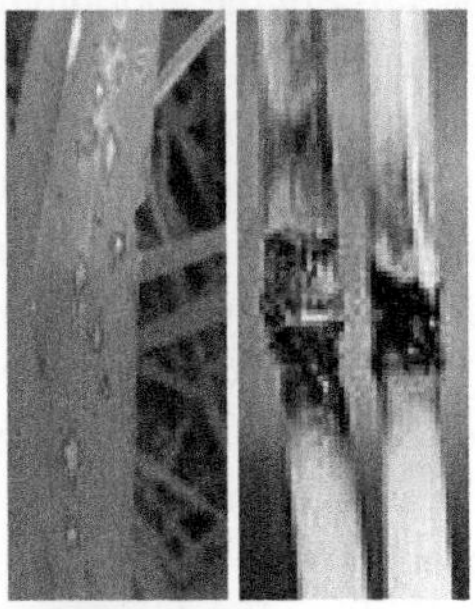
Brown spot	*Helminthosporium oryzae*	The lesions are circular to oval in shape with a grayish center, surrounded by a reddish-brown margin	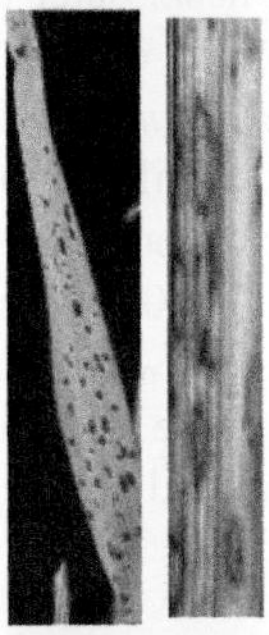
Sheath blight	*Rhizoctonia solani*	Gray lesions on the leaf sheath, similar to rattlesnake skin pattern	
Seedling blight	Fungal consortium	Seedlings rot or die	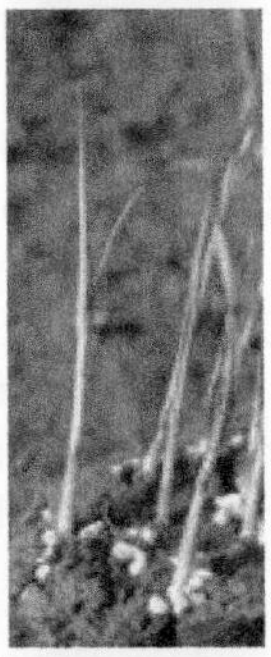
False smut	*Ustilaginoidea virens*	Rice grain transforms into a mass of yellow fruiting bodies	

Disease	Causal Pathogen	Symptoms	Image
Bakanae/foot rot	*Fusarium* sp.	Seedlings dry at early tillering stage; infected plants are several inches taller than normal plants, with yellowish-green leaves and pale green flag leaves	
Bacterial leaf blight	*Xanthomonas oryzae*	Yellowish stripes on leaf blades, which turn white with progression of the disease	
Tungro	Rice tungro virus	Leaves become yellow or orange-yellow with rust-colored spots, which leads to stunting and reduced tillering	

Ref. Mukherjee et al. 2016

TABLE 10.2

Microbes with Prophylactic Effect against Important Rice Pathogens

Microbe	Pathogens	Reference
Fungus		
Trichoderma sp.	*Fusarium* sp. *Phytopthara* sp. *Scelerotia* sp. *Rhizoctonia solani*	Parab et al. 2009 Mukherjee et al. 2013 Woo et al. 2014
Penicillium sp.	*Pyricularia oryzae*	Paraiso et al. 2019
Harpophora oryzae	*Pyricularia oryzae*	Su et al. 2013
Piriformospora indica	*Pyricularia oryzae*	Nassimi and Taheri 2017
Yeast	*Pyricularia oryzae*	Kunyosying et al. 2018
Bacteria		
Pseudomonas sp.	*Rhizoctonia solani*	Laha and Venkataraman 2001
Pseudomonas fluorescens	*Xanthomonas oryzae*	Vidhyasekaran et al. 2001
Bacillus sp.	*Rhizoctonia solani*	Yu et al. 2017
B. subtilis	*Rhizoctonia solani*	Jayaraj et al. 2004
Bacillus amyloliquefaciens and *B. pumilus*	*Magnaporthe oryzae*	Sha et al. 2020
Bacillus spp.	*Pyricularia oryzae*	Rais et al. 2017
Bacillus firmus	*P. oryzae*, *Kocuria rhizophila* *Sclerotium* sp.	Chaiharn et al. 2009
Ochrobactrum anthropi	*Fusarium oxysporum*	Chaiharn et al. 2009
Streptomyces sp.	*Magnaporthe oryzae*	Law et al. 2017
Streptomyces sp.	*Pyricularia oryzae Rhizoctonia solani*	Prabavathy et al. 2006

10.6 Conclusion and Future Perspectives

The microbes associated with the rice plant have both beneficial and harmful effects. Disease-causing microbes have been explored well, but microbes, especially the fungi, with beneficial effects are still to be explored fully. Many of these beneficial microbes not only inhibit the growth of pathogens but also promote plant growth by directly secreting or inducing the secretion of plant hormones. Some of the symbiont fungi have also been found to provide tolerance against abiotic stress. These beneficial effects of the microbes need to be fully explored using advanced omics tools, and harnessing them for promoting plant growth and preventing diseases could be useful in enhancing production and ensuring sustainability in the agriculture sector.

Advanced omics tools in rice microbiome research deciphered the multifaceted network of biochemical, genetic, and metabolic relations among rice plant–associated microbiomes—the surrounding environment. Future meta-omics studies on the microbiome of wild land rices as descendants of modern cultivars can decode their potential and help the future generation of breeders to understand the resistance mechanism of wild varieties against deadly pathogens and stress conditions. Large-scale experiments should be conducted to integrate data from genomics, proteomics, transcriptomics, ionomics, and metabolomics research to identify agronomically important traits and metabolic pathways. Further multi-omics research has to be accelerated to characterize rice-derived phyto-compounds and other active bio-compounds for human health. Genome-wide association studies should be carried out to isolate effective genes for various abiotic and biotic stress tolerances, to decode the molecular mechanisms of tolerance, and to establish the relationship between mutant phenotypes and genotypes. Genes that are expressed under certain conditions or tissues and the identification of superfluous genes are still the challenges that need to be addressed through integrative omics technology, integrative databases, and advanced tools such as modeling and machine learning.

REFERENCES

Arjun JK, Harikrishnan K (2011) Metagenomic analysis of bacterial diversity in the rice rhizosphere soil microbiome. *Biotechnology, Bioinformatics and Bioengineering* 1(3): 361–367. ISSN 2249-9075.

Chaiharn M, Chunhaleuchanon S, Lumyong S (2009) Screening siderophore producing bacteria as potential biological control agent for fungal rice pathogens in Thailand. *World Journal of Microbiology and Biotechnology* 25: 1919–1928.

Ding LJ, Cui HL, Nie SA, Long XE, Duan GL, Zhu YG (2019) Microbiomes inhabiting rice roots and rhizosphere. *FEMS Microbiology Ecology* 95(5): fiz040. https://doi.org/10.1093/femsec/fiz040

Edwards J, Johnson C, Santos-Medellin C, et al. (2015) Structure, variation, and assembly of the root-associated microbiomes of rice. *Proceedings of the National Academy of Sciences of the United States of America* 112: E911–E20.

Eyre AW, Wang M, Oh Y, Dean RA (2019) Identification and characterization of the core rice seed microbiome. *Phytobiomes Journal.* https://doi.org/10.1094/PBIOMES-01-19-0009-R

FAO (2004) Rice and human nutrition. www.fao.org/rice2004/en/f-sheet/factsheet3.pdf Accessed on 22/1/2021.

Gaiero JR, McCall CA, Thompson KA, Day NJ, Best AS, Dunfield KE (2013) Inside the root microbiome: Bacterial root endophytes and plant growth promotion. *American Journal of Botany* 100(9): 1738–1750.

Hallmann J, Quadt-Hallmann A, Mahaffee WF et al (1997) Bacterial endophytes in agricultural crops. *Canadian Journal of Microbiology* 43: 895–914.

Hardoim PR, van Overbeek LS, van Elsas JD (2008) Properties of bacterial endophytes and their proposed role in plant growth. *Trends Microbiology* 16: 463–471.

Hunter MC, Smith RG, Schipanski ME, Atwood LW, Mortensen DA (2017) Agriculture in 2050: Recalibrating targets for sustainable intensification. *BioScience* 67: 386–391.

Ikeda S, Sasaki K, Okubo T, Yamashita A, Terasawa K, Bao Z, Liu D, Watanabe T, Murase J, Asakawa S, Eda S, Mitsui H, Sato T, Minamisawa K (2014) Low nitrogen fertilization adapts rice root microbiome to low nutrient environment by changing biogeochemical functions. *Microbes and Environment* 29(1): 50–59.

Imchen M, Kumavath R, Vaz ABM, Góes-Neto A, Barh D, Ghosh P, Kozyrovska N, Podolich O, Azevedo V (2019) 16S rRNA gene amplicon based metagenomic signatures of rhizobiome community in rice field during various growth stages. *Frontiers Microbiology* 20 September 2019 | https://doi.org/10.3389/fmicb.2019.02103

Jayaraj J, Yi H, Lian GG, et al (2004) Blattapplikation von *Bacillus subtilis* AUBS1 reduziert die Blattscheidendürre und induziertAbwehrmechanismen in Reis [Foliar application of *Bacillus subtilis* AUBS1 reduces sheath blight and triggers defense mechanisms in rice]. *Journal of Plant Diseases and Protection* 111(2): 115–125.

Kim H, Lee YH (2020) The rice microbiome: A model platform for crop holobiole. *Phytobiomes Journal* 4: 5–18.

Kunyosying D, To-anun C, Cheewangkoon R (2018) Control of rice blast disease using antagonistic yeasts. *International Journal of Agricultural Technology* 14(1): 83–98.

Laha GS, Venkataraman S (2001) Sheath blight management in rice with biocontrol agents. *IndianPhytopath* 54 (4): 461–464.

Law JW, Fei, Ser HL, Khan TM, Chuah LH, Pusparajah P, Chan KG, Goh BH, Learn-Han Lee (2017) The potential of Streptomyces as biocontrol agents against the rice blast fungus. *Magnaportheoryzae (Pyriculariaoryzae). Frontiers in Microbiology.* https://doi.org/10.3389/fmicb.2017.00003.

Li Y, Wang X (2013) Root-induced changes in radial oxygen loss, rhizosphere oxygen profile, and nitrification of two rice cultivars in Chinese red soil regions. *Plant Soil* 365: 115–126.

Liechty Z, Santos-Medellín C, Edwards J, Nguyen B, Mikhail D, Eason S, Phillips G, Sundaresan V (2020) Comparative analysis of root microbiomes of rice cultivars with high and low methane emissions reveals differences in abundance of methanogenic archaea and putative upstream fermenters. *mSystems* 5: e00897–19. https://doi.org/10.1128/mSystems.00897-19.

Masuda Y, Itoh H, Shiratori Y, Senoo K (2018) Metatranscriptomic insights into microbial consortia driving methane metabolism in paddy soils. *Soil Science and Plant Nutrition* 64 (4)

Midha S, Bansal K, Sharma S, Kumar N, Patil PP, Chaudhry V, Patil PB (2016) Genomic resource of rice seed associated bacteria. *Frontiers in Microbiology.* https://doi.org/10.3389/fmicb.2015.01551.

Moronta-Barrios F, Gionechetti F, Pallavicini A, Marys E, Venturi V (2018) Bacterial microbiota of rice roots: 16S-Based taxonomic profiling of endophytic and rhizospheric diversity, endophytes isolation and simplified endophytic community. *Microorganisms* 6(1): 14.

Mukherjee AK, Karthikeyan SR, Raghu S, Yadav M, Aravindan S, Lenka S, Bag MK, Dhua U, Adak T, Jena M (2016) Diagnostic guide for rice diseases. ICAR-NRRI pocket diary no. 4. https://icar-nrri.in/wp-content/uploads/2018/07/3.-Diagnostic-guide-for-rice-diseases.pdf.

Mukherjee AK, Sampath Kumar A, Kranthi S, Mukherjee PK (2013) Biocontrol potential of three novel Trichoderma strains: Isolation, evaluation and formulation. *Biotechnology* 3. doi: 10.1007/s13205-013-0150-4.

Nalley L, Tsiboe F, Durand-Morat A, Shew A, Thoma G (2016) Economic and environmental impact of rice blast pathogen (*Magnaporthe oryzae*) alleviation in the United States. *PLoS One* 11(12): e0167295.

Nassimi Z and Taheri P. 2017. Endophytic fungus *Piriformospora indica* induced systemic resistance against rice sheath blight via affecting hydrogen peroxide and antioxidants. *Biocontrol Science and Technology* 27(2): 252–267.

Parab PB, Diwakar MP, Sawant UK (2009) Exploration of *Trichoderma harzianum* as antagonist against *Fusarium* spp. causing damping off and root rot disease and its sensitivity to different fungicides. *Journal of Plant Disease Sciences* 4(1): 52–56.

Paraiso L, Dela Pena F and Capacao R (2019) Fungal epiphytes for biological control of rice blast fungus (*Pyriculariaoryzae*). *Plant Pathology & Quarantine* 9(1): 64–71.

Peng J, Wegner CE, Bei Q, Liu P, Liesack W (2018) Metatranscriptomics reveals a differential temperature effect on the structural and functional organization of the anaerobic food web in rice field soil. *Microbiome* 6: 169.

Potshangbam M, Devi SI, Sahoo D et al (2017) Functional characterization of endophytic fungal community associated with *Oryza sativa* L. and *Zea mays* L. *Frontiers in Microbiology* 8: 325.

Prabavathy VR, Mathivanan N, Murugesan K (2006) Control of blast and sheath blight diseases of rice using antifungal metabolites produced by *Streptomyces* sp. PM5. *Biological Control* 39(3): 313–319.

Rais A, Jabeen Z, Shair F, Hafeez FY, Hassan MN (2017) *Bacillus* spp., a bio-control agent enhances the activity of antioxidant defense enzymes in rice against *Pyricularia oryzae PLoS One* 12(11): e0187412.

Raj G, Shadab M, Deka S, Das M, Baruah J, Bharali R, Talukdar NC (2019) Seed interior microbiome of rice genotypes indigenous to three agroecosystems of Indo-Burma biodiversity hotspot. *BMC Genomics* 20: 924.

Reinhold-Hurek B, Hurek T (2011) Living inside plants: Bacterial endophytes. *Current Opinion in Plant Biology* 14: 435–443.

Roman-Reyna V, Pinili D, Borja FN, Quibod IL, Groen SC, Alexandrov N, Mauleon R, Oliva R (2020) Characterization of the leaf microbiome from whole-genome sequencing data of the 3000 rice genomes project. *Rice* 13: 72.

Santos-Medellín C, Edwards J, Liechty Z et al (2017) Drought stress results in a compartment-specific restructuring of the rice root-associated microbiomes. *MBio* 8: e00764–17.

Sebastien M, Margarita MM, Haissam JM (2015) Biological control in the microbiome era: Challenges and opportunities. *Biological Control* 89: 98–108.

Sessitsch A, Hardoim P, Doring J et al (2012) Functional characteristics of an endophyte community colonizing rice roots as revealed by metagenomic analysis. *Molecular Plant-Microbe Interactions* 25: 28–36.

Sha Y, Zeng Q, Sui S (2020) Screening and application of *Bacillus* strains isolated from nonrhizospheric rice soil for the biocontrol of rice blast. *Plant Pathology Journal* 36(3): 231–243.

Sindelar K, Bastola D (2020) Life in the phyllobiome: Functional adaptations in *Novosphingobium* sp. ‘Leaf2’, a leaf-borne alphaproteobacteria. https://digitalcommons.unomaha.edu/cgi/viewcontent.cgi?article=2473&context=srcaf Accessed on 9/8/2020.

Su ZZ, Mao LJ, Li N, Feng XX, Yuan ZL, Wang LW, Lin FC, Zhang CL (2013) Evidence for biotrophic lifestyle and biocontrol potential of dark septate endophyte *Harpophora oryzae* to rice blast disease. *PLoS One* 8(4): 1–14.

University of Queensland (2020) www.qub.ac.uk/courses/postgraduate-research/phd-opportunities/the-rice-paddy-microbiome-and-its-impact-on-plant-mineral-nutrition.html accessed on 23/12/2020.

Vidhyasekaran P, Kamala N, Ramanathan A, et al (2001) Induction of systemic resistance by *Pseudomonas fluorescens* Pf1 against *Xanthomonas oryzae* pv. Oryzae in rice leaves. *Phytoparasitica* 29(2): 155–166.

Vukanti RVNR (2020) Structure and function of rhizobiome. In: Varma A, Tripathi S, Prasad R (eds) *Plant Microbe Symbiosis*. Springer, Cham. https://doi.org/10.1007/978-3-030-36248-5_13

Wang M, Eyre AW, Thon MR, Oh Y, Dean RA (2020) Dynamic changes in the microbiome of rice during shoot and root growth derived from seeds. *Frontiers in Microbiology* 11: 559728. doi: 10.3389/fmicb.2020.559728

Wang W, Zhai Y, Cao L, Tan H, Zhang R (2016) Endophytic bacterial and fungal microbiota in sprouts, roots and stems of rice (*Oryza sativa* L.). *Microbiological Research* 188–189: 1–8.

Woo SL, Ruocco M, Vinale F, Nigro M, Marra R, Lombardi N, Pascale A, Lanzuise S, Manganiello G, Lorito M (2014) Trichoderma-based products and their widespread use in agriculture. *The Open Mycology Journal* 8: 71–126.

Yu Y, Jiang CH, Wang C, Chen LJ, Li HY, Xu Q, Guo JH (2017) An improved strategy for stable biocontrol agents selecting to control rice sheath blight caused by *Rhizoctoniasolani Microbiological Research* 203: 1–9.

Yuan H, Zhu Z, Liu S et al. (2016) Microbial utilization of rice root exudates: 13C labeling and PLFA composition. *Biology and Fertility of Soils* 52: 615–627.

Zhao Z, Ge T, Gunina A et al. (2018) Carbon and nitrogen availability in paddy soil affects rice photosynthate allocation, microbial community composition, and priming: Combining continuous 13C labeling with PLFA analysis. *Plant Soil*. doi: 10.1007/s11104-018-3873-5.

11

Ecosystem Services of Microbes

Joystu Dutta*
Department of Environmental Science, Sant Gahira Guru University, Sarguja
Tirthankar Sen*
Indian Institute of Technology, Guwahati

CONTENTS

11.1 Ecosystem Services: An Overview

Ecosystems are complex emergent systems that are composed of a variety of abiotic and biotic components. These include humans and microbes as well as a lot of other biotic and abiotic factors. All these biotic and abiotic components interact with each other in various ways, resulting in complex interaction patterns spanning several orders of biological organization. For example, plant–microbe interactions in the soil result in an increased quantity of nutrients and growth-promoting factors being accessible to growing plants. This, over time, may result in an overall increase in the agricultural productivity and yield of the soil.

DOI: 10.1201/9781003042570-16

A discourse into how microbes have, since time immemorial, shaped the course of human evolution and civilization is beyond the scope of this chapter. Instead, this chapter will focus on giving readers a high-level understanding of the following:

- How microbes, directly and indirectly, have promoted the growth and development of mankind.
- How new high-throughput omics-based approaches are providing new insights into the complex interplay of microbes occurring 'behind the scenes' of complex natural and ecological processes.
- How our improved understanding of the complex interplay between the biotic and abiotic components of the biosphere is shaping new approaches to revolutionize biotechnology, bioprocess engineering, and healthcare.

The Millennium Ecosystem Assessment, 2005 (Reid et al. 2005), report distinguishes four categories of ecosystem services—*supporting*, *provisioning*, *regulating*, and *cultural*, as represented in Figure 11.1. Supporting services are regarded as the basis for the services of the other three categories and include fundamental complex natural processes such as nutrient cycling, soil formation, organic matter transformation, and so on. Provisioning services are ecosystem services that describe the material or energy outputs from ecosystems. In other words, provisioning ecosystem services refer to products obtained from ecosystems. They include food, raw materials (such as minerals, wood, etc.), freshwater, clean breathable air, and medicinal resources (such as pharmaceuticals, chemical models, test and assay organisms, bioactive phytochemicals, etc.). Regulating services are the services that ecosystems provide by regulating the quality of resources present within the ecosystem. Examples of this include the regulation of the quality of air, water, and soil in the ecosystem that ensures both the healthy growth of the inhabitants of the ecosystem as well as disease control. In other words, they are the benefits obtained from the regulation of ecosystem services. They include local climate and air quality regulation, carbon sequestration, carbon storage, moderation of extreme events, wastewater treatment, erosion prevention and maintenance of soil fertility, pollination, and biological control of pests and pathogens. Cultural services refer to the nonmaterial benefits obtained from ecosystems and will not be covered at depth in this chapter due to their obtuse and indirect relationship with microbes and their metacommunity-level interactions.

11.2 Microbial Ecology and Supporting Ecosystem Services

Microorganisms impact the entire biosphere and play a pivotal role in the regulation of biogeochemical systems. From hydrothermal vents to volcanic craters, from polar ice caps to frozen glacial lakes, from tropical forests to the arid deserts, from the islands of Kiribati to the mighty Himalayan mountains, microorganisms are omnipresent (Bowler et al. 2009; Konopka 2009).

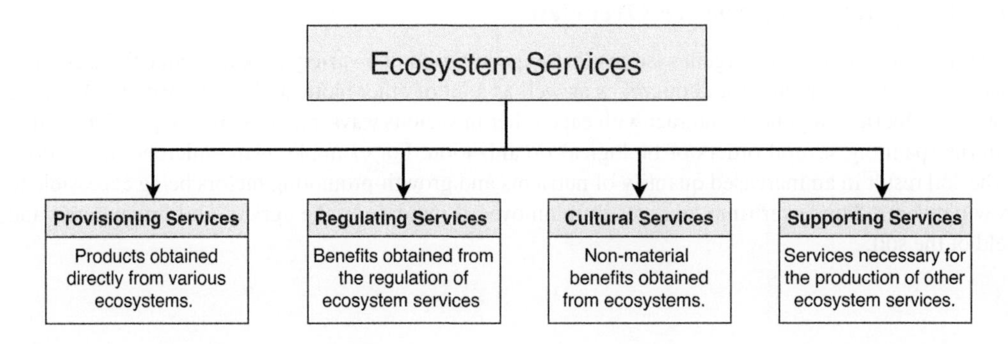

FIGURE 11.1 The four categories of ecosystem services as given by the Millennium Ecosystem Assessment, 2005.

The global microbial biomass is enormous and constitutes a significant carbon sink (Reddy 2004). Microorganisms help in carbon fixation and are known to influence key metabolic processes such as nitrogen fixation, methane metabolism, and sulfur metabolism (Delong 2009). They are also known to have significant control over global biogeochemical cycling (Delong 2009). The understanding of the ecology of microorganisms and their specific role in ecosystem processes is crucial, as most of mankind's current challenges and existential threats have their solutions deeply knitted within the ecological world of microorganisms. Advanced techniques such as community fingerprinting and metagenomics are now being used extensively to temporally monitor the growth and behavior of complex microbial communities. They can be further employed to assess biodiversity. It can thus be concluded that microbes form the backbone of all ecosystems.

11.3 Provisioning Ecosystem Services

Provisioning services are ecosystem services are responsible for the natural materials and products that are obtained and extracted from ecosystems. Provisioning services can also be said to describe the material or energy outputs from ecosystems. The provisioning ecosystem services described in this section are illustrated in Figure 11.2.

11.3.1 Human Health and the Microbiome

The entire set of microbes living symbiotically on and within various locations of the human body is collectively called the human microbiome (Kho and Lal 2018). These microbes include protozoa, archaea, eukaryotes, viruses, and bacteria, bacteria being the most predominant (Kho and Lal 2018). Our oral cavity, genitalia, respiratory tract, skin, and gastrointestinal systems are examples of microbe-occupied habitats in our body (Lloyd-Price et al. 2016). The human microbiota is expected to consist of approximately 10^{13}–10^{14} microbial cells (Sender et al. 2016). Around 3.8×10^{13} bacterial cells are estimated to be inhabiting the colon, the organ with the highest microbial density (Sender et al. 2016). The diverse gut microbiota is mainly composed of three major bacterial phyla: Firmicutes, Bacteroidetes, and Actinobacteria (Tap et al. 2009). It is estimated that the gut microbiota harbors 50× to 100× more genes as compared to the host's genome. The genomes of the microbes comprising the gut microbiome expand the host (human) genome's functional capabilities and play a critical role in a variety of metabolic and other roles (Hooper and Gordon 2001).

As the role of the gut microbiome in the healthy development and maintenance of our overall health is being actively explored, we are coming to know more and more about the beneficial roles and aspects of the gut microbiome. For example, studies using rodents raised in a germ-free environment have demonstrated that the gut microbiota appears to influence proper emotional behavior development, stress and

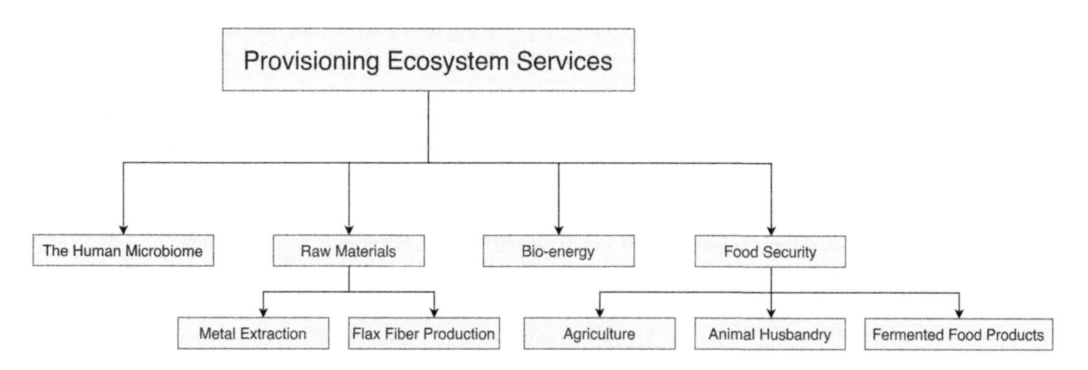

FIGURE 11.2 Provisioning ecosystem services.

pain-modulation systems and even brain neurotransmitter systems (Mayer et al. 2015). Active research is currently going on in the domain of the gut/brain axis. A lot of high-quality research in this domain concerns the bidirectional interactions between the central nervous system (CNS), the enteric nervous system (ENS) and the gastrointestinal (GI) tract. The microbiota also functions in association with the host's immune system to confer protection against pathogen colonization and invasion (Carding et al. 2015). There is growing evidence that an imbalance (dysbiosis) of the gut microbiota can trigger the pathogenesis of both intestinal disorders (inflammatory bowel disease, irritable bowel syndrome, celiac disease) as well as extra-intestinal disorders (allergy, asthma, metabolic syndrome, cardiovascular disease, obesity) (Carding et al. 2015). Due to the difficulty of examining non-cultivable microbes as well as the lack of population-scale datasets, the properties of the human microbiome and its concomitant host–microbe interactions have been largely unexplored (Kho and Lal 2018). It was demonstrated for the first time in 2018 that the microbial signatures of women suffering from triple negative and triple positive breast cancer are distinct and significantly different from the microbial signatures found in ER-positive and HER 2–positive breast cancer patients (Banerjee et al. 2018). In another exciting paper (H.-Y. Liu et al. 2017), the authors examined the microbiological relationships between oropharyngeal swabs and sputum and found remarkable similarities between the two. Therefore, based on the insights obtained from the study, oropharyngeal swabs can potentially be used as a viable alternative to sputum samples for patients with exacerbated chronic obstructive pulmonary disease symptoms. McKenney et al. (2018) draws exciting parallels between natural ecosystem services and their functional counterparts in the human gut microbiome. The authors further utilized the Millennium Ecosystem Assessment, 2005, framework to outline a 'microbiome services' framework that incorporated supporting, provisioning, regulating, and cultural services but all in context of how the gut microbes execute these roles in service (or disservice) to the human host (McKenney et al. 2018).

11.3.2 Raw Materials

11.3.2.1 Metal Extraction

Biomining is the application of our understanding of microbiology and microorganisms to facilitate the extraction and recovery processes of precious and base metals from their primary ores and concentrates. It is now a promising and fast-progressing field of biotechnology (Rawlings and Johnson 2007). The microorganisms that catalyze the biomining processes are required to grow in inorganic, aerobic, and highly acidic environments. Although the particulars of the energy source(s) used by these microorganisms are highly variable and situation dependent, they generally oxidize reduced sulfur or ferrous iron (or both) (Rawlings and Johnson 2007).

Rawlings and Johnson (2007) discuss the theoretical, pragmatic, and engineering challenges associated with construction and application of microbial consortia to process different mineral ores and concentrates in non-sterile, industrial-scale, on-site operations. Apart from the application of microbial consortia for the solubilization of metals from their respective minerals, bioaugmentation with iron-reducing bacteria (IRB) is generally considered an effective method to enhance the activity of iron reduction. A bioaugmentation study conducted by Pan et al. (2017) involving Fe (II)-poor sediments and an iron-reducing bacterial consortium suggested that bacterial consortia might be a better choice for bioaugmentation purposes as compared to pure strains because of the less stringent environmental condition requirement.

11.3.2.2 Flax Fiber

Flax fiber is extracted from the bast beneath the stem of the flax plant. It is soft, lustrous, flexible, and stronger but less elastic than cotton fiber. Flax dew-retting is a key step in the industrial extraction of fibers from flax stems. This process depends on the production of a variety of hydrolytic enzymes produced by a consortium of microorganisms. To obtain deeper insights about the microbial communities involved in this process, Djemiel et al. (2017) applied a high-throughput sequencing technique on plant and soil samples obtained over a period of nearly two months. Their work was the first exhaustive attempt to explore the microbial community dynamics involved in the flax dew-retting process, thus providing a valuable benchmark for future studies aiming to evaluate the effects of other parameters such as spatial and temporal variability in this complex yet industrially important process.

11.3.3 Bioenergy

Human civilization's ever-increasing demand for energy, our fast-depleting fossil fuel reserves, and the impending doom of global warming are projected to bring about catastrophic consequences in the near future. Leveraging the potential of versatile microorganisms to process biomass and other biological waste products into renewable fuels is a promising approach to diminish this existential challenge. Interest in the production of various biofuels using microorganisms has steadily increased in recent years (Liao et al. 2016). This is partially accelerated by enhancing our understanding of the metabolic idiosyncrasies of the diverse range of microorganisms inhabiting the planet. A lot of these microbial biochemical strategies can be utilized to produce biofuels from various substrates. For example, a considerable range of bacterial species can metabolize sugars into ethanol. Cellulose-degrading microbes can utilize plant-driven substrates. Cyanobacteria and microalgae can photosynthetically reduce atmospheric CO_2 into biomass, which can be utilized as biofuels. Methanotrophs can capture methane and utilize it to produce methanol (Liao et al. 2016). Some bacterial species such as *Geobacter sulfurreducens* and *Shewanella oneidensis* can directly transfer electrons from their microbial outer membrane onto conductive surfaces (Kracke et al. 2015). Biohydrogen and bioelectricity can be generated by utilizing this unique property (Kumar and Kumar 2017).

Biogas is a renewable fuel primarily composed of methane and carbon dioxide produced by the natural decomposition of organic matter. Anaerobic digestion is an effective way of producing biogas while at the same time recycling waste organic biomass residues (Surendra et al. 2014). The methanogenesis stage of the anaerobic digestion bioprocess is performed exclusively by distinct archaeal groups. This lack of functional redundancy and diversity of methanogenic archaea often pushes methanogenesis into unfavorable bioprocess conditions such as trace element deprivation (Demirel 2014). The work done by Wintsche et al. (2018) provides a much better understanding of how methanogens cope with trace element limitation while sustaining their growth and metabolism, which ultimately results in bioprocess stabilization. Microalgae are a promising substrate for biogas production due to their ability to utilize CO_2 to synthesize biomass and their high productivity compared to other biomasses. However, biogas production from microalgae cannot still be scaled to an industrially viable level, mainly because the microalgae cell wall is difficult to degrade by hydrolytic bacteria. Attempts have been made by Córdova et al. (2018) to tackle this limitation and improve biogas production by studying the meta-transcriptome of anaerobic sludge–associated microbial communities involved in the bio-methanization of *Chlorella sorokiniana* in both the presence as well as absence of enzymatic pretreatment.

11.3.4 Food Security

11.3.4.1 Agriculture

Agriculture is one of the most significant and impactful developments in the history of modern man. The role of microbes in plant growth and agriculture is extremely complex and varied and has been extensively studied for decades. Agricultural microbiologists worldwide have undertaken species-specific research to further understand the role of microbes in maintaining optimal crop health, thus further contributing to the development of better agricultural practices to feed the global population. Li et al. (2017) isolated and characterized *Pseudomonas* spp. from the sugarcane rhizosphere to evaluate their plant-growth–promoting (PGP) traits and nitrogenase activity. Their study was the first that provided rare insights into the bacterial genus *Pseudomonas* and its role in the sugarcane rhizosphere.

The impact the sustained application of fertilizers has on topsoil microbial communities due to changes in nutrient availability has been extensively explored. However, the effects of the same on communities and their associations with soil nutrients in the subsoil are still unclear (Gu et al. 2017). A study conducted by Gu et al. (2017) revealed that long-term fertilization played a significant role in determining the composition and interactions of the bacterial and archaeal communities inhabiting the soil.

Although triple superphosphate is an excellent soluble phosphorus source, its prohibitive cost makes the long-term use of crude rock phosphate a more attractive alternative in developing countries. However, the influence rock phosphate exerts on the plant-associated microbiota is inadequately understood. Silva et al. (2017) compared the long-term effects triple superphosphate and rock phosphate fertilization had on the structure of microbial communities inhabiting the maize rhizosphere using next-generation sequencing (NGS). The study indicated a few candidate microorganism groups possibly involved in rock phosphate solubilization, which could further be evaluated for their potential to serve as inoculants in greenhouses and agricultural applications. Similar work by Correa-Galeote et al. (2018) studied the role of soil cultivation history in the structuring of the endophytic bacterial communities of maize plants. The authors found evidence for the fact that cultivation history is an important driver of the endophytic colonization of maize. Metagenomics-based approaches are also enriching our understanding of food crops such as paddy. Bai et al. (2017) employed high-throughput sequencing and microarray-based techniques to better understand the variability of soil microorganisms in various soil types and their potential economic significance.

Decoding the mechanistic underpinnings of microbial community interactions in agriculture would be an economic boon for farmers living in regions with problematic soil conditions where conventional modes of agriculture are cumbersome and non-productive.

11.3.4.2 Animal Husbandry

The role of the gut microbiome in human health and wellness has already been discussed. However, the gut microbiome is not exclusive to humans. Different animal species possess different microbiomes, each equally or more complex than the human microbiome. Moreover, animal microbiomes influence the health of livestock, pets, disease vectors, and species that uphold ecosystems (Trinh et al. 2018). An improved understanding of these different and functionally diverse microbiomes as well as their interaction patterns can help us optimize livestock rearing and consequently the economic benefits therein.

B. Yang et al. (2018) observed that alfalfa supplementation to starter diets during the pre-weaning period of Hu lambs (a fat-tailed sheep with origins in Mongolia) increased rumen papillae length and rumen weight, decreased the incidence of feed plaques, and consequently led to increased feed intake, average daily gain, and carcass weight during the pre- and post-weaning periods. B. Yang et al. (2018) characterized the basis of their observations by studying the effects of alfalfa intervention on the rumen microbial colonization of Hu lambs during the early life, thus providing strategies to improve rumen function through manipulation of the rumen microbiota during early life.

To effectively carry out the pre-intestinal digestion of plant materials, ruminants possess a characteristic four-chambered stomach. The four chambers of the stomach are the rumen, reticulum, omasum, and abomasum, respectively. The rumen is the largest and the most well studied among these four compartments. Lei et al. (2018) studied the microbial profiles in the reticulum, omasum, and abomasum of baby goats, thus providing interesting insights into the microbiota dynamics of ruminant stomachs. The insights obtained from the study are expected to allow livestock owners to better design juvenile ruminant starter diets during their early stages of development.

11.3.4.3 Fermented Food Products

The application of microbes in the preservation of foods as well as the preparation of fermented food products dates back through thousands of years of human civilization. The earliest evidence of wheat-and-barley–based alcohol preparation dates to around 13,000 years ago. Recent developments in metagenomics are providing us with a high-resolution view of the microbial processes underlying the production of food products, flavors, and beverages.

Cheeses harbor a diverse microbiome. The cheese microbiota plays a critical role in determining its flavor, odor, appearance, and mouthfeel, as well as other nutritional properties (Yeluri Jonnala et al. 2018). It is composed primarily of a resident microflora that interacts with deliberately inoculated starter and adjunct cultures. The cheese microbiome mainly consists of Firmicutes (lactic acid bacteria, staphylococci), Actinobacteria (coryneform bacteria), Proteobacteria, Bacteroidetes, yeasts, and molds. High-throughput sequencing (HTS) is gaining particular attention by virtue of its potential to provide insights into the total microbial community of the cheese. In their review, Yeluri Jonnala et al. (2018) highlighted several studies in which the cheese microbiota was studied using HTS approaches. The authors paid particular attention to those variants of cheese that have the greatest relevance to industry. The authors also noted that while there are very many advantages associated with the use of HTS, some barriers, such as the detection of DNA sequences from non-viable bacteria, cost effectiveness, and experience in bioinformatic analysis, need to be overcome before these benefits can be applied extensively across the dairy industry.

Cheese and dairy products are not the only fermented products where metagenomic approaches are finding applications. *Zygosaccharomyces bailii* is a yeast species widely present in a variety of fermented foods such as wine, tea, and vinegar. It is also the dominant species in Maotai-flavor liquor fermentation (Xu et al. 2017). Despite its wide range of applications, the metabolic underpinnings of *Z. bailii* in the context of liquor fermentation are poorly understood. Xu et al. (2017) conducted a genome sequencing and comparative transcriptome analysis of a strain of *Z. bailii* to unravel its metabolic mechanisms at the relatively high temperature used in liquor fermentation. The results shed new light on the biology of *Z. bailii* and provide insights into how to utilize it more efficiently in food fermentation processes.

11.4 Regulating Ecosystem Services

Regulating ecosystem services maintain the environmental conditions favorable for the existence and propagation of life within the ecosystem. Regulating ecosystem services maintain the proper cycling of nutrients and reproduction of healthy organisms within the ecosystem. The regulating ecosystem services described in this section are illustrated in Figure 11.3.

11.4.1 Heavy Metal Bioremediation

Heavy metal toxicity is an immense environmental and human health concern because of their bioaccumulation potential and natural nonbiodegradability (Igiri et al. 2018). The chemistry underlying the toxic effects heavy metals have on biological systems can involve various mechanisms. These may include the poisoning of critical enzymatic reactions, reactive oxygen species (ROS) production, ion balance dysregulation, DNA strand breaks, and protein structure damage (Igiri et al. 2018; Ramasamy et al. 2007). The critical ecosystem service of heavy metal detoxification is performed by a multitude of microbes inhabiting the planet alongside us. Microorganisms adopt a plethora of different strategies to survive and even thrive in the presence of heavy metals at concentrations that would be unsustainable for humans. Some of the significant mechanistic pathways microbes utilize to resist the deleterious effects of heavy metals are illustrated in Figure 11.4 (Igiri et al. 2018; Ramasamy et al. 2007).

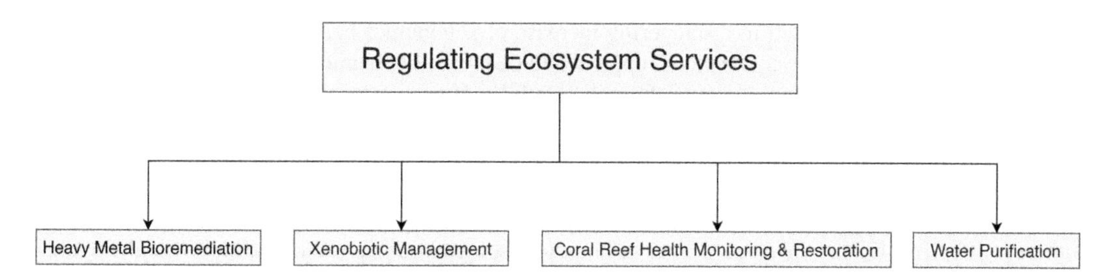

FIGURE 11.3 Regulating ecosystem services.

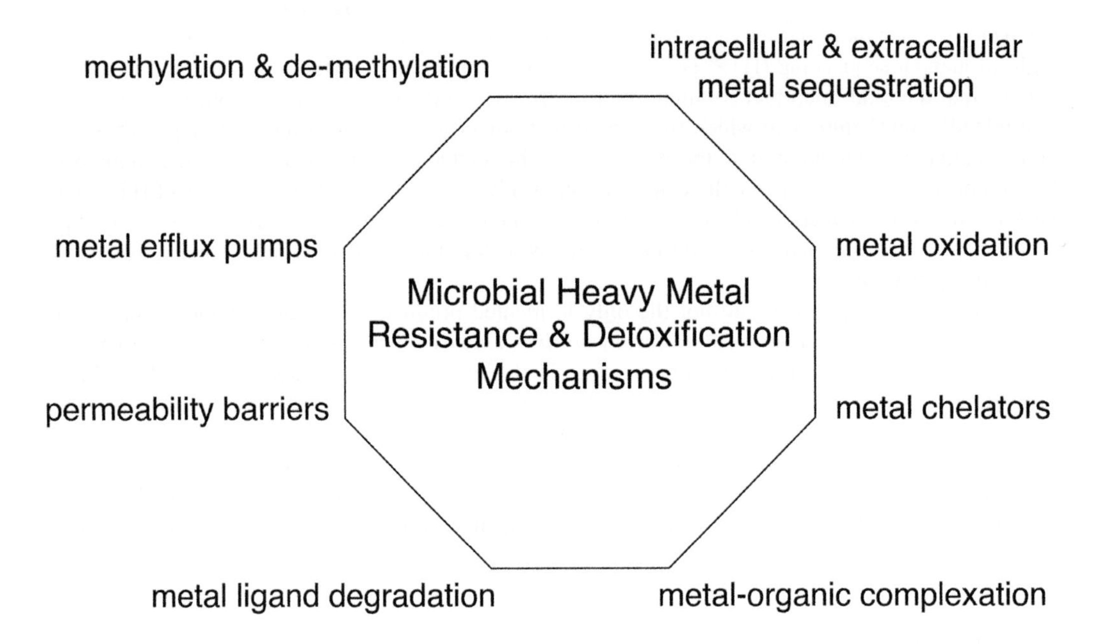

FIGURE 11.4 Some major mechanistic pathways microbes utilize to resist the deleterious effects of heavy metals.

Recent advances in metagenomics have allowed us to peer much deeper into the mechanistic underpinnings of the diverse set of biochemical strategies microbes employ to bioremediate heavy metals. These 'omics' technologies have also allowed us to deeply appreciate the complex community dynamics under complex ecological processes. These technologies have allowed us, for the first time, to peer into the obscure world of organisms recalcitrant to isolation and culture-based propagation. Hemmat-Jou et al. (2018) observed a diverse bacterial and archaeal consortium from samples collected from the Bama Mine located in the Isfahan province, Iran, using a metagenomic approach. Another study (Costa et al. 2015) analyzed the heavy metal–rich tropical sediment of one of the world's largest mining regions, the Mina Stream (located in the Iron Quadrangle in Brazil). The authors uncovered interesting insights regarding the structure of prokaryotic communities inhabiting the tropical freshwater sediment and their possible overlapping roles in biogeochemical cycles and heavy metal biotransformation.

11.4.2 Xenobiotic Management

Xenobiotics are 'chemicals to which an organism is exposed that are extrinsic to the normal metabolism of that organism' (Croom 2012). According to Qadir et al. (2017), 'environmental pollutants, hydrocarbons, food additives, oil mixtures, pesticides, other xenobiotics, synthetic polymers, carcinogens, drugs, and antioxidants are the major groups of xenobiotics'. The recent decades of rapid scientific and industrial development have resulted in a staggering increase of xenobiotics being dumped into the various ecosystems of our planet. The conventional approach to disposing of chemical wastes is to transfer them to waste disposal sites such as (but not exclusively) landfills. However, these methods are unsustainable and inefficient due to a variety of risks, including leakage and leaching as well as contaminant handling and transport (Vidali 2001).

A much more efficient long-term approach is to chemically transform the pollutants such that they are no longer a threat to the biosphere. Thermal and chemical technologies to destroy or detoxify pollutants, although effective, have their own caveats such as technological complexity, scaling difficulties, and distributed public consensus and approval (particularly applicable for waste disposal technologies such as incineration) (Vidali 2001).

Bioremediation techniques are typically more efficient and economical than the traditional approaches to waste disposal due to lower risk and resource requirements. Bioremediation is done using naturally occurring microbes to degrade or detoxify substances hazardous to human health and/or the environment. A wide variety of taxonomically diverse microorganisms can transform environmentally hazardous chemicals through reactions that are a part of their metabolic processes (Singh et al. 2016; Vidali 2001). The microorganisms may be indigenous or extraneous (Vidali 2001). Xenobiotic bioremediation may be in situ or ex situ depending on the degree of saturation and aeration of an area, as illustrated in Figure 11.5.

It has been observed that xenobiotic degrading microorganisms do not generally exist in monocultures in the environment. Instead, many bacteria are found to exist in close association with one another. These naturally occurring microbial consortia exhibit a biodegradation performance much superior to that exhibited by a single species (Mikesková et al. 2012). In their study of soil bacterial community shifts in response to polycyclic aromatic hydrocarbon contamination, Storey et al. (2018) highlighted the need to better characterize unculturable soil microorganisms that are thought to have significant roles in bioremediation.

Metagenomic approaches are slowly allowing scientists to explore, comprehend, and visualize the structural, functional, and temporal aspects of the microbial consortia present in the environment in response to different environmental pollutants.

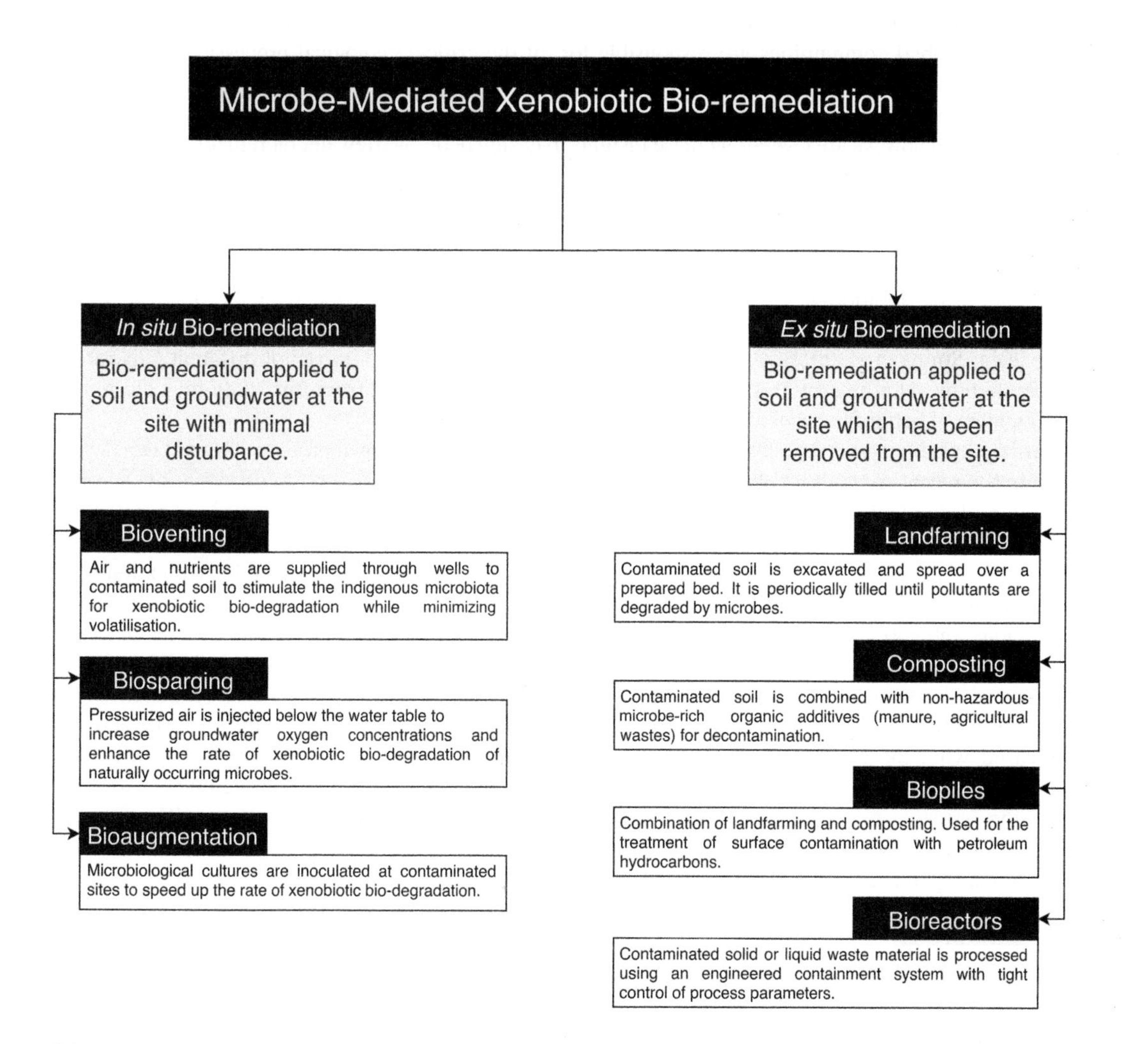

FIGURE 11.5 In-situ vs. ex-situ bioremediation.

11.4.3 Coral Reef Health Monitoring and Restoration

Coral reefs are often referred to as the rainforests of the sea. Approximately 25% of the ocean's fishes are dependent on coral reefs for their food, shelter, safety, and reproduction ("Value of Corals | Coral Reef Systems" 2020).

Despite the undeniable importance of coral reefs around the world, they are undergoing massive extinction due to a combination of anthropogenic activities (overfishing, pollution) and natural causes (infectious diseases, ocean warming, acidification) (De'Ath et al. 2009, 2012). For example, the genus *Vibrio* is known to harbor strains pathogenic to marine organisms. *Vibrio coralliilyticus*, a causative agent of the white plague disease (WPD), is among the best-characterized coral pathogens from which the main Abrolhos reef builders, *Mussismilia* species, have suffered massively (Garcia et al. 2013). Chimetto Tonon et al. (2017) developed a rapid and reproducible qPCR assay-based detection system for *Vibrio*, thus providing a reliable tool for the monitoring of coral pathogens. Levin et al. (2017) integrated the growing wealth of *Symbiodinium* NGS data to design tailored genetic engineering strategies and develop a testable expression construct. The authors also discussed how the application of genetic engineering could enhance the stress tolerance of *Symbiodinium* and, in turn, coral reefs.

11.4.4 Water Purification

Microbes play a critical role in the purification and nutrient cycling of aquatic ecosystems. Diverse and complex microbial communities are responsible for all the critical ecological processes occurring in water bodies such as oceans, lakes, rivers, streams, estuaries, and wetlands. In this section, we will look into the critical roles microbial communities play in driving supporting ecosystem services in aquatic ecosystems and how we are leveraging these insights to design better wastewater bioremediation systems.

Nitrogen is an important nutrient for aquatic organisms. An excess of nitrogen in freshwater ponds leads to eutrophication and the death of fish and shrimps (Camargo and Alonso 2006). Denitrification is a major pathway for removing nitrogen from water bodies (Chen et al. 2017). It can help reduce the negative impacts of nitrogen pollution via the conversion of nitrate to nitrogen gas. Previous studies have investigated how a multitude of factors such as temperature, pH, dissolved oxygen level, organic carbon species, and organic carbon to nitrogen ratio (C/N) can impact the efficiency of denitrification (Kraft et al. 2014; Strong et al. 2011). Chen et al. (2017) studied the effect and impact of different C/N ratios on nitrate removal efficiency in pond sediments. The insights obtained from this study may aid in the development of better freshwater nitrogen pollution reduction systems.

Partial nitritation/anammox (PN/A) is a novel biological nitrogen removal process that has been developed in the recent years. It involves utilizing 1) ammonium oxidizing bacteria to carry out the aerobic oxidation of ammonium to nitrite and 2) anammox bacteria to carry out the anaerobic conversion of ammonium and nitrite to nitrogen gas (Yang et al. 2018). This process saves 60% of the aeration energy and 100% of the external carbon source requirements of conventional nitrification/denitrification processes (Yang et al. 2018). Gonzalez-Martinez et al. (2018) subjected a partial-nitritation biofilter to a continuous antibiotic loading of azithromycin, norfloxacin, trimethoprim, and sulfamethoxazole. This study is especially interesting, as wastewater treatment systems have been implicated in the emergence and spread of antibiotic resistance, a major environmental and public health concern (Barancheshme and Munir 2019). The authors reported that due to the antibiotic load in the water, the bacterial activity decreased, while the activity of archaea and fungi increased (Gonzalez-Martinez et al. 2018). The results linked the performance of the bioreactor, the bacterial community dynamics, and the microbial activity in the system to reach an evaluation of the overall impact of antibiotics on the system. The insights obtained from this study have the potential to be of immense value for the treatment of highly antibiotic contaminated effluents (Gonzalez-Martinez et al. 2018).

11.5 Cultural Services

Metagenomics is bringing forth a revolution in science, technology, and healthcare. These approaches are being adopted to preserve our heritage, culture, and history. They have been applied to understand the nature of the microflora, colonizing storeroom objects in the largest museum in China, the Tianjin Museum, that causes aesthetic changes on the surfaces of objects such as discoloration, biofilm formation, and eventually complete structural integrity failures (Cappitelli et al. 2010; Sterflinger and Pinzari 2012; Z. Liu et al. 2018). In another similar study, Z. Liu et al. (2017) explored the nature of the fungal communities colonizing a wooden tomb dating back to the Western Han Dynasty (206 B.C.–25 A.D.) and uncovered useful insights for informing future contamination mitigation efforts for cultural artifacts. Vilanova and Porcar (2020) discuss how metagenomics can be central for identifying the microbial key players on artwork surfaces. This knowledge can then be utilized to drive artifact conservation efforts using a range of techniques. One of these approaches involves seeding the artifact with a consortium of selected microorganisms because of their direct or indirect effect in stabilizing and preserving valuable art objects.

11.6 Conclusion

The roles of microbes are central to the functioning of almost all ecological processes. Over the past decade, our understanding of metagenomics and the complex interaction patterns between different microbes and between microbes and other ecosystem components has increased significantly. We are starting to realize how truly inextricable the underlying microbiology and community dynamics of the various ecosystem services are. Although categorizing ecosystem services as supporting, provisioning, regulating, or cultural provides us with a convenient heuristic, when we consider the roles of microbes behind these ecosystem services, these categories start to converge into each other. This issue is more pronounced when we look at fast-developing knowledge of microbial interaction patterns used for human benefit.

This chapter focuses on the microbiological and metagenomic aspects of some ecosystem services that represent other similar ecosystem services and the research efforts being directed to understand them better. In summary, this chapter is just an introduction to the interesting yet esoterically complex world of microbes, microbial interactions, metagenomes, and the emergent phenomena that arise due to these interactions.

11.7 Acknowledgments

The authors acknowledge the contributions of their respective institutions. The authors would also like to acknowledge Dr Abhijit Mitra, Director of Research, Techno India University, West Bengal, and Dr Amrita Kumari Panda, Assistant Professor, Department of Biotechnology, Sant Gahira Guru University, Sarguja, Ambikapur, Chhattisgarh, for their constant motivation and encouragement.

REFERENCES

"Value of Corals I Coral Reef Systems." 2020. Accessed August 2. https://scripps.ucsd.edu/projects/coralreefsystems/about-coral-reefs/value-of-corals/.

Bai, Ren, Jun-Tao Wang, Ye Deng, Ji-Zheng He, Kai Feng, and Li-Mei Zhang. 2017. "Microbial Community and Functional Structure Significantly Varied among Distinct Types of Paddy Soils but Responded Differently along Gradients of Soil Depth Layers." *Frontiers in Microbiology* 8 (May). Frontiers Media S.A.: 945. doi:10.3389/fmicb.2017.00945.

Banerjee, Sagarika, Tian Tian, Zhi Wei, Natalie Shih, Michael D. Feldman, Kristen N. Peck, Angela M. DeMichele, James C. Alwine, and Erle S. Robertson. 2018. "Distinct Microbial Signatures Associated with Different Breast Cancer Types." *Frontiers in Microbiology* 9 (May). Frontiers Media S.A.: 951. doi:10.3389/fmicb.2018.00951.

Barancheshme, Fateme, and Mariya Munir. 2019. "Development of Antibiotic Resistance in Wastewater Treatment Plants." In *Antimicrobial Resistance—A Global Threat*. IntechOpen. doi:10.5772/intechopen.81538.

Bowler, Chris, David M. Karl, and Rita R. Colwell. 2009. "Microbial Oceanography in a Sea of Opportunity." *Nature* 459 (7244): 180–4.

Camargo, Julio A., and Álvaro Alonso. 2006. "Ecological and Toxicological Effects of Inorganic Nitrogen Pollution in Aquatic Ecosystems: A Global Assessment." *Environment International.* Elsevier Ltd. doi:10.1016/j.envint.2006.05.002.

Cappitelli, Francesca, Giovanna Pasquariello, Gianfranco Tarsitani, and Claudia Sorlini. 2010. "Scripta Manent? Assessing Microbial Risk to Paper Heritage." *Trends in Microbiology.* Elsevier. doi:10.1016/j.tim.2010.09.004.

Carding, Simon, Kristin Verbeke, Daniel T. Vipond, Bernard M. Corfe, and Lauren J. Owen. 2015. "Dysbiosis of the Gut Microbiota in Disease." *Microbial Ecology in Health & Disease* 26. Co-Action Publishing. doi:10.3402/mehd.v26.26191.

Chen, Rong, Min Deng, Xugang He, and Jie Hou. 2017. "Enhancing Nitrate Removal from Freshwater Pond by Regulating Carbon/Nitrogen Ratio." *Frontiers in Microbiology* 8 (SEP). Frontiers Media S.A.: 1712. doi:10.3389/fmicb.2017.01712.

Chimetto Tonon, Luciane A., Janelle R. Thompson, Ana P. B. Moreira, Gizele D. Garcia, Kevin Penn, Rachelle Lim, Roberto G. S. Berlinck, Cristiane C. Thompson, and Fabiano L. Thompson. 2017. "Quantitative Detection of Active Vibrios Associated with White Plague Disease in *Mussismilia braziliensis* Corals." *Frontiers in Microbiology* 8 (November). Frontiers Media S.A.: 2272. doi:10.3389/fmicb.2017.02272.

Córdova, Olivia, Rolando Chamy, Lorna Guerrero, and Aminael Sánchez-Rodríguez. 2018. "Assessing the Effect of Pretreatments on the Structure and Functionality of Microbial Communities for the Bioconversion of Microalgae to Biogas." *Frontiers in Microbiology* 9 (June). Frontiers Media S.A.: 1388. doi:10.3389/fmicb.2018.01388.

Correa-Galeote, David, Eulogio J. Bedmar, and Gregorio J. Arone. 2018. "Maize Endophytic Bacterial Diversity as Affected by Soil Cultivation History." *Frontiers in Microbiology* 9 (March). Frontiers Media S.A.: 484. doi:10.3389/fmicb.2018.00484.

Costa, Patrícia, Mariana Reis, Marcelo Ávila, Laura Leite, Flávio Araújo, Anna Salim, Guilherme Oliveira, Francisco Barbosa, Edmar Chartone-Souza, and Andréa Nascimento. 2015. "Metagenome of a Microbial Community Inhabiting a Metal-Rich Tropical Stream Sediment." *PloS One* 10 (March): e0119465. doi:10.1371/journal.pone.0119465.

Croom, Edward. 2012. "Chapter Three—Metabolism of Xenobiotics of Human Environments." In *Progress in Molecular Biology and Translational Science*, edited by Ernest Hodgson, vol. 112, 31–88. Academic Press. https://doi.org/10.1016/B978-0-12-415813-9.00003-9.

De'Ath, Glenn, Katharina E. Fabricius, Hugh Sweatman, and Marji Puotinen. 2012. "The 27-Year Decline of Coral Cover on the Great Barrier Reef and Its Causes." *Proceedings of the National Academy of Sciences of the United States of America* 109 (44). National Academy of Sciences: 17995–99. doi:10.1073/pnas.1208909109.

De'Ath, Glenn, Janice M. Lough, and Katharina E. Fabricius. 2009. "Declining Coral Calcification on the Great Barrier Reef." *Science* 323 (5910). American Association for the Advancement of Science: 116–19. doi:10.1126/science.1165283.

Demirel, Burak. 2014. "Major Pathway of Methane Formation from Energy Crops in Agricultural Biogas Digesters." *Critical Reviews in Environmental Science and Technology* 44 (3). Taylor and Francis Inc.: 199–222. doi:10.1080/10643389.2012.710452.

Delong, Edward F. 2009. "The Microbial Ocean from Genomes to Biomes." *Nature* 459 (7244): 200–6.

Djemiel, Christophe, Sébastien Grec, and Simon Hawkins. 2017. "Characterization of Bacterial and Fungal Community Dynamics by High-Throughput Sequencing (HTS) Metabarcoding during Flax Dew-Retting." *Frontiers in Microbiology* 8 (October). Frontiers: 2052. doi:10.3389/FMICB.2017.02052.

Garcia, Gizele D., Gustavo B. Gregoracci, Eidy de O. Santos, Pedro M. Meirelles, Genivaldo G.Z. Silva, Rob Edwards, Tomoo Sawabe, et al. 2013. "Metagenomic Analysis of Healthy and White Plague-Affected *Mussismilia braziliensis* Corals." *Microbial Ecology* 65 (4) (Springer): 1076–86. doi:10.1007/s00248-012-0161-4.

Gonzalez-Martinez, Alejandro, Alejandro Margareto, Alejandro Rodriguez-Sanchez, Chiara Pesciaroli, Silvia Diaz-Cruz, Damia Barcelo, and Riku Vahala. 2018. "Linking the Effect of Antibiotics on Partial-Nitritation Biofilters: Performance, Microbial Communities and Microbial Activities." *Frontiers in Microbiology* 9 (February). Frontiers Media S.A.: 354. doi:10.3389/fmicb.2018.00354.

Gu, Yunfu, Yingyan Wang, Sheng'e Lu, Quanju Xiang, Xiumei Yu, Ke Zhao, Likou Zou, Qiang Chen, Shihua Tu, and Xiaoping Zhang. 2017. "Long-Term Fertilization Structures Bacterial and Archaeal Communities along Soil Depth Gradient in a Paddy Soil." *Frontiers in Microbiology* 8 (August). Frontiers Media S.A.: 1516. doi:10.3389/fmicb.2017.01516.

Hemmat-Jou, M. H., A. A. Safari-Sinegani, A. Mirzaie-Asl, and A. Tahmourespour. 2018. "Analysis of Microbial Communities in Heavy Metals-Contaminated Soils Using the Metagenomic Approach." *Ecotoxicology* 27 (9): 1281–91. doi:10.1007/s10646-018-1981-x.

Hooper, Lora V., and Jeffrey I. Gordon. 2001. "Commensal Host-Bacterial Relationships in the Gut." *Science*. American Association for the Advancement of Science. doi:10.1126/science.1058709.

Igiri, Bernard E., Stanley I. R. Okoduwa, Grace O. Idoko, Ebere P. Akabuogu, Abraham O. Adeyi, and Ibe K. Ejiogu. 2018. "Toxicity and Bioremediation of Heavy Metals Contaminated Ecosystem from Tannery Wastewater: A Review." Edited by Valerio Matozzo. *Journal of Toxicology* 2018. Hindawi: 2568038. doi:10.1155/2018/2568038.

Kho, Zhi Y., and Sunil K. Lal. 2018. "The Human Gut Microbiome—A Potential Controller of Wellness and Disease." *Frontiers in Microbiology*. Frontiers Media S.A. doi:10.3389/fmicb.2018.01835.

Konopka, Allan. 2009. "What Is Microbial Community Ecology?" *The ISME Journal* 3 (11): 1223–30.

Kracke, Frauke, Igor Vassilev, and Jens O. Krömer. 2015. "Microbial Electron Transport and Energy Conservation—The Foundation for Optimizing Bioelectrochemical Systems." *Frontiers in Microbiology*. Frontiers Media S.A. doi:10.3389/fmicb.2015.00575.

Kraft, Beate, Halina E. Tegetmeyer, Ritin Sharma, Martin G. Klotz, Timothy G. Ferdelman, Robert L. Hettich, Jeanine S. Geelhoed, and Marc Strous. 2014. "The Environmental Controls That Govern the End Product of Bacterial Nitrate Respiration." *Science* 345 (6197). American Association for the Advancement of Science: 676–79. doi:10.1126/science.1254070.

Kumar, Ravinder, and Pradeep Kumar. 2017. "Future Microbial Applications for Bioenergy Production: A Perspective." *Frontiers in Microbiology* 8 (March). Frontiers Research Foundation: 450. doi:10.3389/fmicb.2017.00450.

Lei, Yu, Ke Zhang, Mengmeng Guo, Guanwei Li, Chao Li, Bibo Li, Yuxin Yang, Yulin Chen, and Xiaolong Wang. 2018. "Exploring the Spatial-Temporal Microbiota of Compound Stomachs in a Pre-Weaned Goat Model." *Frontiers in Microbiology* 9 (August). Frontiers Media S.A.: 1846. doi:10.3389/fmicb.2018.01846.

Levin, Rachel A., Christian R. Voolstra, Shobhit Agrawal, Peter D. Steinberg, David J. Suggett, and Madeleine J. H. van Oppen. 2017. "Engineering Strategies to Decode and Enhance the Genomes of Coral Symbionts." *Frontiers in Microbiology* 8: 1220. doi:10.3389/fmicb.2017.01220.

Li, Hai-Bi, Rajesh K. Singh, Pratiksha Singh, Qi-Qi Song, Yong-Xiu Xing, Li-Tao Yang, and Yang-Rui Li. 2017. "Genetic Diversity of Nitrogen-Fixing and Plant Growth Promoting Pseudomonas Species Isolated from Sugarcane Rhizosphere." *Frontiers in Microbiology* 8 (July). Frontiers Media S.A.: 1268. doi:10.3389/fmicb.2017.01268.

Liao, James C., Luo Mi, Sammy Pontrelli, and Shanshan Luo. 2016. "Fuelling the Future: Microbial Engineering for the Production of Sustainable Biofuels." *Nature Reviews Microbiology*. Nature Publishing Group. doi:10.1038/nrmicro.2016.32.

Liu, Hai-Yue, Shi-Yu Zhang, Wan-Ying Yang, Xiao-Fang Su, Yan He, Hong-Wei Zhou, and Jin Su. 2017. "Oropharyngeal and Sputum Microbiomes Are Similar Following Exacerbation of Chronic Obstructive Pulmonary Disease." *Frontiers in Microbiology* 8 (June). Frontiers Media S.A.: 1163. doi:10.3389/fmicb.2017.01163.

Liu, Zijun, Yu Wang, Xiaoxuan Pan, Qinya Ge, Qinglin Ma, Qiang Li, Tongtong Fu, Cuiting Hu, Xudong Zhu, and Jiao Pan. 2017. "Identification of Fungal Communities Associated with the Biodeterioration of Waterlogged Archeological Wood in a Han Dynasty Tomb in China." *Frontiers in Microbiology* 8 (August). Frontiers Media S.A.: 1633. doi:10.3389/fmicb.2017.01633.

Liu, Zijun, Yanhong Zhang, Fengyu Zhang, Cuiting Hu, Genliang Liu, and Jiao Pan. 2018. "Microbial Community Analyses of the Deteriorated Storeroom Objects in the Tianjin Museum Using Culture-Independent and Culture-Dependent Approaches." *Frontiers in Microbiology* 9 (April). Frontiers Media S.A.: 802. doi:10.3389/fmicb.2018.00802.

Lloyd-Price, Jason, Galeb Abu-Ali, and Curtis Huttenhower. 2016. "The Healthy Human Microbiome." *Genome Medicine*. BioMed Central Ltd. doi:10.1186/s13073-016-0307-y.

Mayer, Emeran A., Kirsten Tillisch, and Arpana Gupta. 2015. "Gut/Brain Axis and the Microbiota." *Journal of Clinical Investigation*. American Society for Clinical Investigation. doi:10.1172/JCI76304.

McKenney, E. A., K. Koelle, R. R. Dunn, and A. D. Yoder. 2018. "The Ecosystem Services of Animal Microbiomes." *Molecular Ecology* 27 (8). Blackwell Publishing Ltd: 2164–72. doi:10.1111/mec.14532.

Mikesková, H., C. Novotný, and K. Svobodová. 2012. "Interspecific Interactions in Mixed Microbial Cultures in a Biodegradation Perspective." *Applied Microbiology and Biotechnology*. Springer. doi:10.1007/s00253-012-4234-6.

Pan, Yuanyuan, Xunan Yang, Meiying Xu, and Guoping Sun. 2017. "The Role of Enriched Microbial Consortium on Iron-Reducing Bioaugmentation in Sediments." *Frontiers in Microbiology* 8 (March). Frontiers Research Foundation: 462. doi:10.3389/fmicb.2017.00462.

Qadir, Abdul, Muhammad Zaffar Hashmi, and Adeel Mahmood. 2017. "Xenobiotics, Types, and Mode of Action." In *Xenobiotics in the Soil Environment: Monitoring, Toxicity and Management*, edited by Muhammad Zaffar Hashmi, Vivek Kumar, and Ajit Varma, 1–7. Soil Biology. Cham: Springer International Publishing. doi: 10.1007/978-3-319-47744-2_1.

Ramasamy, K., Kamaludeen, and Sara Parwin Banu. 2007. "Bioremediation of Metals: Microbial Processes and Techniques." In *Environmental Bioremediation Technologies*, edited by Shree N. Singh and Rudra D. Tripathi, 173–87. Berlin, Heidelberg: Springer Berlin Heidelberg. doi:10.1007/978-3-540-34793-4_7.

Rawlings, Douglas E., and D. Barrie Johnson. 2007. "The Microbiology of Biomining: Development and Optimization of Mineral-Oxidizing Microbial Consortia." *Microbiology* 153 (2). Microbiology Society: 315–24. doi:10.1099/MIC.0.2006/001206-0.

Reddy, K. Ramesh, and Ronald D. DeLaune. 2004. *Biogeochemistry of Wetlands: Science and Applications*, 116. Boca Raton: Taylor & Francis. ISBN 978-0-203-49145-4. Retrieved 25 May 2013.

Reid, Walter V., Harold A. Mooney, Angela Cropper, Doris Capistrano, Stephen R. Carpenter, Kanchan Chopra, Partha Dasgupta, Thomas Dietz, Anantha Kumar Duraiappah, and Rashid Hassan. 2005. *Ecosystems and Human Well-Being-Synthesis: A Report of the Millennium Ecosystem Assessment*. Washington, DC: Island Press.

Sender, Ron, Shai Fuchs, and Ron Milo. 2016. "Revised Estimates for the Number of Human and Bacteria Cells in the Body." *PLoS Biology* 14 (8). Public Library of Science: e1002533. doi:10.1371/journal.pbio.1002533.

Silva, Ubiana C., Julliane D. Medeiros, Laura R. Leite, Daniel K. Morais, Sara Cuadros-Orellana, Christiane A. Oliveira, Ubiraci G. de Paula Lana, Eliane A. Gomes, and Vera L. dos Santos. 2017. "Long-Term Rock Phosphate Fertilization Impacts the Microbial Communities of Maize Rhizosphere." *Frontiers in Microbiology* 8 (July). Frontiers Media S.A.: 1266. doi:10.3389/fmicb.2017.01266.

Singh Anjali, Shivani Chaudhary, Bhawna Dubey, and Prasad Vishal. 2016. "Microbial-Mediated Management of Organic Xenobiotic Pollutants in Agricultural Lands." In *Plant Responses to Xenobiotics*, edited by Sheo Mohan, Singh Rajeev Pratap, and Singh Anita Prasad, 211–30. Singapore: Springer Singapore. doi:10.1007/978-981-10-2860-1_9.

Sterflinger, Katja, and Flavia Pinzari. 2012. "The Revenge of Time: Fungal Deterioration of Cultural Heritage with Particular Reference to Books, Paper and Parchment." *Environmental Microbiology* 14 (3). John Wiley & Sons, Ltd: 559–66. doi:10.1111/j.1462-2920.2011.02584.x.

Storey, Sean, Mardiana Mohd Ashaari, Nicholas Clipson, Evelyn Doyle, and Alexandre B de Menezes. 2018. "Opportunistic Bacteria Dominate the Soil Microbiome Response to Phenanthrene in a Microcosm-Based Study." *Frontiers in Microbiology* 9: 2815. doi:10.3389/fmicb.2018.02815.

Strong, P. J., B. McDonald, and D. J. Gapes. 2011. "Enhancing Denitrification Using a Carbon Supplement Generated from the Wet Oxidation of Waste Activated Sludge." *Bioresource Technology* 102 (9). Elsevier: 5533–40. doi:10.1016/j.biortech.2010.12.025.

Surendra, K. C., Devin Takara, Andrew G. Hashimoto, and Samir Kumar Khanal. 2014. "Biogas as a Sustainable Energy Source for Developing Countries: Opportunities and Challenges." *Renewable and Sustainable Energy Reviews*. Elsevier Ltd. doi:10.1016/j.rser.2013.12.015.

Tap, Julien, Stanislas Mondot, Florence Levenez, Eric Pelletier, Christophe Caron, Jean Pierre Furet, Edgardo Ugarte, et al. 2009. "Towards the Human Intestinal Microbiota Phylogenetic Core." *Environmental Microbiology* 11 (10). John Wiley & Sons, Ltd: 2574–84. doi:10.1111/j.1462-2920.2009.01982.x.

Trinh, Pauline, Jesse R. Zaneveld, Sarah Safranek, and Peter M. Rabinowitz. 2018. "One Health Relationships between Human, Animal, and Environmental Microbiomes: A Mini-Review." *Frontiers in Public Health* 6 (August). Frontiers Media SA: 235. doi:10.3389/fpubh.2018.00235.

Vidali, M. 2001. "Bioremediation. An Overview." *Pure and Applied Chemistry* 73 (7). Berlin, Boston: De Gruyter: 1163–72. https://doi.org/10.1351/pac200173071163.

Vilanova, Cristina, and Manuel Porcar. 2020. "Art-Omics: Multi-Omics Meet Archaeology and Art Conservation." *Microbial Biotechnology* 13 (2). John Wiley and Sons Ltd: 435–41. doi:10.1111/1751-7915.13480.

Wintsche, Babett, Nico Jehmlich, Denny Popp, Hauke Harms, and Sabine Kleinsteuber. 2018. "Metabolic Adaptation of Methanogens in Anaerobic Digesters upon Trace Element Limitation." *Frontiers in Microbiology* 9 (March). Frontiers Media S.A.: 405. doi:10.3389/fmicb.2018.00405.

Xu, Yan, Yan Zhi, Qun Wu, Rubing Du, and Yan Xu. 2017. "*Zygosaccharomyces bailii* Is a Potential Producer of Various Flavor Compounds in Chinese Maotai-Flavor Liquor Fermentation." *Frontiers in Microbiology* 8 (December). Frontiers Media S.A.: 2609. doi:10.3389/fmicb.2017.02609.

Yang, Bin, Jiaqing Le, Peng Wu, Jianxin Liu, Le L. Guan, and Jiakun Wang. 2018. "Alfalfa Intervention Alters Rumen Microbial Community Development in Hu Lambs During Early Life." *Frontiers in Microbiology* 9 (March). Frontiers Media S.A.: 574. doi:10.3389/fmicb.2018.00574.

Yang, Yandong, Liang Zhang, Jun Cheng, Shujun Zhang, Xiyao Li, and Yongzhen Peng. 2018. "Microbial Community Evolution in Partial Nitritation/Anammox Process: From Sidestream to Mainstream." *Bioresource Technology* 251 (March 1): 327–33. https://doi.org/10.1016/j.biortech.2017.12.079.

Yeluri Jonnala, Bhagya R., Paul L. H. McSweeney, Jeremiah J. Sheehan, and Paul D. Cotter. 2018. "Sequencing of the Cheese Microbiome and Its Relevance to Industry." *Frontiers in Microbiology* 9 (May). Frontiers Media S.A.: 1020. doi:10.3389/fmicb.2018.01020.

12

Commercial Exploitation of Microbial Communal Services to Enrich Plant Microbiome

Adhikesavan Harikrishnan*, Ramasamy Shanmugavalli
Department of Chemistry, School of Arts and Sciences, Vinayaka Mission Research Foundation-Aarupadai Veedu (VMRF-AV) Campus, Paiyanoor, Chennai, India
Oshin K, Vijaykumar Veena*
Department of Biotechnology, REVA University, Kattigenahalli, Yelahanka, Bengaluru, India
Basavegowda Lakshmi
Department of Chemistry, REVA University, Kattigenahalli, Yelahanka, Bengaluru, India

12.1 Introduction

Microbially derived enzymes and metabolites have commercial applications in the market. Around 50% of commercial enzymes are obtained from the microbial origin of plants (Borrelli and Trono 2015). This is projected to reach >6 billion USD by 2022, with predicted annual growth of 6%, and the demand for enzyme catalysts is growing tremendously. In addition, metabolites are very exclusively used in human health. In 2013, around 40% of new chemical entities of microbial products such as mimics or derivatives were approved by the Food and Drug Administration (FDA) (Katz and Baltz 2016). It is also expected that the market can reach >250 billion USD in 2023, of which 40% are of pharmaceutical importance, with an annual growth rate of 8% from 2017. Similarly, biostimulant markets are increasing by 10.9% annually in 2022. In agriculture, microbial metabolites like spinosyn from *Saccharopolyspora* are used as insecticides, and bacterial biosurfactants are expected to have important agricultural applications. Despite large commercial applications, major microbial communities (99%) are unculturable with conventional techniques. However, biosynthetic dark matter can be discovered by high-throughput sequencing techniques to extract the novel lead structure and compounds. Since microbial communities have the ability to adapt to different ecological niches, the different chemical compositions of plant hosts and the plant microbiome have unique diverse metabolic and enzymatic potential (Aleti et al. 2017; Hassani et al. 2018). Using sequence comparison of certain plant microbiomes, several gene clusters responsible for secondary metabolites have been identified (Aleti

DOI: 10.1201/9781003042570-17

et al. 2015). These plant microbes have specific functions, such as diverse metabolites, toxin degradation, and signaling molecules to adapt to stress conditions.

12.1.1 Plant Microbial Communities

Microbial communities represent defined taxa, physiologically established and well documented to have a critical role in nature, including viruses, bacteria, fungi, algae, and protozoans. Studies and microbial community preservation are needed not only for research and development of various applications but also to maintain the ecosystem (Vero et al. 2019). Microbial communities find vast applications in four major domains of life: food, environment, industry, and medicinal fields. In the food domain, microbes are used as probiotics and starter cultures and for the production of functional compounds and traditional fermented products. In the environment domain, microbes are used for sustainability and protection of the ecosystem, such as biofuel, wastewater treatment, and purification of soil and air. In the medicinal field, microbes are used for drugs, antibiotics, and vaccine production. Industrial applications of microbes include the production of high-value products such as biofuels, bioplastics, and bioelectricity. In 1904, Lorenz-Hiltner explained the importance of the plant microbiome and developed several techniques, including microscopy and analytical tools, to study plant microbiome interactions (Caporaso et al. 2012; Jansson et al. 2012).

The plant microbiome has a myriad of mutualistic, pathogenic, and beneficial interactions (Sanchez-Canizares et al. 2017). However, their combinations and abundance of microbial taxa among the plants and other microbes rely on their multitrophic interactions and metabolic interdependency for existence. Plant microbiomes are broadly classified into aboveground and belowground associations with the plant. Above the ground, the microbes in aerial plant parts (leaves, stems, and reproductive organs) are considered the phyllosphere/phyllobiome, whereas belowground plant part–associated microbes constitute the rhizosphere/rhizobiome. The plant microbiome is further subdivided into epiphytes and endophytes. Epiphytes are associated with the exterior surfaces of the plant, while endophytes are microbes that can penetrate layers of plants with symbiotic relations either intercellularly or intracellularly (Bringel and Couee 2015; Hacquard et al. 2015).

Conventional agriculture uses tillage, chemical fertilizers, pesticides, and fungicides with monoculture that degrades soil microbiomes. These results are threatening to agroecosystems, especially soil quality and human health. Soil microbes that are beneficial include plant growth–promoting rhizobacteria (PGPR), actinomycetes, and mycorrhizal fungi (Gianinazzi et al. 2010), which interact and associate with the plant and elicit the production of metabolites of high economic value. Research has suggested that biostimulant-based microbial inoculants have increased crop yield (Bhardwaj et al. 2014). However, commercially available microbial inoculants are commonly limited to a few or one taxonomic groups of microbes (Hart et al. 2018). But the soil microbiome is very complex and diverse and gives regulatory and supportive service to plants.

12.2 Concept of Plant Microbiome Interactions

Plant microbiomes are now focused on the important microbial communities that colonize plants' internal tissues without any disease-causing effects, referred to as endophytes. These types of microbiomes have interactions with their plant host and elicit a response to phytopathogens, pests, and abiotic stress to their host. Further, these microbes have been documented for the synthesis of secondary metabolites, phytohormones, improved stress tolerance, and plant immunity (Igiehon and Babalola 2018; Omomowo and Babalola 2019).

The major organ in the plant where the abundance of microbiome association is observed is the root system, and it has very complex microbial communities (Hacquard et al. 2015). However, the complex interactions within the rhizobiome either directly or indirectly have an influence on the growth and development of plants. Hence, microbes are the key players in nutrient cycling, acquisition, and absorption in plants (Figure 12.1) (Mishra et al. 2012). Further, the rhizobiome also provides microbe-derived

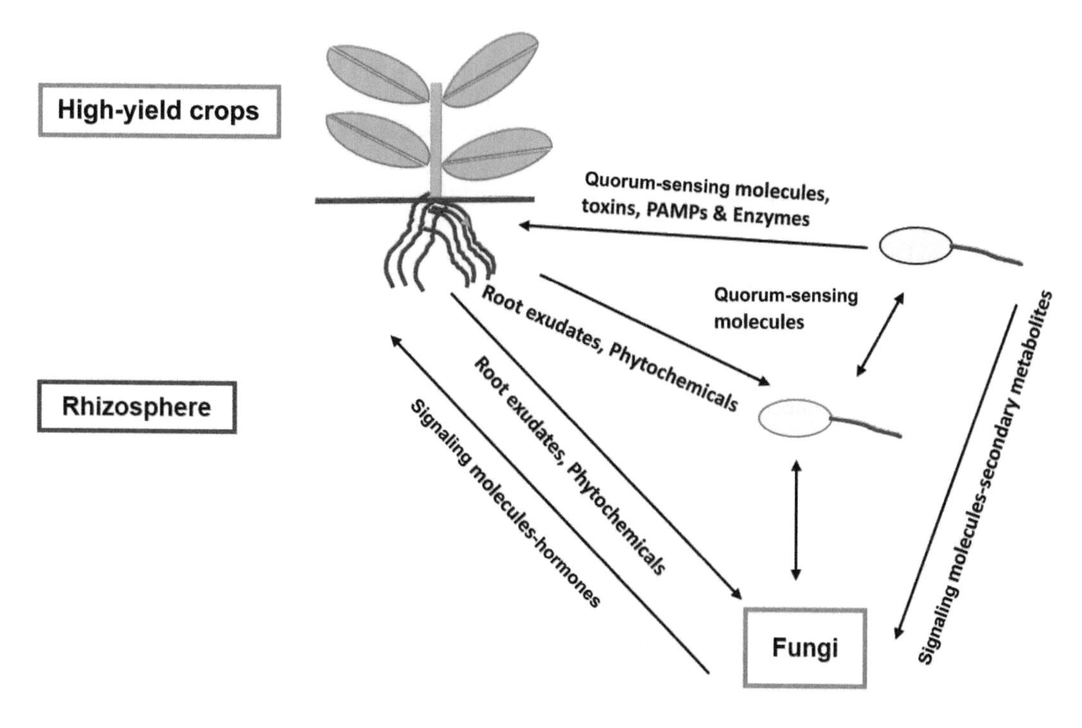

FIGURE 12.1 Plant microbiome interactions at the rhizobiome and mechanisms of natural existence.

compounds to host plants that influence their traits as well as inducing stress tolerance in the host (Rout et al. 2013). Hence, rhizobiomes are referred to as the secondary genome of the plants.

The major focus seems to be the rhizobiome, as several parameters and multiple interaction changes in microbiome structure affect the plant's health and development. However, their interactions are not clearly understood. Exploiting these potential benefits from interactions of the rhizobiome could lead to suitable solutions in the restoration of the ecosystem and increasing agricultural crop production (Gaba et al. 2014). In certain disease conditions, the microbiome has a negative impact on plant productivity and accounts for severe loss in rice, wheat, potato, soya bean, and maize. These plant microbiomes are responsible for a wide variety of metabolites, proteins, and small RNA that are involved in plant growth promotion and to adapt to environmental changes. It is being identified that highly diverse endophytes of medicinal plants are responsible for aiding seed germination and stress-reducing properties. Enormous numbers of plant growth–promoting endophytes are being isolated and utilized to improve the quantity and qualitative aspects of plants. Utilization of the plant microbiome is necessary because of the high demands of the expanding population, reduction of agriculture productivity, land degradation, and climatic changes (Gaba et al. 2014). Understanding plant microbiome interactions and finding developmental methods to analyze microbiome composition may enhance plant-beneficial interactions and improve current climatic conditions.

12.3 Elucidation of Plant Microbiome for Enriched Plants and Their Impact

Diverse microbial communities are associated with different plant parts. Unraveling the diverse microbial communities and their types of interactions with plants is important. Any disturbance in the healthy microbiome can result in a disease outbreak. These microbiomes have profound effects on seedling vigor, seed germination, development of immunity, and nutrition (Mendes et al. 2013). Different plant functional traits such as roots and leaves are due to plant microbial interactions. Inoculation of plant microbiomes that promote plant growth such as *Pseudomonas* strain, *Pseudomonas* sp., *Bacillus subtilis, Pantoea* sp.,

Pseudomonas putida, Azospirilium, Azotobacter, Phyllobacterium brassicacearum, P. brassicacearum, Glomus species, *Rhizobium leguminosarum, Mesorhizobium ciceri*, and *R. phaseoli* have been reported to affect plant traits. Microbial inoculation has affected plant traits such as increased weight of cormlets; enhanced plant growth and nutrition; increased anthocyanin; increased radical length; increased shoot length, morphology, and dry and fresh weight of plants; increased flowers, fruit, and fruit size; and increased plant biomass. Major crops studied were *Fragariax ananassa* (Lingua et al. 2013), *Crocus sativus, Cynara scolymus* (Parray et al. 2013), *Arabidopsis, Crocus sativus* (Lone et al. 2016), *Solanum tuberosum*, and *Allium cepa* (Shuab et al. 2014, 2017).

The plant microbiome even maintains healthy conditions by bioremediation of soil and air. This is evidenced by the rhizobiome of *H. cannabinus* consisting of *Enterobacteriaceae, Pseudomonadaceae*, and *Comamonadaceae* being able to bioremediate metals (Chen et al. 2018). The plant microbiome has the ability to degrade, detoxify, and sequester air pollutants to clean environments (Andreote et al. 2018). An interesting result of inoculated arbuscular mycorrhizal fungi (AMF) in leguminous trees was enhancement of phytoremediation of polluted soil with lead and aiding the growth of natural flora and fauna for ecological sustainability (Yang et al. 2016).

12.4 Novel Plant Microbial Communities to Improve Economic Importance

The plant microbiome, especially the rhizobiome, is a nutrient-enriched area with a highly microbially competitive environment. These microbiomes produce enormous secondary metabolites to occupy niches and establish in the rhizosphere (Pierson et al. 2010; Kim et al. 2011). This chemical warfare in the rhizosphere consists of antibiotics, toxins, enzymes, and siderophores (Bais et al. 2006). Utilizing omics approaches, these microbiomes possess larger gene clusters that are involved in products, as observed in cultured microbes such as *Bacillus amyloliquefaciens* and *Pseudomonas fluorescens* (Chen et al. 2007; Paulsen et al. 2005). Secondary metabolites that include phenazine compounds, 2,4-diacetylphloroglucinol (DAPG), hydrogen cyanide (HCN), and oomycin-A are involved in disease suppression in soil (Pierson et al. 2010; Kim et al. 2011). However, these metabolites are also used for pharmacologically significant lead molecules for anticancer, antioxidant, and anti-inflammatory molecules for synthetic methods (Veena et al. 2015). Further, these molecules are also involved in communications establishing signaling and metabolism in plants for microbial interaction in the rhizosphere (Kim et al. 2011). However, microbial reprogramming of the plants has been altered by root exudates and selectively enriches the plant rhizobiome (Bulgarelli et al. 2013). Hence, the plant microbiome establishes microbial communities that help in niche exclusion and inclusion (Lareen et al. 2016).

It is suggested that microbial communities have highlighted the significance of multipartite interactions to impacts on host plants. The tripartite symbiotic association of bacteria, fungi, and host plants is very important for the plant productivity (Bais et al. 2006). Legume seedlings inoculated with AMF and rhizobium showed a 15-fold increase in plant productivity. Further, stable beneficial microbial populations are selectively recruited and stabilized by host plants through secretion of essential metabolites (Doornbos et al. 2012). However, enriching microbial communities to increase crop production in terms of plant growth and disease suppression and to withstand biotic and abiotic stress seems to be a major challenge. But the use of microbes in agriculture is not a new concept. For example, *Bacillus thuringiensis* is commercially exploited for crop protection (Sanchis and Bourguet 2008), and PGPRs such as *Bacillus pumilus, Bacillus subtilis*, and *Curtobacterium flaccumfaciens* are commonly used as biological control of plant pathogens in addition to for productivity.

In many crop plants, seed inoculation of *Azospirullum* strains has enhanced seed yield by 8.9% and dry weight of the wheat by 17.8% as biofertilization of soil by nitrogen fixation (Namvar and Khandan 2015). Similar results are documented in increased grain yield and oil content in rapeseed using dual inoculation of *Azospirillum* and *Azotobacter* spp. through the production of indole acetic acid, gibberellins, polyamines, and amino acids and increasing nutrient availability to plants (Bashan et al. 2010). In order to apply any strategies to improve microbial community persistence, it is necessary to understand and elucidate the biotic factors such as genotype of plants; plant age; microbe–microbe interactions; plant–microbe interactions; and abiotic factors such as soil type, soil physics, environmental factors, nutrients, and water availability that determine the stabilization of microbial communities and dynamics among microbial community composition in field conditions (Rehman et al. 2019).

12.5 Multi-Omics Approaches in Plant Microbiome and Their Interactions

The microbiome defines the microbial communities occupying a reasonably well-defined habitat. Ninety-nine percent of unculturable microbial communities are a major factor for ecosystem services and maintenance of ecological functions. Multi-omics and sequencing technologies have provided insight into the examination of microbiomes that impact plant gene expression (Schenk et al. 2012). Metagenomics approaches are used to identify and characterize the complex DNA mixture from microbial habitats. Omics approaches of meta-transcriptomics, meta-proteomics, and meta-metabolomics have provided the functional characteristics of plant microbiota (Bell et al. 2014; Bressan et al. 2009). Advances in sequencing technologies and bioinformatics tools have improved the affordability with increased accuracy to study the plant microbiome (Figure 12.2) (Table 12.1).

Transcriptomics and genomics studies have suggested that soil microbes have been involved in the alteration of the developmental process in plants (Wolf 2013). Studying the microbiome through metagenomes yields the taxonomic profile of a microbial community. Hence, metagenomic studies of the microbiome and their genomes aggregate with respect to environmental and host factors. In order to understand the function of microbial communities, meta-transcriptomics are utilized that involves sequencing of the complete transcriptome of the plant microbiome. Metagenomics addresses the microbial diversity in different conditions, while transcriptomics reveals the type of genes collectively expressed under specific conditions, but metabolomics can reveal the by-products under various conditions. These microbial metabolites are essential for the maintenance of plant health in their environmental conditions. In order to study the role of microorganisms in the ecosystem processes, statistical modeling is used (Zhang et al. 2019). The value of environmental variables and microbial community structure either alone or both in

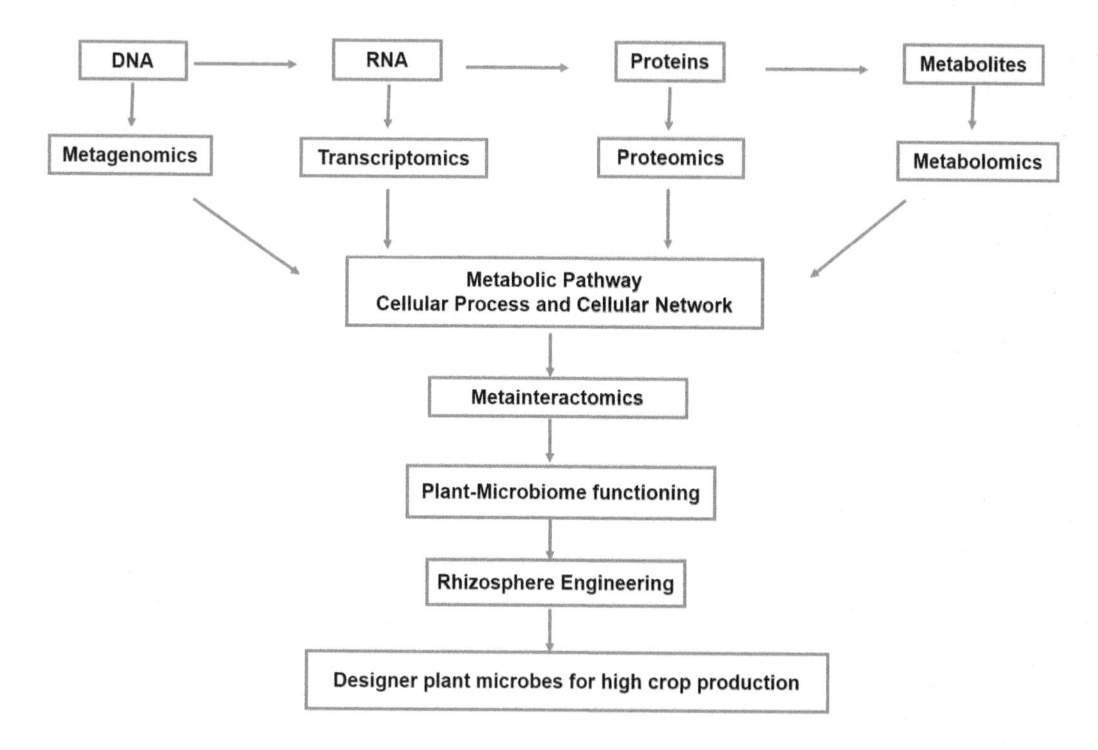

FIGURE 12.2 Multi-omics approaches in elucidation of plant microbiome interaction and designer plant technology.

TABLE 12.1

Assessment of Plant Microbiomes Using Metagenomic Approaches (Adapted from Rehman et al. 2019)

Microbial Communities Assessed	Techniques	References
Rhizobiome of *Arachis hypogea*	Amplicon sequence of conserved gene of 16S rRNA	Su et al. 2015
Microbial communities in hydrocarbon-contaminated soils	Amplicon sequence of conserved gene of 16S rRNA	Bell et al. 2014
Rhizobiome of soya bean	Metagenomic sequencing	Mendes et al. 2013
Fungal microbial communities of soya growth	Metagenomic sequencing	Sugiyama et al. 2014
Grassland plant microbiome and soil edaphic	Metagenomic sequencing	LeBlanc et al. 2015
Root surface microbes and assemblage of plant microbiome Chaparo et al. 2014	Metatranscriptome	Ofek-Lalzar et al. 2014
Phyllobiome and rhizobiome of paddy and sugarcane Lin et al. 2013	Metaproteomics	Knief et al. 2012
Fungal-associated tomato plants	Metabolomic profiling	Tschaplinki et al. 2014

conjunction with ecosystem process using global datasets can elucidate the crucial role of microbes. This research has suggested that the environment seems to be a strong driver of most ecosystem processes, and microbial community structure adds to our understanding of processes under certain circumstances (Graham et al. 2014). Omics approaches have been applied for analysis of microbial communities in other areas, including the cheese-making process; fiber product aspects in lamb and goat; and formulation of newer, efficient production of products with industrial and environment importance (Marco et al. 2019; Starke et al. 2019).

Using functional genomics approaches, the efficient researcher has identified the set of genes that are responsible for stress management; production of bioactive compounds to improve crops; defense against phytopathogens and nitrogen-fixing; and salt-tolerant microbial communities in crops such as sugarcane, rice, and maize (García-Fraile et al. 2015). These ecosystem services for the biofertilization of soil and plants with eco-friendly agriculture to reduce chemical fertilizers and pesticides were also evaluated by omics approaches (Gaba et al. 2014). Researchers have also used metagenomics to develop an ecological model for microbial communities at the lab scale, referred to as a denitrifying sulfur conversion–associated enhanced biological phosphorous removal (DS-EBPR) system for operating the removal of C, N, P, and S from water samples to avoid eutrophication in water systems. This DS-EBPR system showed 11 complete drafts of genomes of sulfate-reducing and sulfide-oxidizing bacteria that constituted the majority (39.4%). Among them, ten bacteria were new potential polyphosphate- and sulfur-accumulating organisms that had not been cultivated earlier. These interesting findings highlighted the new model and enhanced the understanding of the various phosphate- and sulfur-reducing microbial communities in efficient waste water treatment (Bedmar et al. 2013). It is also observed that the influence of host microbial communities on the plant at transcriptional-level analysis was combined with network analysis. This report suggests novel genes and metabolic pathways are needed for specific microbial partner/microbial consortium interactions with the host plant. Also, the effect of specific microbial associations at the rhizosphere and their nutritional requirements were identified. Utilizing plant multi-omics approaches, it was possible to elucidate plant–soil systems and the role of rhizospheric microbiomes to influence various types of conditions. As an interesting example, exometabolomics techniques have suggested various metabolites from microbiomes are needed for interaction with the plant host (Swenson et al. 2000). Further, these data suggested that microbial exo-metabolites in the niches of the rhizosphere are linked with the composition of the microbial structure (Baran et al. 2015). Proteomics analysis was able to determine the various functions of the microbiome that influence the plant host (White et al. 2017). Hence, the integrated omics technologies researchers have developed seems to be more robust in deep understanding of plants and associated microbiomes that influence the soil process and plant biology.

Any systemic alterations in plant physiology or architecture are also useful in studies of how plants can control microbiome assembly and their functionality. An interesting example was interruption of the mutual relationships of microbes that aid in uptake of nutrient through modification of root exudates through gene editing (Ahkami et al. 2017). The results suggested that reduction in carbon sources among soil microbes affected the dynamics of mineralization. In addition, modification in the composition of root exudates and a microbial taxonomic shift in the plant suggested plant selection for specific microbial taxa (Su et al. 2015).

Recent studies have suggested the plant microbiome has a profound role in inducing tolerance to biotic and abiotic stress due to climatic changes. The role of the root microbiome is well established in stress tolerance utilization of next-generation sequencing methods such as 454 pyrosequencing, Illumina sequencing, and so on. Many plant microbiome studies have used amplicon-based microbial ecology to describe the community structure and model plant pathogens or beneficial microbes (Caporaso et al. 2012). The most commonly used microbial tools include 16S rRNA/gyrB gene amplicon sequencing to elucidate the taxonomic structure of microbial communities. However, this does not characterize the interaction of microbes as harmful, neutral, or mutualistic. Accessory genes such as disease resistance (R-genes), virulence factors, plant stress interactions, plant hormones, or nutrient-mobilizing genes of the microbiomes were used to study the interactions (Haldar and Sengupta 2015; Sugiyama et al. 2014; Mendes et al. 2013). However, meta-proteomics and meta-transcriptomics tools can explore species diversity and relate to their functions (Chaparo et al. 2014; Knief et al. 2012; Lin et al. 2013).

Studies of microbial communities have boosted the intensive results in sequencing technologies from culturable and environmental samples. Microbial communities are considered complex-adaptive systems where individuals and populations interact and can give rise to a variety of interactions such as mutualism, antagonism, and parasitism between the host and its environment. In the case of socio-microbiological aspects, there are single-system to complex models related to host–microbe interaction and their microbial consortium. The study of conceptual frameworks from market economy theory has also predicted the evolution of these complex interactions of microbial communities (Singh et al. 2019). Mathematical modeling of microbial communities such as Lotka-Voltera, evolutionary game models, thermodynamic non-linear models, trait-based models, and stoichiometric models can be used. These models are based on either top-down or bottom-up approaches, as defined by population-level models (PLMs) and individual-based models (IBMs). PLMs are best for homogenous environments, while IBMs are the best fit for heterogeneous environments. Using the output data from biomedicine, environmental, and metabolic pathways paved the way for the possible outcome of predictive microbial ecology or synthetic ecology (Figure 12.2) (Haruta and Yamamoto 2018). Using this knowledge, artificial microbial community structures can be designed towards precise and efficient bio-performance at the same time, maintaining the resilience and complexity of the microbial community. In this view, bacteriotherapy and extraterrestrial life support were tightly and efficiently programmed (Diaz et al. 2011).

12.6 Synthetic Microbial Consortium

Comprehensive mechanistic studies on microbial ecology have realized the taxonomic and functional diversity of natural microbial communities in plant and soil health. Furthermore, in microbial communities, there is a wide range of interest in several aspects that result in higher functionality and genetic adaptation in complex ecosystems. In the last few decades, synthetic ecology seems to be diverging from synthetic biology. More precisely, the Synthetic Microbial Consortium (SMC) has been developed to overcome several aspects of current problems. Many consortia-based systems are being designed and adopted for various applications, including production of biofuels and bioproducts and greenhouse management. In the SMC, several attempts have been made to model microbial consortia that are designed and built (Arif et al. 2020; Corato 2020). This research has resulted in a number of experimental systems that continuously grow for testing to learn ecological methods to understand microbial diversity and functions. Several approaches, such as binary, three species, higher-member complex culture systems, and enriched model systems, were developed to elucidate the functions of microbial communities and their wide variety of behaviors in gene regulatory networks, metabolic interactions, and ecological significance. The development of the SMC involves the combination of specific microbial genotypes with desirable traits (Dodd and Ruiz-Lozano 2012; Thijs et al. 2014).

12.7 Designer Plant Microbiome Technology for Sustainable Agriculture

Natural rhizobiomes are complex, diverse, and dynamic and make up the entire food web (Bender et al. 2016). There are commercially available microbial inoculants that have been used to replace fertilizers, but their major limitation is the lack of diversity with one or few microbial taxa (Hart et al. 2018). Hence, the concept of designer plant technology/synthetic biology approaches is introduced to increase biodiversity utilizing unculturable microbes. The SMC can replace or restore the structure and function of plant microbiota. Construction of the SMC consists of the microbial consortium with multiple functions that promote crop productivity and improve ecosystem quality. However, it is ineffective because of incompatibility with the host plant and cannot adapt to local environmental conditions (Hart et al. 2018).

Development of the SMC is based on the synergistic combination of specific microbial genotypes with desirable traits. It consists of PGPRs and mycorrhiza that can target an increase of metabolite concentrations such as essential oils; zein; glucosinolate; sugar; ascorbic and folic acid; volatile compounds; vitamins; anthocyanin; and nutrients such as N, Ca, P, Mg, K, Na, Fe, Mn, Cu, Zn, and B that represent higher nutraceutical values in crops (Berta et al. 2014). Recent advances in NGS platforms allow us to perceive whole microbial communities of crops through metagenomics. Based on metagenomic approaches, the hub species that co-exist with plants can be identified with the help of network topological properties. These hubs are core microbial communities with desired traits to construct the core microbiome's cultural aspects and can be elucidated using web-based platforms, such as Known Media Database (KOMODO), to predict the media components for culturing the core microbiome (Kong et al. 2018) (Figure 12.3).

In bio-stimulant–based SMC biofertilizer development, PGPRs and mycorrhizae were targeted to increase the metabolites and nutrients that enhanced high nutraceutical values for the crops (Berta et al. 2014; Cosme et al. 2014; Bona et al. 2016; Avio et al. 2017). However, the SMC cannot be same for all crops, and the SMC has to be specific by the addition of additives for the benefit of the crop. It is evidenced that indigenous soil and plant microbiomes from high-quality crops confer resistance to stress due to changes and seem to be ideal for construction of SMCs. Thus, the specific SMC of the same plant can be better for growth and quality. Further, the rhizo-microbiome seems to be the hot spot for SMCs due to extensive stress resistance and intensive interactions with host plants (Huang et al. 2018; Kong et al. 2017). SMCs seem promising but have two challenges that need to be addressed. First, the developed SMCs need adaptation to various soils, crops, and soil types with climatic changes that might affect the efficacy and sustainability of SMCs. Second, the ecological risks, including the invasive nature of SMCs and their interactions with indigenous soil microbes, are a critical challenge.

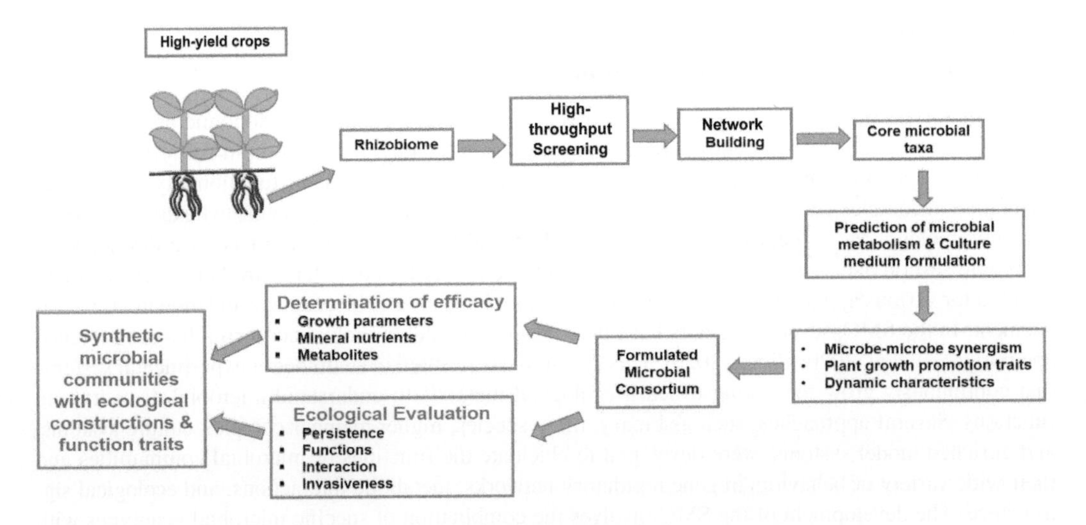

FIGURE 12.3 Illustration of omics for crop quality and development of SMC-based biofertilizers.

12.6 Conclusion

The critical evidence of several studies proved plant microbiome interactions are highly conserved with characteristic traits. Due to the usage of chemicals and fertilizers, there is a loss of the plants' essential microbiome. The majority of the plant microbiome is unculturable but can only be observed using microscopy and sequencing methods. However, multi-omics research on plant microbial interactions includes bacterium–bacterium, bacterium–microeukaryote, and microeukaryote–microeukaryote interactions that are critical for plant development and restoration of the ecosystem. Integrated synthetic biology and omics technology have aided in creating next-generation biofertilizers based on the enrichment of unculturable plant microbiomes. Those SMCs can be developed for specific crops to reduce chemical fertilizers and pesticides with healthy natural ecosystem conservation.

REFERENCES

Ahkami, A.H. White, R.A. Handakumbura, P.P. Jansson, C. 2017. Rhizosphere engineering: Enhancing sustainable plant ecosystem productivity. *Rhizosphere* 3: 233–243.

Aleti, G. Nikolic, B. Brader, G. Pandey, R.V. Antonielli, L. Pfeiffer, S. et al. 2017. Secondary metabolite genes encoded by potato rhizosphere microbiomes in the Andean highlands are diverse and vary with sampling site and vegetation stage. *Sci Rep* 7: 2330.

Aleti, G. Sessitsch, A. Brader, G. 2015. Genome mining: Prediction of lipopeptides and polyketides from *Bacillus* and related firmicutes. *Comp Struct Biotechnol J* 13: 192–203.

Andreote, F.D. Gumiere, T. Durrer, A. 2018. Exploring interactions of plant microbiomes. *Sci Agric* 71: 528–539.

Arif, I. Batool, M. Schenk, P.M. 2020. Plant microbiome engineering: Expected benefits for improved crop growth and resilience. *Trends in Biotechnology* 38(2): 1385–1396.

Avio, L. Sbrana, C. Giovannetti, M. Frassinetti, S. 2017. Arbuscular mycorrhizal fungi affect total phenolics content and antioxidant activity in leaves of oak leaf lettuce varieties. *Sci. Hortic* 224: 265–227.

Bais, H.P. Weir, T.L. Perry, L.G. Gilroy, S. Vivanco, J.M. 2006. The role of root exudates in rhizosphere interactions with plants and other organisms. *Annu Rev Plant Biol* 57: 233–266.

Baran, R. Brodie, E.L. Mayberry-Lewis, J. Hummel, E. Da Rocha, U. N. Chakraborty, R. et al. 2015. Exometabolite niche partitioning among sympatric soil bacteria. *Nat. Commun* 6: 8289.

Bashan, Y. de-Bashan, L.E. 2010. How the plant growth promoting bacterium *Azospirillum* promotes plant growth-A critical assessment. *Adv Agron* 108: 77–136.

Bedmar, E.J. Bueno, E. Correa, D. Torres, G. et al. 2013. Ecology of denitrification in soils and plant-associated bacteria. In *Beneficial Plant-microbial Interactions: Ecology and Applications*, eds M. González and J. Gonzalez-López. Boca Raton, FL: CRC Press, 164–182.

Bell, T.H. Hassan, E.H. Lauron-Moreau, A. Al-Otaibi, F. Hijri, M. et al. 2014. Linkage between bacterial and fungal rhizosphere communities in hydrocarbon-contaminated soils is related to plant phylogeny. *ISME J* 8: 331–343.

Bender, S.F. Wagg, C. Heijden, M.G.A. 2016. An underground revolution: Biodiversity and soil ecological engineering for agricultural sustainability. *Trends Ecol. Evol* 31: 440–452.

Berta, G. Copetta, A. Gamalero, E. Bona, E. Cesaro, P. Scarafoni, A. et al. 2014. Maize development and grain quality are differentially affected by mycorrhizal fungi and a growth-promoting pseudomonad in the field. *Mycorrhiza* 24: 161–170.

Bhardwaj, D. Ansari, M.W. Sahoo, R.K. Tuteja, N. 2014. Biofertilizers function as key player in sustainable agriculture by improving soil fertility, plant tolerance and crop productivity. *Microb. Cell Fact* 13: 66.

Bona, E. Lingua, G. Todeschini, V. 2016. Effect of bioinoculants on the quality of crops. In *Bioformulations: For Sustainable Agriculture*, eds N. K. Arora, S. Mehnaz, and R. Balestrini. New Delhi: Springer India, 93–124.

Borrelli, G.M. Trono, D. 2015. Recombinant lipases and phospholipases and their use as biocatalysts for industrial applications. *Int J Mol Sci* 16: 20774–20840.

Bressan, M. Roncato, M.A. Bellvert, F. Comte, Zahar-Haichar, F. Achouak, W. Berge, O.et al. 2009. Exogenous glucosinolate produced by *Arabidopsis thaliana* has an impact on microbes in the rhizosphere and plant roots. *ISME J* 3(11): 1243.

Bringel, F. Couee, I. 2015. Pivotal roles of phyllosphere microorganisms at the interface between plant functioning and atmospheric trace gas dynamics. *Front. Microbiol* 6: 486.

Bulgarelli, D. Schlaeppi, S.S. Ver, L. van-Themaat, E. Schulze-Lefert, P. 2013. Structure and functions of the bacterial microbiota of plants. *Annu Rev Plant Biol* 64: 807–838

Caporaso, J.G. Lauber, C.L. Walters, W.A. Berg-Lyons, D. Huntley, J. Fierer, N. et al. 2012. Ultra-high throughput microbial community analysis on the Illumina HiSeq and MiSeq platforms. *ISME J* 6(8): 1621.

Chaparo, J.M. Badri, D.V. Vivanco J.M. 2014. Rhizosphere microbiome assemblage is affected by plant development. *ISME J* 8: 790–803.

Chen, X.H. Koumoutsi, A. Scholz, R. Eisenreich, A. Schneider, K. et al. 2007. Comparative analysis of the complete genome sequence of the plant growth promoting bacterium *Bacillus amyloliquefaciens* FZB42. *Nat Biotechnol* 25(9): 1007–1014.

Chen, Y. Ding, Q. Chao, Y. Wei, X. Wang, S. Qiu, R. 2018. Structural development and assembly patterns of the root-associated microbiomes during phytoremediation. *Sci Total Environ* 644: 1591–1601.

Corato, U.D. 2020. Soil microbiota manipulation and its role in suppressing soil-borne plant pathogens in organic farming systems under the light of microbiome-assisted strategies. *Chem. Biol. Technol. Agric* 7: 17.

Cosme, M. Franken, P. Mewis, I. Baldermann, S. Wurst, S. 2014. Arbuscular mycorrhizal fungi affect glucosinolate and mineral element composition in leaves of *Moringa oleifera*. *Mycorrhiza* 24: 565–570.

Diaz, S. Quetier, F. Caceres, D.M. Trainor, S.F. Perez-harguideguy, N. et al. 2011. Linking functional diversity and social actor strategies in a framework for interdisciplinary analysis of nature's benefits to society. *Proceedings of the National Academy of Sciences* 108: 895–902.

Dodd, I.C. Ruiz-Lozano, J.M. 2012. Microbial enhancement of crop resource use efficiency. *Curr Opin Biotechnol* 23: 236–242.

Doornbos, R.F. van-Loon, L.C. 2012. Impact of root exudates and plant defence signalling on bacterial communities in the rhizosphere: A review. *Agron Sustain Dev* 32: 227–243.

Gaba, S. Bretagnolle, F. Rigaud, T. Philippot, L. 2014. Managing biotic interactions for ecological intensification of agroecosystem. *Frontiers in Ecology and Evolution* 2: 29.

García-Fraile, P. Menéndez, E. Rivas, R. 2015. Role of bacterial biofertilizers in agriculture and forestry. *AIMS Bioeng* 2: 183–205.

Gianinazzi, S. Gollotte, A. Binet, M.N. van Tuinen, D. Redecker, D. Wipf, D. 2010. Agroecology: The key role of arbuscular mycorrhizas in ecosystem services. *Mycorrhiza* 20: 519–530.

Graham, E.B. Wieder, W.R. Leff, J.W. Weintraub, S.R. Townsend, A.R. Cleveland, C. C. et al. 2014. Do we need to understand microbial communities to predict ecosystem function? A comparison of statistical models of nitrogen cycling processes. *Soil Biol. Biochem* 68: 279–282.

Hacquard, S. Garrido-Oter, R. Gonzalez, A. Spaepen, S. Ackermann, G. et al. 2015. Microbiota and host nutrition across plant and animal kingdoms. *Cell Host Microbe* 17(5): 603–616.

Haldar, S. Sengupta, S. 2015. Impact of plant development on the rhizobacterial population of *Arachis hypogea*: A multifactorial analysis. *J Basic Microbiol* 55: 922–928.

Hart, M. Antunes, P. Chaudhary, V.B. Abbott, L. 2018. Fungal inoculants in the field: Is the reward greater than the risk? *Funct. Ecol* 32: 126–135.

Haruta, S. Yamamoto, K. 2018. Model microbial consortia as tools for understanding complex microbial communities. *Current Genomics* 19: 723–733.

Hassani, M.A. Duran, P. Hacquard, S. 2018. Microbial interactions within the plant holobiont. *Microbiome* 6: 58.

Huang, L.H. Yuan, M.Q. Ao, X.J. Ren, A.Y. Zhang, H.B. Yang, M.Z. 2018. Endophytic fungi specifically introduce novel metabolites into grape flesh cells in vitro. *PLoS One* 13: e0196996.

Igiehon, N.O. Babalola, O.O. 2018. Below-ground-above-ground plant-microbial interactions: Focusing on soybean, rhizobacteria and mycorrhizal fungi. *Open Microbiol. J* 12: 261–279.

Jansson, J.K. Neufeld, J.D. Moran, M.A. Gilbert, J.A. 2012. Omics for understanding microbial functional dynamics. *Environ Microbiol* 14(1): 1–3.

Katz, L. Baltz, R.H. 2016. Natural product discovery: Past, present, and future. *J Ind Microbiol Biotechnol* 43: 155–176.

Kim, Y.C. Leveau, J. Gardener, M.B.B. Pierson, E.A. Pierson, L.S.III. Ryu, C. 2011. The multifactorial basis for plant health promotion by plant-associated bacteria. *Appl Environ Microbiol* 77: 1548–1555.

Knief, C. Delmotte, N. Chaffron, S. Stark, M. Innerebner, G. et al. 2012. Metaproteogenomic analysis of microbial communities in the phylogenetic in the phyllosphere and rhizosphere of rice. *ISME J*: 1–30.
Kong, Z. Glick, B.R. 2017. The role of plant growth-promoting bacteria in metal phytoremediation. *Adv. Microb. Physiol* 71: 97.
Kong, Z. Hart, M. Liu, H. 2018. Paving the way from the lab to the field: Using synthetic microbial consortia to produce high quality crops. *Frontiers in Plant Sciences* 9: 1467.
Lareen, A. Burton, F. Schafer, P. 2016. Plant root-microbe communication in shaping root microbiomes. *Plant Mol Biol* 90: 575–587.
LeBlanc, N. Kinkel, L.L. Kistler H.C. 2015. Soil fungal communities respond to the grassland plant community richness and soil edaphics. *Micro Ecol* 70: 188–195.
Lin, W. Wu, L. Lin, S. Zhang, A. Zhou, M. et al. 2013. Metaproteogenomic analysis of ratoon sugarcane rhizospheric soil. *BMC Microbiol* 13: 135.
Lingua, G. Bona, E. Manassero, P. Marsano, F. Todeschini, V. Cantamessa, S. et al. 2013. Arbuscular mycorrhizal fungi and plant growth-promoting pseudomonads increases anthocyanin concentration in strawberry fruits (*Fragaria x ananassa* var. Selva) in conditions of reduced fertilization. *Int J Mol Sci* 14(8): 16207–16225.
Lone, R. Shuab, R. Koul, K.K. 2016. AMF association and their effect on metabolite mobilization, mineral nutrition and nitrogen assimilating enzymes in saffron (*Crocus sativus*) plant. *J Plant Nutr* 39(13): 1852–1862.
Marco, D.E. Abram, F. 2019. Using genomics, metagenomics and other omics to access valuable microbial ecosystem services and novel biotechnological applications. *Frontier in Microbiology* 10: 151.
Mendes, R. Krujit, M. de-Bruijn, I. Dekkers, E. vander-Voort, M. et al. 2013. Deciphering the rhizosphere microbiome for disease-suppressive bacteria. *Science* 332: 1097–1100.
Mishra, P. Bisht, S. Mishra, S. Selvakumar, G. Bisht, J. Gupta, H. 2012. Coinoculation of *Rhizobium leguminosarum*-PR1 with a cold tolerant *Pseudomonas* sp. improves iron acquisition, nutrient uptake and growth of field pea (*Pisum sativum* L.). *J Plant Nutr* 35(2): 243–256.
Namvar, A. Khandan, T. 2015. Inoculation of rapeseed under different rates of inorganic nitrogen and sulfur fertilizer: Impact on water relations, cell membrane stability, chlorophyll content and yield. *Arch Agron Soil Sci* 61(8): 1137–1149.
Ofek-Lalzar, M. Sela, N. Goldman-Voronov, M. Green, S.J. Hadar, Y. 2014. Niche and host-associated functional signatures of the root surface microbiome. *Nat Commun* 5: 4950.
Omomowo, O.I. Babalola, O.O. 2019. Bacterial and fungal endophytes: Tiny giants with immense beneficial potential for plant growth and sustainable agricultural productivity. *Microorganisms* 7: 481.
Parray, J.A. Kamili, A.N. Reshi, Z.A.Hamid, R. Qadri, R.A. 2013. Screening of beneficial properties of rhizobacteria isolated from saffron (*Crocus sativus* L) rhizosphere. *Afr J Microbiol Res* 7(23): 2905–2910.
Paulsen, I.T. Press, C.M. Ravel, J. Kobayashi, D.Y. Myers, G.S.A. Mavrodi, D.V. 2005. Complete genome sequence of the plant commensal *Pseudomonas fluorescens* Pf-5. *Nat Biotechnol* 23: 873–878.
Pierson, L.S.III. Pierson, E.A. 2010. Metabolism and function of phenazines in bacteria: Impacts on the behavior of bacteria in the environment and biotechnological processes. *Appl Microbiol Biotechnol* 86: 1659–1670.
Rehman, A. Ijaz, M. Mazhar, K. Ui-allah, S. Ali, Q. Kumar, V. (eds). 2019. Metagenomic approach in relation to microbe-microbe and plant microbiome interactions. *Microbiome in Plant Health and Disease*: 540–572.
Rout, M. Southworth, D. 2013. The root microbiome influences scales from molecules to ecosystems: The unseen majority. *Am J Bot* 100: 1689–1691.
Sanchez-Canizares, C. Jorrin, B. Poole, P.S. Tkacz, A. 2017. Understanding the holobiont: The interdependence of plants and their microbiome. *Curr Opin Microbiol* 38: 188–196.
Sanchis, V. Bourguet, D. 2008. *Bacillus thuringiensis*: Applications in agriculture and insect resistance management. *Agron Sustain Dev* 28: 11–20.
Schenk, P.M. Carvalhais, L.C. Kazan, K. 2012. Unraveling plant-microbe interactions: Can multi-species transcriptomics help? *Trends Biotechnol* 30: 177–184.
Shuab, R. Lone, R. Koul, K.K. 2017. Influence of arbuscular mycorrhizal fungi on storage metabolites, mineral nutrition, and nitrogen-assimilating enzymes in potato (*Solanum tuberosum* L.) plant. *J Plant Nutr* 40(10): 1386–1396.

Shuab, R. Lone, R. Naidu, J. Sharma, V. Imtiyaz, S. Koul. K.K. 2014. Benefits of inoculation of arbuscular mycorrhizal fungi on growth and development of onion (*Allium cepa*) plant. *Am Eurasian J Agric Environ Sci* 14(6): 527–535.

Singh, P.P. Kujur, A. Yadav, A. Kumar, A. Singh, SK. Prakash B. 2019. Mechanisms of plant-microbe interactions and its significance for sustainable agriculture. In *PGPR Amelioration in Sustainable Agriculture Food Security and Environmental Management*. Sawston: Woodhead Publishing, 17–39.

Starke, R. Jehmlich, N. Bastida, F. 2019. Using proteins to study how microbes contribute ecosystem services: The current state and future perspectives of soil metaproteomics. *Journal of Proteomics* 198: 50–58.

Su, J. Hu, C. Yan, X. Jin, Y. Chen, Z. Guan, Q. et al. 2015. Expression of barley SUSIBA2 transcription factor yields high-starch low-methane rice. *Nature* 523: 602–606.

Sugiyama, A. Udea, Y. Zushi, T. Takase, H. Yazaki, K. 2014. Changes in the bacterial community of soya bean rhizospheres during growth in the field. *PLoS One*: e100709.

Swenson, W. Wilson, D.S. Elias, R. 2000. Artificial ecosystem selection. *Proc. Natl. Acad. Sci.* U.S.A. 97: 9110–9114.

Thijs, S. Weyens, N. Sillen, W. Gkorezis, P. Carleer, R. Vangronsveld, J. 2014. Potential for plant growth promotion by a consortium of stress-tolerant 2,4-dinitrotoluene-degrading bacteria: Isolation and characterization of a military soil. *Microb Biotechnol* 7: 294–306.

Tschaplinki, T.L., J.M. Plett, N.L. Engle, A. Deveau, et al. 2014. *Populus trichocarpa* and *Populus deltoides* exhibits different metabolomic response to colonization by the symbiotic fungus *Laccaria bicolor. Mol. Plant-Microbe Interact* 27: 546–556.

Veena, V. Popavath, P.R. Kennedy, R.K. Sakthivel, N. 2015. In vitro antiproliferative, proapoptotic, antimetastatic and anti-inflammatory potential of 2,4-diacteylphloroglucinol (DAPG) by *Pseudomonas aeruginosa* strain FP10. *Apoptosis* 20(10): 1281–1295.

Vero, L.D. Boniotti, M.B. Budroni, M. Buzzini, P. Cassanelli, S. Comunian, R. et al. 2019. Preservation, characterization and exploitation of microbial biodiversity: The perspective of the Italian network of culture collections. *Microorganisms* 7: 685.

White, R.A., Rivas-Ubach, A. Borkum, M.I. et al. 2017. The state of rhizospheric science in the era of multiomics: A practical guide to omics technologies. *Rhizosphere* 3: 212–221.

Wolf, J.B.W. 2013. Principles of transcriptome analysis and gene expression quantification: An RNA-seq tutorial. *Mol. Ecol. Resour* 13: 559–572.

Yang, Y. Liang, Y. Han, X. Chiu, T.Y. Ghosh, A. Chen, H. Tang, M. et al. 2016. The roles of arbuscular mycor rhizal fungi (AMF) in phytoremediation and tree-herb interactions in Pb contaminated soil. *Sci Rep* 6: 20469.

Zhang, Y. Hua, Z. Lu, H. Oehman, A. Guo, J. et al. 2019. Elucidating functional microorganisms and metabolic mechanisms in a novel engineered ecosystem integrating C, N, P and S biotransformation by metagenomics. *Water Research* 148: 219–230.

Index

For Product Safety Concerns and Information please contact our EU representative GPSR@taylorandfrancis.com Taylor & Francis Verlag GmbH, Kaufingerstraße 24, 80331 München, Germany

Printed and bound by CPI Group (UK) Ltd, Croydon, CR0 4YY
05/05/2025
01860979-0001